MEDICAL INTELLIGENCE UNIT

Abdominal Compartment Syndrome

Rao R. Ivatury, M.D.
Division of Trauma, Critical Care and Emergency Surgery
Department of Surgery
Virginia Commonwealth University
VCU Reanimation Engineering Shock Center (VCURES)
Richmond, Virginia, U.S.A.

Michael L. Cheatham, M.D.
Department of Surgical Education
Orlando Regional Medical Center
Orlando, Florida, U.S.A.

Manu L. N. G. Malbrain, M.D.
Intensive Care Unit
ZiekenhuisNetwerk Antwerpen
Campus Stuivenberg
Antwerp, Belgium

Michael Sugrue, M.D.
Department of Trauma
Liverpool Hospital
Liverpool, Sydney, Australia

CRC Press
Taylor & Francis Group
Boca Raton London New York

CRC Press is an imprint of the
Taylor & Francis Group, an **informa** business

ABDOMINAL COMPARTMENT SYNDROME

Medical Intelligence Unit

First published 2006 by Landes Bioscience

Published 2018 by CRC Press
Taylor & Francis Group
6000 Broken Sound Parkway NW, Suite 300
Boca Raton, FL 33487-2742

© 2006 by Taylor & Francis Group, LLC
CRC Press is an imprint of Taylor & Francis Group, an Informa business

First issued in paperback 2019

No claim to original U.S. Government works

ISBN 13: 978-0-367-44633-8 (pbk)
ISBN 13: 978-1-58706-196-7 (hbk)

Visit the Taylor & Francis Web site at
http://www.taylorandfrancis.com

and the CRC Press Web site at
http://www.crcpress.com

While the authors, editors and publisher believe that drug selection and dosage and the specifications and usage of equipment and devices, as set forth in this book, are in accord with current recommendations and practice at the time of publication, they make no warranty, expressed or implied, with respect to material described in this book. In view of the ongoing research, equipment development, changes in governmental regulations and the rapid accumulation of information relating to the biomedical sciences, the reader is urged to carefully review and evaluate the information provided herein.

Library of Congress Cataloging-in-Publication Data

Abdominal compartment syndrome / [edited by] Rao R. Ivatury ... [et al.].
     p. ; cm. -- (Medical intelligence unit)
   Includes bibliographical references and index.
   ISBN 1-58706-196-1
   1. Abdomen--Diseases. 2. Compartment syndrome.   I. Ivatury,
Rao R. II. Series: Medical intelligence unit (Unnumbered : 2003)
   [DNLM: 1. Abdomen--physiopathology. 2. Compartment Syndromes.
3. Hypertension.   WI 900 A1334 2006]
RC944.A163 2006
617.5'5--dc22

                                                    2006012693

To the memory of my parents and to my family:
Leela, Gautam and Arun. Your patience, love and support
is the greatest gift. I am forever grateful.
Rao R. Ivatury

To my wife Susie and my daughters, Kaitlin, Amelia,
Julianne and Melinda. Thank you for your unwavering
love and support.
To my father and mother. Thank you for the example
you have always been for me.
Michael L. Cheatham

To the memory of Dr. Hubert Malbrain, my inspiring father, colleague
and friend.
I'm in gratitude to my devoted wife Bieke, who experienced a husband
"never being there when needed most."
To share my passion, success, failure and dreams with her
is still a pleasant never ending story to me.
I also thank my three sons, Jacco, Milan and Luca for always providing a
quiet writing environment.
I thank my parents who gave me the opportunity to study medicine and
to develop my talents.
Manu L. N. G. Malbrain

To my family: Pauline, Conor, Alan, Gavin and Ryan.
Thanks for being there and allowing me the time
to work on ACS.
Michael Sugrue

CONTENTS

EDITORS

Rao R. Ivatury
Division of Trauma, Critical Care and Emergency Surgery
Department of Surgery
Virginia Commonwealth University
VCU Reanimation Engineering Shock Center (VCURES)
Richmond, Virginia, U.S.A.
Email: rivatury@hsc.vcu.edu
Chapters 2, 9, 23

Michael L. Cheatham
Department of Surgical Education
Orlando Regional Medical Center
Orlando, Florida, U.S.A.
Email: michael.cheatham@orhs.org
Chapters 4, 6, 23

Manu L. N. G. Malbrain
Intensive Care Unit
ZiekenhuisNetwerk Antwerpen
Campus Stuivenberg
Antwerp, Belgium
Email: manu.malbrain@skynet.be
Chapters 3, 4, 5, 6, 23

Michael Sugrue
Department of Trauma
Liverpool Hospital
Liverpool, Sydney, Australia
Email: michael.sugrue@swsahs.nsw.gov.au
Chapters 8, 23

CONTRIBUTORS

Georg Auzinger
Institute of Liver Studies
Kings College Hospital
London, U.K.
Chapter 10

Zsolt Balogh
Department of Traumatology
University of Szeged
Szeged, Hungary
Email: zsoltbalogh@yahoo.com
Chapters 13, 22

Lorenzo Berra
Intensive Care Unit
San Gerardo Hospital
University of Milano-Bicocca
Milan, Italy
Chapter 11

Gianni Biancofiore
Post-Surgical and Transplant
   Intensive Care Unit
Azienda Ospedaliera Pisana
Pisa, Italy
Email: g.biancofiore@med.unipi.it
Chapter 10

Martin Björck
Vascular Surgery
Uppsala University Hospital
Uppsala, Sweden
Email: martin@bjorck.pp.se
Chapter 16

L. D. Britt
Department of General Surgery
Eastern Virginia Medical School
Norfolk Viriginia, U.S.A.
Foreword

Giuseppe Citerio
UO Neuroanestesia
   e Neurorianimazione
Dipartimento di Medicina Perioperatoria
   e Terapie Intensive
H San Gerardo, Monza (Mi), Italy
Email: g.citerio@hsgerardo.org
Chapter 11

Scott D'Amours
Department of Trauma
Liverpool Hospital
Liverpool, Sydney, Australia
Email: Scott.Damours@swsahs.nsw.gov.au
Chapter 8

Dries H. Deeren
Department of Internal Medicine
University Hospital Leuven
Leuven, Belgium
Chapter 5

Lawrence N. Diebel
Department of Surgery
Wayne State University
Detroit, Michigan, U.S.A.
Email: ldiebel@med.wayne.edu
Chapter 9

Davis Elliot
Departments of Surgery and Radiology
Vancouver Hospital
   and Health Sciences Centre
Vancouver, British Columbia, Canada
Chapter 16

Luciano Gattinoni
Department of Anesthesia
   and Intensive Care
University of Milan
Istituto di Anestesia e Rianimazione,
   Fondazione IRCCS
Ospedale Maggiore Policlinico
   Mangiagalli Regina Elena
Milano, Italy
Email: gattinon@policlinico.mi.it
Chapter 20

Claudia E. Goettler
Department of Surgery
Brody School of Medicine
East Carolina University
Greenville, North Carolina, U.S.A.
Email: cgoettle@pcmh.com
Chapter 22

Ali Hallal
Department of Trauma
Liverpool Hospital
Liverpool, Sydney, Australia
Email: ali.hallal@swsahs.nsw.gov.au
Chapter 8

Giselle G. Hamad
Department of Surgery
University of Pittsburgh Medical Center
Pittsburgh, Pennsylvania, U.S.A.
Email: hamadg@upmc.edu
Chapter 15

Michael E. Ivy
Department of Surgery
Hartford Hospital
Hartford, Connecticut, U.S.A.
and
Department of Surgery
University of Connecticut
    School of Medicine
Farmington, Connecticut, U.S.A.
Email: mivy@harthosp.org
Chapter 14

Felicity Jones
Department of Trauma
Liverpool Hospital
Liverpool, Sydney, Australia
Chapter 3

Mark J. Kaplan
Division of Trauma
    and Surgical Critical Care
Albert Einstein Medical Center
Philadelphia, Pennsylvania, U.S.A.
Email: kaplanm@einstein.edu
Chapter 22

Andrew Kirkpatrick
Departments of Surgery, Critical Care
    Medicine and Radiology
Vancouver Hospital
    and Health Sciences Centre
Vancouver, British Columbia, Canada
Email: andrew.kirkpatrick@
    calgaryhealthregion.ca
Chapter 16

Karel A. Kolkman
Department of Surgery
Rijnstate Hospital Arnhem
Arnhem, The Netherlands
Email: KKolkman@rijnstate.nl
Chapter 21

M. Ann Kuhn
Department of Surgery
Section of Pediatric Surgery
The University of Oklahoma
    College of Medicine
Oklahoma City, Oklahoma, U.S.A.
Email: Ann-Kuhn@ouhsc.edu
Chapter 17

Ari Leppäniemi
Department of Surgery
Meilahti Hospital
University of Helsinki
Helsinki, Finland
Email: ari.leppaniemi@hus.fi
Chapters 2, 16

John C. Mayberry
Department of Surgery
Oregon Health and Science University
Portland, Oregon, U.S.A.
Email: mayberrj@ohsu.edu
Chapter 18

Ingrid R. A. M. Mertens zur Borg
Department of Anesthesiology
Erasmus Medical Center
Rotterdam, The Netherlands
Email: i.mertenszurborg@erasmusmc.nl
Chapters 7, 21

Ernest E. Moore
Department of Surgery
Denver Health Medical Center
Denver, Colorado, U.S.A.
Email: Ernest.Moore@dhha.org
Chapter 12

Frederick A. Moore
Department of Surgery
University of Texas Medical School
   at Houston
University of Texas Health Science
   Center
Houston, Texas, U.S.A.
Email: Frederick.A.Moore@uth.tmc.edu
Chapters 13, 22

David J. J. Muckart
Department of Surgery
Nelson R. Mandela School of Medicine
University of KwaZulu/Natal
Durban, Republic of South Africa
Email: MuckartD@ukzn.ac.za
Chapter 2

Savvas Nicolaou
Trauma Program
Foothills Medical Centre
Calgary, Alberta, Canada
and
Department of Radiology
Vancouver Hospital
   and Health Sciences Centre
Vancouver, British Columbia, Canada
Chapter 16

Claudia I. Olvera
Department of Critical Care Medicine
The American British Cowdray
   Medical Centre
Mexico City, Mexico
Email: Claudia_olvera@hotmail.com
Chapters 7, 19

Michael J. A. Parr
Intensive Care Units
Liverpool and Campbelltown Hospitals
University of New South Wales
Sydney, Australia
Email: Michael.Parr@swsahs.nsw.gov.au
Chapter 19

Andrew B. Peitzman
Department of Surgery
University of Pittsburgh Medical Center
Pittsburgh, Pennsylvania, U.S.A.
Email: peitzmanab@upmc.edu
Chapter 15

Christopher D. Raeburn
Department of Surgery
University of Colorado
   Health Sciences Center
Denver, Colorado, U.S.A.
Email: christopher.raeburn@uchsc.edu
Chapter 12

Michael F. Rotondo
Department of Surgery
Brody School of Medicine
East Carolina University
Greenville, North Carolina, U.S.A.
Email: MFRotond@PCMH.com
Chapter 22

Anastazia Salazar
Departments of Surgery
   and Transplantation
Vancouver Hospital
   and Health Sciences Centre
Vancouver, British Columbia, Canada
Chapter 16

Moshe Schein
Surgical Specialists of Keokuk
Keokuk Area Hospital
Keokuk, Iowa, U.S.A.
Email: mschein1@mindspring.com
Chapter 1

C. William Schwab
Department of Surgery
University of Pennsylvania
Philadelphia, Pennsylvania, U.S.A.
Chapter 22

R. Stephen Smith
Department of Surgery
University of Kansas School
   of Medicine - Wichita
Wichita, Kansas, U.S.A.
Email: rsmith3@kumc.edu
Chapter 2

David W. Tuggle
Department of Surgery
Section of Pediatric Surgery
The University of Oklahoma
   College of Medicine
Oklahoma City, Oklahoma, U.S.A.
Email: David-Tuggle@ouhsc.edu
Chapter 17

Franco Valenza
Universita degli Studi di Milano
Istituto di Anestesia e Rianimazione
Ospedale Maggiore di Milano
Milano, Italy
Email: Franco.valenza@unimi.it
Chapter 20

Serge J. C. Verbrugge
Department of Anesthesiology
Erasmus Medical Centre
Rotterdam, The Netherlands
Chapters 7, 21

Jean-Louis Vincent
Professor, Intensive Care
University Hospital Erasme
Brussels, Belgium
Foreword

Julia Wendon
Institute of Liver Studies
Kings College Hospital
London, U.K.
Email: julia.wendon@kcl.ac.uk
Chapter 10

FOREWORD

Abdominal compartment syndrome is a condition that, until relatively recently, has been poorly appreciated, despite the important potential implications for all organ systems. Recent interest has helped clarify the local and systemic effects of increased intra-abdominal pressure and heightened awareness of the importance of early recognition and treatment. This book, written by international experts in the field follows the first World Congress on Abdominal Compartment Syndrome held in Noosa, Australia in December 2004, and provides an excellent update on this very topical subject. Comprehensively constructed, the book covers all key aspects of the syndrome from definitions and diagnosis to monitoring and management, and is supported throughout by appropriate and ample illustrations. Many of the detrimental consequences of abdominal compartment syndrome are reversible with appropriate and timely intervention, and although many questions remain unanswered regarding this relatively newly recognized condition, this book provides a welcome and detailed source of current information and opinion on abdominal compartment syndrome, which will be helpful to all health care practitioners, and ultimately to our patients.

Jean-Louis Vincent, M.D., Ph.D., F.C.C.M.
Professor, Intensive Care
University Hospital Erasme
Brussels, Belgium

Abdominal compartment syndrome is now a well-known entity. It is defined as intra-abdominal hypertension with associated renal, pulmonary, or hemodynamic compromise. However, our profession was late in recognizing abdominal compartment syndrome and the devastating complications. Although Henricus first implicated abdominal compartment syndrome as a cause of death in 1890, Kron et al highlighted this syndrome in 1984. They reported a series of 11 patients with elevated intra-abdominal pressure and oliguria following aneurysm surgery. Unfortunately, only recently has there been a widespread awareness of this syndrome. Trauma patients represent a heterogenous group who are at risk for the development of both primary and secondary abdominal compartment syndrome with the latter occurring when there is no primary intra-abdominal injury. Early identification of patients who are prone to developing this syndrome is essential.

After a highly successful first World Congress on abdominal compartment syndrome, Dr. Rao Ivatury and the other international experts have produced the most definitive publication on this topic. This well-written book covers every aspect of the abdominal compartment syndrome, including diagnosis, complications, management, and prevention. To date, there is nothing comparable. This is a landmark contribution to the literature.

L. D. Britt, M.D., M.P.H., F.A.C.S.
Chairman, Brickhouse Professor of Surgery
Department of General Surgery
Eastern Virginia Medical School
Norfolk Viriginia, U.S.A.

PREFACE

First suggested in 1863 by Marey, abdominal compartment syndrome (ACS) is a constellation of the physiologic sequelae of increased intra-abdominal pressure (IAP), termed intra-abdominal hypertension (IAH). Recent observations suggest an increasing frequency of this complication in all types of patients, neonates to the elderly and in diverse clinical conditions, surgical to non-surgical. Even chronic elevations of IAP seem to effect the various organ systems in the body. Despite its obvious clinical implications, not enough attention is paid to IAP and IAH. ACS still is not uniformly appreciated or diagnosed. Only a few medical and surgical intensivists believe in the concept of IAH and actively attempt its prevention and treatment. The result, as is strongly substantiated by retrospective and prospective data, is a successful reduction in organ failures and mortality.

The literature on IAH and ACS has exponentially increased in the last decade. Several unanswered questions, however, cloud our understanding of the pathophysioology of elevated IAP. To name a few, what is the ideal method of measuring IAP? What level of IAP requires abdominal decompression? Is it a level at which the classic manifestations of ACS become evident? Or is it a level at which subtle changes in physiology precede the development of ACS? Is it the IAP that is important or is abdominal perfusion pressure (APP) the critical determinant? Are IAH and ACS synonymous?

This book is an overview of the current knowledge on IAH and ACS. The authors have been chosen for their original work in the field. They write with conviction from personal experience. We have preserved to the fullest extent possible their own concepts in their own writing style. As a consequence, some repetition of basic concepts is unavoidable and may be excused. We sincerely hope that this work will stimulate interest and attract clinicians and scientists to the fascinating field of IAH and its sequelae.

Finally, our heartfelt appreciation to the authors for their labor of love in contributing to this work, despite their over-filled calenders. We sincerely thank Cynthia Conomos and Sara Lord at Landes Bioscience for their expertise and commitment to the publication of this book. We are profoundly thankful to Ms. Charmaine Miranda for her labor of love in bringing to life the WSACS.

Rao R. Ivatury
Michael L. Cheatham
Manu L. N. G. Malbrain
Michael Sugrue

CHAPTER 1

Abdominal Compartment Syndrome:
Historical Background

Moshe Schein*

Abstract

The brief history of IAH and ACS are typical of any medical innovation: described, forgotten, rediscovered, and faced with skepticism and ridicule. Eventually, after being scientifically proven and reproven and supported by "clinical leaders" and widely published in reputable journals—it is accepted as "truth". This chapter summarizes phases in the history of IAH and ACS from the mid 19th century until today.

At a first glance, a comprehensive history of the abdominal compartment syndrome (ACS) is an impossible task. It is like writing the history of vomiting or urinating or any other basic human (and mammalian) physiological activity. For as long as humans existed, abdomens ballooned with gas, fluid, blood, pus—resulting in all the physiological changes that we today call intra-abdominal hypertension (IAH) and ACS.

Obviously, IAH and ACS always existed but were not understood as such. For example: in his book *Surgical Errors and Safeguards*[1] Max Thorek wrote: "It is of paramount importance to recognize acute dilation of the stomach at once. Practically all unrecognized cases die." He mentioned that (according to Hamilton Bailey) "the condition can be diagnosed even before the patient vomits...if the pulse is rising...and if the urine is scanty in amount." Clearly, what was described was yet another example of the ACS—the syndrome which remained, until very recently, elusive and almost unknown.

Even we—so called modern surgeons—until a decade ago observed patients with full-blown ACS, failing to understand what we see. We saw massively distended patients dying after operations for ruptured abdominal aneurysm and blamed their death on myocardial infarction or respiratory failure; we watched patients with massive abdomens in the early days of severe acute pancreatitis—blaming their cardiorespiratory demise on pancreatic toxins; we rushed to resuture dehisced abdomens—pushing everything back, and tightly closing with "retention sutures"—wondering why these patients spend weeks on the ventilator and then die.

What Is History?

At what point in time should we start looking at the history of the abdominal compartment syndrome and where should we stop? Various dictionaries define history as "A branch of knowledge that records and explains *past events*", or "the aggregate of *past events*". History is also defined as "the discipline that records and interprets past events involving human beings", or "the continuum of events occurring in succession leading from the past to the present and even into the future."

*Moshe Schein—Surgical Specialists of Keokuk, Keokuk Area Hospital, Keokuk, Iowa 52632, U.S.A. Email: mschein1@mindspring.com

Abdominal Compartment Syndrome, edited by Rao R. Ivatury, Michael L. Cheatham, Manu L. N. G. Malbrain and Michael Sugrue. ©2006 Landes Bioscience.

Since the present, the future—and even the recent past—of all aspects of ACS are covered in great details elsewhere in this book I will start this historical overview at mid 19th century—at the time when people started measuring intra-abdomianl pressure (IAP) and study its physiological consequences. I still do not know how late into the 20th should I carry this narrative on.

Early Rays of Light

Haven Emerson's (1874-1957) comprehensive treatise "intra-abdominal pressures"[2] (1911) spares us the arduous task of finding yet older and hard to retrieve manuscripts. Emerson's manuscript was typical of its day when people had more to write about the history of things than about what is actually new. His started his manuscript with 10 pages of "historical sketch" covering the second half of the 19th century. His opening testament may be true even today:

" The standard text-books of obstetrics, gynecology and surgery treat of the matter so rarely, and when it is mentioned, so inaccurately, that no information is to be had from them…Most of the text-books of physiology fail to mention intra-abdominal pressure at all."

Emerson provided a detailed historical review of which I will mention only selected key points:

- It is difficult to define with certainty who was the first to write about the physiology of IAP. Contemporary reviews bestow such honor on Marey of Paris who in his paper "Physiologie médicale de la circulation du sang" (1863) wrote that the "effects that respiration produces on the thorax are the inverse pf those present in the abdomen." However, according to Emerson,[2] Marey "describes no tests and gives no records or figures."
- Braune of Germany (1865) appears to be the first to measure IAP through the rectum.[2]
- Another pioneer mentioned by Coombs[3] and contemporary reviews is Paul Bert who in 1870 published a volume on "Physiologie comparée de la respiration." Based on experiments in anesthetized animals, measuring thoracic and abdominal pressures through tubes inserted in the trachea and rectum, respectively, Bert described elevation of IAP on inspiration and the descent of the diaphragm.[2]
- Schroeder of Germany (1886) noted the slightly increased IAP in pregnancy, hypothesizing that there must be some adaptation of abdominal wall tension to the increasing size of the uterus.[2]
- Schatz of Germany (1872) used balloon tube connected to a mercury manometer to measure pressures within the gravid uterus. According to him IAP is positive and during pregnancy IAP rises slightly, though not in proportion to the increase in the size of the uterus, until the last month, when usually the abdominal muscles are stretched beyond their ability to respond, and there is then a fall of pressure below normal. He also noted that the pressure in the inferior vena cava must be at least as high as IAP to avoid obliteration of its lumen, and that a moderately positive IAP—increased in the erect position—assists the return of blood flow, and the flow of chyle from the abdominal viscera.[2]
- Wendt of Germany (1873) measured IAP through the rectum, noting that the higher the abdominal pressure the less the secretion of urine.[2]
- Oderbrecht of Germany (1875) measured pressures within the urinary bladder, concluding that IAP is normally positive.[2]
- Wegner of Germany (1877) noted that the normal positive IAP aids absorption of fluids from the peritoneal surface.[2]
- Quinke of Germany (1878) noted in patients with ascites the obstructive effects of high IAP on venous return from abdominal viscera.[2]
- Mosso and Pellacani of Italy (1882) measured positive IAP through the urinary bladder.[2]
- Senator of France (1883) noted that IAP is much diminished by weakness of the abdominal walls.[2]
- Heinricius of Germany (1890) found that IAP's between 27 to 46 cm. water were fatal to animals owing to prevention of respiration, decreasing cardiac diastolic distention and a low blood pressure. He also contended that rapid abdominal distention at low pressure is of much harm while gradually established high pressure may be well tolerated.[2]

- During the late 19th century, and the early years of the 20th, many other authors confirmed, or refuted, the above observations in multiple experimental and clinical observations. The latter included a few fallacies which led to confusion: that the "normal" IAP is subatmospheric, that IAP varies in the different regions of the abdomen and that trans-visceral measurements are not accurate.[2]

Emerson himself (1911)[2] conducted numerous experiments in dogs—showing what was to be rediscovered again and again: that contraction of the diaphragm is the chief factor in the rise of IAP during inspiration; that anesthesia and muscle paralysis—with loss of muscle tone—decreases the IAP; that elevated IAP increases peripheral vascular resistance; that excessive IAP can cause death from cardiac failure even before terminal asphyxia develops ("Pressure as high as 45 cm. Aq. Will kill a small animal")... Emerson understood that elevated IAP decreases blood pressure because of diminished venous return to the heart as well as depressed cardiac contractility. He then provided amazingly relevant clinical correlation, which subsequently has been totally ignored by many generations of surgeons:

"(in) excessive IAP, the difficulty in breathing is even more marked, and this often plays an important role in the circulatory emergencies in infectious disease where meteorism, abdominal distention and interference with the descent of the diaphragm may determine cardiac failure."

Emerson understood that the cardiovascular collapse associated with "distention of the abdomen with gas or fluid, as in typhoid fever, ascites, or peritonitis" are caused by "overloading the resistance in the splanchnic area" and that "relief of the laboring heart is constantly seen after removal of ascitic fluid."

Thus we see that almost 100 years ago, before the world was engulfed by the chaos of WW-I, ample evidence existed concerning the adverse physiological effects of high IAP on cardiac, respiratory and renal function. There were also those who understood the clinical implications of such knowledge: that high IAP due to ascites, ileus and peritonitis results in morbidity and mortality. Such early rays of light however failed to penetarte the opaque minds of contemporary clinicians and researchers, and the significance of IAP has almost dissapeared druing the ensuing Dark Age.

The Dark Age

This long era of gloom lasted for over 50 years with only scattered but totally ignored attempts to shed old or new light on IAP.

- Thorington and Schmidt (1923) studied urinary output and blood pressure changes in experimental ascites.[4]
- Overholt (1931) seemed to be the first to introduce the issue in an American surgical journal[5]—showing and postulating what essentially has been already known.
- Bellis and Wangensteen (1939) demonstrated changes in venous flow in the abdomen and extremities associated with abdominal distention[6]
- Bradely and Bradely (1947) showed decreased glomerular filtration rate and renal plasma flow with increased IAP.[7]
- Gross (1948) introduced the so-called "staged abdominal repair" in the management of omphalocele, thus acknowledging the importance of avoiding abdominal closure under excessive tension.[8]
- Olerud (1953) studied the effects of increased IAP on portal circulation.[9]

But it was M.G Baggot, an anesthesist from Dublin (currently retired in Granite City, Illinois) who really saw the light.[10] Already in 1951 he suggested that forcing distended bowel back into the abdominal cavity of limited size might kill the patient. He had conceived that the factor leading to the high mortality rate associated with abdominal wound dehiscence is not the dehiscence itself but the emergency procedure to correct it—that produces high IAP. He termed such abdominal dehiscence "abdominal blow-out" and concluded that the ensuing death is due to respiratory dysfunction. Baggot coined also the term "acute tension pneumoperitoneum" believing that excessive free air trapped in the abdomen during its closure increases IAP. Significantly, he recommended avoiding abdominal closure under tension, leaving instead such abdomens open, using a technique described during WW-II by the great British surgeon Sir Heneage Ogilvie.[11] But typically, the destiny of lone prophets—however truthful they are—is to be ridiculed and ignored. And such was Baggot's fate.

Dawn

Early sunrays, heralding the dawn, pierced the dark horizon in 1970 when Sönderberg and Westin correlated IAP directly measured during laparoscopy, to that measured through the urinary bladder.[12] In 1972 Shenansky and Gillenwater showed how increased IAP generated by applying abdominal counterpressure (i.e., the MAST suit) depresses cardiac and renal function.[13]

Early experience with laparoscopy led to recognition of the adverse effects of pneumoperitroneum-associated increase in IAP: Ivankovich et at described cardiovascular collapse during gynecological laparoscopy[14] and studied the physiology of the phenomenon.[15] Then in 1976, Lenz et al, studying cardiovascular changes during laparoscopy, pointed out the dangers of pneumoperitoneum in patients with cardiovascular dysfunction, anemia or hypovolemia.[16] Simultaneously, (1976), Richardson and Trinkle studied hemodynamic and respiratory alterations with increased intra-abdominal pressure.[17]

Even earlier, during the 1960s and later in the 1970s, supporting evidence to the clinical significance of elevated IAP was provided in studies in patients with ascites, the amount of which correlated with cardiorespiratory morbidity—the later reversed by paracentesis.[18-21] The same was true by the growing number of papers supporting leaving the abdomen open in newborns with omphalocele and gastroschisis.[22,23]

Thus the sun has slowly risen to melt the frozen brains—at least the few which were susceptible and welcoming its warm rays.

Sunrise

The early 1980s produced a few key studies: Kashtan et al (1981) rediscovered the hemodynamic effects of increased IAP;[24] Harman et al (1982),[25] as well as Richards et al (1983)[26] demonstrated how elevated IAP adversely affects renal function and how abdominal decompression improves it, and Le Roith et al (1982) studied the effects of increased IAP on plasma antidiuretic hormone levels.[27]

It was however the paper by Kron, Harman and Nolan (1984) which is considered by many as a "benchmark" in the clinical perception of intra-abdominal hypertension. In this combined clinical and experimental study the authors showed that IAP could be used as a criterion for life-saving abdominal reexploration and decompression.[28] Interestingly, many claim that the term Abdominal Compartment Syndrome (ACS) was first used by Kron et al.[28] However, I could not find such term used within their paper. This reflects the dangers of citing from secondary sources rather than reading the primary, original ones.

Sporadic communications continued to appear: Smith et al (1985) reported reversal of postoperative anuria by decompressive laparotomy;[29] Barnes et al (1985) studied cardiovascular responses to elevated IAP;[30] Caldwell and Ricotta (1987) measured such changes in visceral blood flow;[31] Jacques and Lee (1988) reported improvement in renal function after relief of raised IAP due to retroperitoneal hematoma;[32] Cullen et al (1989) reported on surgical decompression of the abdomen in critically ill patients to reverse cardiovascular, renal and respiratory compromise.[33] Around the same time people started measuring IAP through the urinary catheter in their intensive care units[34]—reinventing a method reported by Oderbrecht more than 100 years earlier.[2]

Then in 1989 the term abdominal compartment syndrome wad finally coined. It may well be that different people begun to use such term before but the first publication mentioning it was that by Fietsam et al from the William Beaumont Hospital, Royal Oak, Michigan.[35] The authors wrote:

"In four patients with ruptured abdominal aortic aneurysms increased intra-abdominal pressure developed after repair. It was manifested by increased ventilatory pressure, increased central venous pressure, and decreased urinary output associated with massive abdominal distension not due to bleeding. This set of findings constitutes an intra-abdominal compartment syndrome caused by massive interstitial and retroperitoneal swelling…four patients received more than 25 liters of fluid resuscitation (electrolyte and blood) during and within 16 hours after operation and had massive abdominal distension. Decompressive laparotomies were performed in the Intensive Care…Opening the abdominal incision was associated with dramatic improvements in central venous pressure, urinary output, ventilatory pressure, arterial carbon dioxide tension, and oxygenation."

Thus, by the end of the 1980s people knew how to measure IAP, what damage elevated IAP can produce and how to treat it. They also defined the clinical syndrome and named it. But commonly people look at truth but refuse to acknowledge it and so it took a few more years for the concept to penetrate our surgical minds.

Morning

The 1990s brought with it a plethora of studies—reflecting a growing recognition of IAH and ACS. As most studies published during this decade will be cited elsewhere in this book I will only mention key developments.

- The introduction of laparoscopic surgery produced numerous studies on the physiology of penumoperitoneum—demonstrating and emphasizing the dangers of increased IAP on various organs and systems.
- Growing popularity of "damage control" strategies in abdominal trauma increased surgeons' and intensivists' awareness of IAH/ACS and the benefits of leaving the abdomen open and/or its decompression.
- Cautious enthusiasm with the use of "laparostomy" in severe abdominal infections led to the recognition that various nontraumatic abdominal catastrophes maybe associated with IAH/ACS and could benefit of abdominal nonclosure.
- Two "collective reviews" of ACS appeared in 1995[36] and 1996[37]—opening the gate to numerous publications, recognizing IAH/ACS in a large number of surgical—abdominal and extra-abdominal, traumatic and nontraumatic scenarios—and providing an ever growing list of complications and consequences.

High Noon

It is my understanding that high noon is yet to come and it will come only when the majority of clinicians will understand and recognize the clinical significance of IAH and ACS. But looking around me I still see surgeons closing distended and tense abdomens with retention sutures under excessive tension, I see surgeons closing hugely swollen abdomens after repair of ruptured AAA, I see abdomens ballooning in the intensive care unit with surgeons and intensivist ignoring the potential benefits of abdominal decompression. Typically, a while ago, when we submitted for publication our observations of IAH/ACS in early severe pancreatitis it has been rejected by reputable surgical journal: the distinguished journal-reviewers commented that the existence of such syndrome is "controversial and unproven." (The manuscript eventually appeared elsewhere.[38])

The brief history of IAH and ACS are characteristic of any medical innovation: described, forgotten, rediscovered, and faced with skepticisms and ridicule. Eventually, after being scientifically proven and reproven and supported by "clinical leaders" and widely published in reputable journals—it is accepted as "truth." But it takes a few generations.

"New ideas seldom have the simplicity of a switched on light bulb." (–Thomas Starzl).

Commentary

Michael Sugrue

Dr. Schein has firmly encapsulated the evolution of the history of intra-abdominal hypertension and the ACS.

Since Wendt's elegant work of 1876 we have become slow to adopt the evidence supporting the causal relationship between intra-abdominal hypertension, the ACS and organ dysfunction. The great work at the early part of the twentieth century by key researchers such as Haven Emerson, Helen Coombs and many others, in defining the correct pattern of intra-abdominal pressure, has been clearly resurrected in this chapter and should not be forgotten in our forward push with ACS. One of the hallmark pieces of research from the 20th century, was probably by Bradley and Bradley. In 1947 they reported from Massachusetts Memorial Hospital fascinating invasive human experiments demonstrating that raised intra-abdominal pressure had a direct effect on renal function. It is only in the last two decades that there has been an exponential rise in the realization in the clinical importance of ACS. Robert Fietsam from Beaumont Hospital in Michigan should be credited with coining the term 'Abdominal Compartment Syndrome' in his report of four patients following aortic surgery.

We owe a great debt to the visionaries who over the last 150 years have driven forward the cause and research relating to the abdominal compartment syndrome.

References

1. Thorek M. Surgical errors and safeguards. Philadelphia: JB Lippincott Company, 1934; 370.
2. Emerson H. Intra-abdominal pressures. Arch Intern Med 1911; 7:754-784.
3. Coombs HC. The mechanisms of the regulation of intra-abdominal pressure. Am J Physiol 1920; 61:159-63.
4. Thorington JM, Schmidt CF. A study of urinary output and blood-pressure changes resulting in experimental ascites. Am J Med Sci 1923; 165:880-90.
5. Overholt RH. Intraperitoneal pressure. Arch Surg 1931; 22:691-703.
6. Bellis CJ, Wangensteen OH. Venous circulatory changes in the abdomen and lower extremities attending abdominal distention. Proc Soc Exp Biol Med 1939; 4:490-498.
7. Bradely SE, Bradely GP. The effect of increased intra-abdominal pressure on renal function in man. J Clin Invest 1947; 26:1010-1015.
8. Gross RE. A new method for surgical treatment of large omphaloceles. Surgery 1948; 24:277-292.
9. Olerud S. Experimental studies on portal circulation at increased intra-abdominal pressure. ACTA Physio Scand 1953; 30(Supp 109):4-93.
10. Baggot MG. Abdominal blow-out: A concept. Current Research Anesthesia Analgesia 1951; 30:295-8.
11. Ogilvie WH. The late complications of abdominal war wounds. Lancet 1940; 2:253-256.
12. Sönderberg G, Westin B. Transmission of rapid pressure increase from the peritoneal cavity to the bladder. Scan J Urol Nephrol 1970; 4:155-165.
13. Shenansky JH, Gillenwater JY. The renal hemodynamic and functional effects of external counterpressure. Surg Gynecol Obstet 1972; 134:253-258.
14. Ivankovich AS, Albrecht RF, Zahed B et al. Cardiovascular collapse during gynecological laparoscopy. Ill Med J 1974; 145:58-61.
15. Motev M, Ivankovich AD, Bieniarz J et al. Cardiovascular effects and acid base and blood gas changes during laparoscopy. Amer J Obstet Gynecol 1973; 116:1002-1012.
16. Lenz RJ, Thomas TA, Wilkins DG. Cardiovascular changes during laparoscopy. Anaesthesia 1976; 31:4-12.
17. Richardson JD, Trinkle JK. Hemodynamic and respiratory alterations with increased intra-abdominal pressure. J Surg Res 1976; 20:401-404.
18. Gordon ME. The acute effects of abdominal paracentesis in Laennec's cirrhosis upon changes of electrolytes and water, renal function and hemodynamics. Am J Gastroenterol 1960; 33:15-37.
19. Suazzi M, Polese A, Magrini F et al. Negative influence of ascites in the cardiac function of cirrhotic patients, Am J Med 1975; 59:165-170.
20. Cruikshank DP, Buschsbalm HJ. Effects of rapid paracentesis, cardiovascular dynamics and body fluid composition. JAMA 1973; 225:1361-1362.
21. Knauer CM, Love HM. Hemodynamics in cirrhotic patient during paracentesis. N Engl J Med 1967; 276:491-496.

22. Ravitch MM. Omphalocle: Secondary repair with the aid of pneumoperitoneum. Arch Surg 1969; 99:166-170.
23. Allen RG, Wrenn Jr EL. Silo as a sac in the treatment of omphalocele and gastroschisis. J Ped Surg 1969; 4:3-8.
24. Kashtan J, Green JF, Parson EQ et al. Hemodynamic effects of increased abdominal pressure. J Surg Res 1981; 30:249-255.
25. Harman PK, Kron IL, McLachan DH et al. Elevated intra-abdomial pressure and renal function. Ann Surg 1982; 196:594-597.
26. Richards WO, Scovill W, Shin B et al. Acute renal failire associated with increased intra-abdominal pressure. Ann Surg 1983; 197:183-187.
27. Le Roith D, Bark H, Nyska M et al. The effect of abdominal pressure on plasma antidiuretic hormone levels. J Surg Res 1982; 32:65-69.
28. Kron IL, Harman PK, Nolan SP. The measurement of intra-abdominal pressures as a criterion for abdominal reexploration. Ann Surg 1984; 199:28-30.
29. Smith JH, Merrell RC, Raffin TA. Reversal of postoperative anuria by decompressive celiotomy. Arch Intern Med 1985; 145:553-4.
30. Barnes GE, Laine GA, Giam PY et al. Cardiovascular responses to elevation of intra-abdominal hydrostatic pressure. Am J Physiol 1985; 248:R208-13.
31. Caldwell CB, Ricotta JJ. Changes in visceral blood flow with elevated intraabdominal pressure. J Surg Res 1987; 43:14-20.
32. Jacques T, Lee R. Improvement of renal function after relief of raised intra-abdominal pressure due to traumatic retroperitoneal haematoma. Anaesth Intensive Care 1988; 16:478-82.
33. Cullen DJ, Coyle JP, Teplick R et al. Cardiovascular, pulmonary, and renal effects of massively increased intra-abdominal pressure in critically ill patients. Crit Care Med 1989; 17:118-21.
34. Iberti TJ, Lieber CE, Benjamin E. Determination of intra-abdominal pressure using a transurethral bladder catheter: Clinical validation of the technique. Anesthesiology 1989; 70:47-50.
35. Fietsam Jr R, Villalba M, Glover JL et al. Intra-abdominal compartment syndrome as a complication of ruptured abdominal aortic aneurysm repair. Am Surg 1989; 55:396-402.
36. Schein M, Wittmann DH, Aprahamian CC et al. The abdominal compartment syndrome: The physiological and clinical consequences of elevated intra-abdominal pressure. J Am Coll Surg 1995; 180:745-53.
37. Burch JM, Moore EE, Moore FA et al. The abdominal compartment syndrome. Surg Clin North Am 1996; 76:833-42.
38. Gecelter G, Fahoum B, Gardezi S et al. Abdominal compartment syndrome in severe acute pancreatitis: An indication for a decompressing laparotomy? Dig Surg 2002; 19:402-4.

CHAPTER 2

Definitions

David J. J. Muckart, Rao R. Ivatury,* Ari Leppäniemi
and R. Stephen Smith

Historical Background

David J. J. Muckart and Rao Ivatury

Introduction

Within any human body compartment a rise in pressure above physiological limits is detrimental. At pressures which still permit axial vessel flow, capillary perfusion may cease to exist resulting in cell death. The physiological sequelae and development of signs and symptoms is dependent upon a number of factors, namely perfusion pressure, rate and magnitude of intracompartmental pressure rise, compliance of the compartment, and the reason for the change in compartmental pressure. A rapid exponential rise in intracranial pressure within the rigid skull in a hypotensive patient is rapidly fatal if urgent intervention is delayed, whereas chronic hydrocephalus in a child may have little, if any, effect on vital organ function. The same principles pertain to the abdomen. During laparoscopy with muscle relaxation IAP may rise acutely to 20 mm Hg or more without any overt interference in organ function.[4] Conversely, in a swine model Simon et al showed that if IAP was elevated to 20 mm Hg following a period of haemorrhagic shock and resuscitation, there was a marked decrease in pulmonary function.[5] Control animals with similar elevations in IAP but without prior haemorrhage had minimal changes in PaO_2/FiO_2 ratio. In humans, ACS as a result of rapid accumulation of intra-abdominal fluid has been documented following massive crystalloid resuscitation in the absence of intra-abdominal injury,[6-9] whereas chronic ascites of up to 15 litres is well tolerated.[10] From the above there appear to be certain preconditions for the development of a pathological rise in IAP. The change must be acute and there must be a prior insult which need not necessarily be intra-abdominal.

In order to minimise the detrimental effects of rising pressures within any compartment it is imperative that the condition is recognised well before the compartment syndrome is established. There can be few more relevant areas for the Dickensian philosophy "Prevention is better than cure"[11] for the latter may prove impossible.

As with most syndromes there exists a prodromal phase before obvious signs and symptoms become manifest. Within the abdomen this represents the period of IAH. In the past this term has been used interchangeably and was synonymous with ACS. A distinction should be made between IAH and ACS and they must be regarded as different phases of a developing pathological process,[12,13] perhaps comparable to the distinctions between the grades of the systemic inflammatory response syndrome (SIRS). IAH should be reserved for the scenario of an increase in IAP before overt signs appear. It is crucial to realise, however, that in the prodromal

*Corresponding Author: Rao R. Ivatury—VCURES, Virginia Commonwealth University, 1521 West Hospital, P.O. Box 980454, Richmond, Virginia, U.S.A. Email: rivatury@hsc.vcu.edu

Abdominal Compartment Syndrome, edited by Rao R. Ivatury, Michael L. Cheatham, Manu L. N. G. Malbrain and Michael Sugrue. ©2006 Landes Bioscience.

phase splanchnic hypoperfusion may occur long before the classic manifestations of ACS become evident.[12,14]

Prodromal symptoms are often "soft" and may be ascribed to a variety of underlying conditions. This is especially true in the critically ill patient who, for whatever reasons, requires intensive care admission following laparotomy. Mild elevations in airway pressure, hypotension, an increase in central venous pressure (CVP) or pulmonary artery occlusion pressure (PAOP), and borderline urine output are all features of an elevated IAP but may equally be ascribed to cardiac insufficiency following severe abdominal sepsis or major trauma. This raises a number of important questions with regard to defining IAH and ACS. What is normal IAP? What pressure defines the change from normotension to hypertension? How common is IAH? At what point does IAH develop into ACS? If the foregoing can be answered then perhaps IAH and ACS can be defined and the point at which the physiological changes to the pathological be identified.

What Is Normal Intra-Abdominal Pressure?

IAP is highly variable in normal individuals[15-22] and depends on body mass index[2] and position.[1] Although readings as high as 80 mm Hg have been noted, the mean is around 6.5 mm Hg with a range from subatmospheric to 16 mm Hg. Abdominal surgery has a positive influence on IAP but this rarely exceeds 15 mm Hg in uncomplicated cases. Emergency surgery has a more marked effect on IAP but this pertains to the underlying pathology rather than to the act of surgery itself. Positive pressure ventilation will obviously increase IAP compared to spontaneous breathing.

What Pressure Defines Hypertension?

With reference to the vascular system, although the classification ranges from mild to malignant, hypertension of any degree mandates intervention to prevent progressive organ dysfunction. The more severe the hypertension, the more urgent is the need. If this principle is applied to the abdomen, the pressure at which organ dysfunction becomes apparent may be used to define IAH. Therein lies the problem. Depending on the tools used, objective evidence of organ dysfunction and perfusion abnormalities may be discovered before they are clinically apparent. When measuring the common haemodynamic parameters, an IAP of up to 10 mm Hg has no significant effect on cardiac output (CO), blood pressure, CVP, or PAOP.[23] With regard to the PAOP however, this must be viewed in the light of a rise in pleural pressure. The true transmural PAOP equals the end-expiratory PAOP minus pleural pressure and this falls at an IAP of 10 mm Hg.[3] As IAP approaches 15 mm Hg a significant rise in pleural pressure occurs, CO falls by 20%, as does venous return.[3,16,24] Paradoxically CVP and PAOP may be elevated as a result of increased intrathoracic pressure. At an IAP of 20 mm Hg CO drops significantly to 70% of baseline.

Haemodynamic parameters alone, however, do not reflect the extent of the problem. In a series of animal experiments, Diebel et al measured the effect of increasing IAP on blood flow to the intestine and liver.[25,26] During incremental rises in IAP, baseline mean arterial pressure was held constant by intravascular volume loading. The combined results of two studies are shown in Figure 1.

The most notable finding of these studies was that although a significant drop in CO was seen only at an IAP of 40 mm Hg, significant falls in blood supply to the liver and intestine were discovered at pressures of 10 and 20 mm Hg respectively. In addition, gut mucosal pH dropped significantly to 7.16 at 20 mm Hg and 6.89 at 40 mm Hg IAP. These levels represent an oxygen supply of 50% and 2% of normal respectively. From these data it is clear that before overt clinical signs of the effects of a rise in IAP become apparent, the intra-abdominal organs are becoming ischaemic. Furthermore, the animals had not been subjected to a prior insult and as mentioned earlier, this would compound the problem. In humans, an abnormal gastric pHi has a strong correlation with an IAP above 15 mm Hg, organ dysfunction, and a poor outcome.[27] In the majority of published studies on IAP authors have arbitrarily selected a variety of cut points to define IAH.[2,9,12,27-32] These range from 12 to 25 mm Hg. One study, however, used receiver operating characteristic (ROC) curves to define objectively the IAP at which detrimental effects occur.[33] Looking at postoperative complications following elective and emergency aortic aneurysm repair, ROC curves showed a level of 15 mm Hg IAP as the most sensitive and specific point to define IAH. In combination with the measurements of IAP in the normal population, these studies provide strong evidence that a level of 15 mm Hg should be used as the lower limit to define IAH. Although this may include patients without any detrimental effect, values higher than this would fail to capture those patients at risk of morbidity from IAH. This level is not absolutely exclusive and in a minority of patients a value of 12 mm Hg may determine IAH, especially in the presence of hypotension. It has been suggested that abdominal perfusion pressure defined as mean arterial pressure minus intra-abdominal pressure may be a more reliable index.[34] Values of less than 50 mm Hg indicate a high risk of detrimental effects of raised IAP.

How Common Is Intra-Abdominal Hypertension?

The problem in answering this question, as indicated earlier, is that studies have used different criteria to define IAH. Sugrue et al[28] in a study of 100 patients admitted to a single intensive care unit following laparotomy used a level of 20 mm Hg to define IAH. Of the 88 patients eligible for final analysis 29 (33%) had elevated pressures. If a lower level of 15 mm Hg had been used, 38 (43%) patients would have been defined as having excessive IAP. Of the 57 patients undergoing emergency surgery 23 (40%) had pressures greater than 20 mm Hg, whereas of the 31 patients who underwent elective surgery only 6 (19%) fulfilled the criterion. Although not achieving statistical significance this must be viewed as a clinically significant finding. In a series of 70 patients suffering life-threatening abdominal injury, 23 (33%) were regarded as having IAH using a level of 25 mm Hg as the definition.[12] This study may be criticised, however, for containing a specific group of patients at high risk of this complication and does not necessarily portray the true incidence. As suggested by the authors, however, the critical level of IAH that needs treatment has not been established. In a prospective study of 405 patients admitted to intensive care[31] the overall incidence of IAH was 17.5% using a cut point of 12 mm Hg. Of patients admitted following emergency surgery 39% fulfilled the criterion compared to only 6% of those undergoing elective surgery. The incidence in medical patients was 20%. Surprisingly, despite the use of three different levels of IAP to define IAH, the incidence is very similar especially in patients who have undergone emergency surgery.

At What Point Does IAH Develop into ACS?

Not all patients with IAH will require abdominal decompression and unnecessary laparotomies with bag closure confer considerable morbidity.[34,35] Conversely, failure to decompress the abdomen in the presence of ACS is uniformly fatal whereas timely intervention is associated with marked improvement in organ function and a 59% overall survival rate.[36] Within the last statement rests the theoretical distinction between IAH and ACS, namely that IAH in combination with organ dysfunction represents ACS (Fig. 2). In practice that pathological point is harder to elucidate, hence the indistinct margins defining IAH and the modest change in organ dysfunction. The shaded area illustrating IAH may undergo shifts to the right or left depending on the clinical scenario. As with prediction of outcome in the critically ill, the extremes are obvious, but with those in the middle range prediction of survival or death is difficult. Patients with an IAP of less than 15 mm Hg and organ dysfunction explicable by their underlying pathology are unlikely to benefit from abdominal decompression.

Those with organ dysfunction and an IAP above 20 mm Hg should undergo decompressive surgery.[37] But what of those between 15-20 mm Hg with mild elevations of airway pressure explicable by mild acute lung injury, a moderate rise in CVP, and borderline renal function? This defines the group with IAH but potential ACS. A judgement call between risk and benefit is required. On balance, from the available accumulating evidence, such patients should be decompressed and the diagnosis confirmed or refuted retrospectively. It is within this narrow

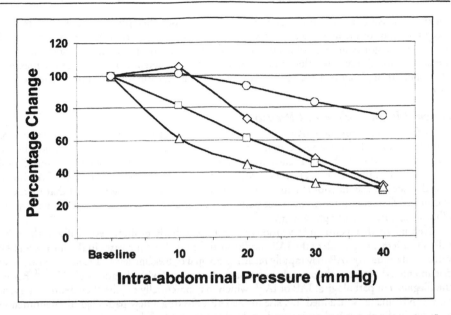

Figure 1. Effect of increasing intra-abdominal pressure on cardiac output (circles), hepatic artery flow (triangles), superior mesenteric artery flow (diamonds), and intestinal mucosa (squares).

no-man's land of IAH between normal IAP and ACS that our efforts should be concentrated in an attempt to clarify definitions and thereby treatment options. Reliance on standard haemodynamic parameters is too crude but measurement of splanchnic perfusion too difficult in the clinical scenario to be currently applicable.

Consensus Definitions

Ari Leppäniemi and R. Stephen Smith

Introduction

The abdominal cavity is surrounded partially by rigid structures such as the pelvis, spine and the costal arches, and partially by flexible tissues – the abdominal wall and the diaphragm. Theoretically, IAP values follow hydrostatic laws where the degree of the flexibility of the abdominal wall and the specific gravity of its contents would determine the pressure at a given point in a given position (prone, supine).[3,38] However, the movements of the diaphragm and the rib cage, resting tone and contractions of the abdominal wall musculature, obesity, and the variations of the content of the intestines (air, liquid, fecal mass) add degrees of physiological variability which limit the usefulness of a strict mathematical description of the IAP. In addition, the techniques used to measure the IAP,[1] discussed in another chapter, add a further degree of uncertainty, and therefore, any subsequently presented numerical values serve only as approximations, and should be considered as such.

One source of confusion in the literature regarding pathological IAP derives from the use of two different units of pressure measurement; mm Hg and cm H_2O. One mm Hg is equal to 1.36 cm H_2O, and conversely, one cm H_2O is equal to 0.74 mm Hg.[39] In this chapter, all pressure values are expressed in mm Hg with the corresponding cm H_2O value presented in brackets when appropriate.

There are inherent variations and fluctuations in the IAP. When comparing values from continuous measurements to measurements taken during short intervals, the question of the relevance of using one maximal value of IAP to guide our therapeutic strategies, instead of using the mean or median of a set of measurements arises. With the lack of a consensus, and because the majority of institutions use maximal IAP values from individual bladder pressure measurements, all pressure values subsequently referred to herein correspond to the maximal IAP values from "standardized" noncontinuous bladder measurements, unless stated otherwise.

Normal Intra-Abdominal Pressure

In the strictest sense, only IAP values ranging from sub-atmospheric to zero mm Hg can be considered normal.[20] However, certain physiological characteristics, such as morbid obesity, can be associated with chronic increased IAP to which the patient has adapted, and the clinical significance of mildly or moderately elevated values needs to be assessed in view of the initial "steady state" of the individual patient. For example, it has been demonstrated that increased abdominal diameter in morbidly obese patients is associated with elevated IAP in the absence of other significant pathophysiology.[22]

Even minor therapeutic interventions or changes in body position, especially in the critically ill patients, might affect the IAP condition and cause brief increases in the measured IAP values. Subsequently, IAP may rapidly return to normal or baseline levels. Previous studies have documented that recent abdominal operations are associated with elevations of IAP.[20,40] Before the diagnosis of pathological IAP or IAH, which may potentially require therapeutic intervention, can be made, a sustained increase in the IAP reflecting a new pathological phenomenon or entity in the abdominal cavity needs to be demonstrated.[9,41-43]

The intra-abdominal pressure (IAP) is the pressure concealed within the abdominal cavity, it varies with respiration and is normally below 10 mm Hg. It should always be measured at end-expiration in the complete supine position.

Pathological Intra-Abdominal Pressure

Obviously, pathological IAP is a continuum ranging from mild increases in the IAP without clinically significant adverse effects to a substantial elevation of the IAP with grave consequences to almost all organ systems in the body. Although the use of a single IAP parameter to define IAH could be questioned, it is important that a consensus on this point be reached in the future. An accepted benchmark for the identification of IAH will facilitate the accurate interpretation of data derived from different institutions and individual studies.

Currently, the definition of IAH in the literature varies most commonly between 12 and 25 mm Hg.[2,12,15,18,27,42,44-49] Some studies have shown deleterious effects on organ function after increases in IAP as low as 10 or 15 mm Hg, respectively.[3,31,33,38,50,51] A recent, and so far the only, multicenter study aimed at establishing the prevalence, etiology and predisposing factors associated with IAH in a mixed population of intensive care patients was conducted. IAH was defined as a maximal IAP value of 12 mm Hg or more in at least one measurement.[2]

Until a universally accepted consensus on the definition of IAH is established, and in order to exclude brief, temporary elevations of IAP with are not clinically significant, the authors of this chapter propose that IAH be defined as a peak IAP value of ≥ 12 mm Hg which is recorded by minimum of two standardized pressure measurements that are conducted 1-6 hours apart.

After establishing a minimum threshold for defining IAH, stratification of the pathological IAP values is needed to calibrate and quantify the "threat" of the insult to produce clinically significant manifestations. Ultimately such a stratification system could be used as an indication for various therapeutic interventions. In 1996, Burch et al[10] presented a four-level grading system upon which treatment could be based. Converting from the units of cm H_2O used in

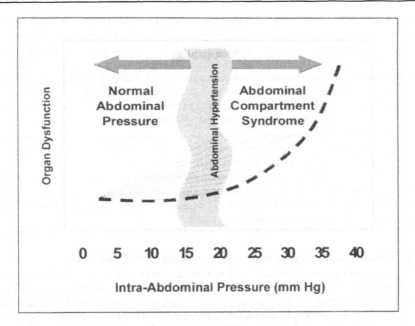

Figure 2. Distinctions between normal intra-abdominal pressure, IAH, and ACS.

the original scale to values in mm Hg, Grade I corresponds to a bladder pressure of 7.5-11 mm Hg (10-15 cm H_2O), Grade II to 11-18 mm Hg (15-25 cm H_2O), Grade III to 18-25 mm Hg (25-35 cm H_2O), and Grade IV > 25 mm Hg (> 35 cm H_2O). They recommended that treatment for Grade I and II IAH was mostly conservative, while treatment for Grade III and IV IAH involved operative intervention.

Definitions of Abdominal Compartment Syndrome

To separate IAH from the clinical Abdominal Compartment Syndrome (ACS), Ivatury et al[39] characterized ACS by the presence of a tensely distended abdomen, elevated intra-abdominal and peak airway pressures, inadequate ventilation with hypoxia and hypercarbia, impaired renal function, and a documented improvement of these features after abdominal decompression. ACS was seen as a late manifestation of uncontrolled IAH. Similar characteristics in different combinations and with additions of persistently low pH_i, labile blood pressure, diminished cardiac output, tachycardia with or without hypotension, or oliguria have subsequently been used by other authors.[52-54]

A more accurate definition of the ACS will enhance the comparison of clinical material from different centers and will be helpful in future clinical trials. To reach this milestone it will require a definition which combines a numerical value identified with increased IAP with the significant clinical consequences of the prolonged IAH, i.e., the development of disturbances in the different organ systems. Meldrum et al[44] defined ACS as IAP > 20 mm Hg complicated by one of the following: peak airway pressure > 40 cm H_2O, oxygen delivery index <600 mL O_2/min/m^2 or urine output < 0.5 mL/kg/hour.

Considering the multiple possible combinations of various degrees of organ dysfunction and failure, and the need to have a clinically useful definition of ACS to further guide therapeutic decisions, one of the existing organ failure grading systems that is currently and widely used to assess the status of critically ill patients in the intensive care environment, could be incorporated into the definition of ACS. In a recent study by Malbrain et al,[2] ACS was defined as IAP ≥ 20 mm Hg with failure of one or more organ systems. They defined organ failure as a Sequential Organ Failure Assessment (SOFA) organ subscore ≥ 3.[55]

The SOFA score includes the sum of six organ system scores (respiratory, cardiovascular, renal, coagulation, liver and neurologic) ranging from 0 (normal) to 4 (severe derangement) for each organ system. The SOFA score is calculated using the worst values of the day. By using the SOFA (or another similar) scoring system to define end-organ failure associated with IAH, one accepts the fact that a SOFA score of 3 for one organ system is not equivalent to a SOFA score of 3 of another organ system as far as outcome prediction is concerned. In addition to the calibration bias, the SOFA score does not account for organ systems which are not included in the score, of which the most important is the gastrointestinal system.

Until a consensus agreement on a definition of ACS is reached, we submit the following to be used in future clinical studies: ACS is defined as a peak IAP value of ≥ 20 mm Hg recorded during a minimum of two standardized measurements that are performed 1-6 hours apart with associated single or multiple organ system failure which was not previously present (as assessed by the daily SOFA or equivalent scoring system; organ failure is defined as a SOFA organ system score of ≥ 3).

Primary and Secondary Abdominal Compartment Syndrome

With the increasing recognition of ACS as a significant contributor to the development of multiple organ failure in critically ill patients, and the multitude of conditions associated with ACS, it is useful to categorize ACS according to the underlying pathology. In trauma patients, primary ACS has been defined as a recognized complication of damage control laparotomy, and secondary ACS as a condition reported in patients without abdominal injury who require aggressive fluid resuscitation.[48,56] In the intensive care environment, primary ACS has been considered as surgical (e.g., ruptured abdominal aortic aneurysm, abdominal trauma) and secondary ACS as medical (e.g., pneumonia with septic shock, toxin release, capillary leak and massive fluid overload).[3] Occasionally a combination of the two may occur, for example when a patient develops sepsis and capillary leakage with fluid overload after initial surgical stabilization for trauma.[57] This overlap of clinical conditions and potential etiologies has added to the confusion regarding the definitions. Additional difficulty arises when patients develop ACS after previous surgical treatment for IAH.[6,46,58,59]

Recognizing the importance of the presence or absence of preexisting intra-abdominal pathology and the crucial role of early abdominal surgery, and with the lack of a widely accepted definition of primary and secondary ACS, the authors submit the following definitions as a basis for further debate.

Primary (surgical) ACS is defined as a condition associated with injury or disease in the abdomino-pelvic region that requires early surgical or angioradiological intervention, or that develops following abdominal surgery (such as abdominal organ injuries that require surgical repair or damage control surgery, secondary peritonitis, bleeding pelvic fractures or other causes of massive retroperitoneal hematomas, liver transplantation). Patients that undergo an initial trial of nonoperative management for solid organ injuries who subsequently develop ACS are included in the Primary (surgical) category.

Secondary (medical) ACS refers to conditions that do not require early surgical or angioradiological intervention (such as sepsis and capillary leak, severe acute pancreatitis, major burns and other conditions requiring massive fluid resuscitation).

Tertiary ACS refers solely to the condition where ACS develops following prophylactic or therapeutic surgical treatment of primary or secondary ACS (e.g., persistence of ACS after decompressive laparotomy or the development of a new ACS episode following the definitive closure of the abdominal wall after the previous utilization of temporary abdominal wall closure).

Summary

Every definition of a clinical situation or syndrome fails to include all possible conditions and variations of an inherently complex phenomenon. Nevertheless, in order to approach scientific accuracy in comparing different clinical reports, and to plan for future clinical trials, definitions are required which are comprehensive, detailed, simple, practical and acceptable to the majority of the scientific community working in the particular field. This chapter does not, and cannot, provide bullet-proof definitions for all issues associated with increased IAP, but puts forward arguments and suggestions that may serve as a springboard for further consensus building endeavors.

Commentary

Manu L. N. G. Malbrain

Definitions of intra-abdominal hypertension (IAH) or abdominal compartment syndrome (ACS) stand or fall with the accuracy and reproducibility of the intra-abdominal pressure (IAP) method used.[1] Not only the absolute cut-off but also the use of mean, median or maximal IAP values will influence the prevalence and incidence of IAH and ACS.[2] Throughout the years different cut-offs have been suggested for IAH and ACS and some have interchanged the terms IAH and ACS. Others suggested terms as primary or secondary ACS, but with ever changing definitions. To date it is therefore very difficult to interpret the literature data. At the cradle of a new era and in response to a louder and louder cry for consensus definitions this chapter aims at providing some uniformisation so that the data and results from future studies can be easily compared.[3]

Most syndromes are preceded by a prodromal phase during which a number of nonspecific symptoms and signs appear. The ACS is no exception to this general rule, and IAH represents the prodromal phase of ACS. A clear distinction must be made between these two terms and attempts made to define them more clearly. From animal experiments and human observational studies it would appear that IAH is best defined as an IAP between 15 and 20 mm Hg in the absence of overt evidence of haemodynamic or organ dysfunction. Intra-abdominal pressures above 20 mm Hg should be defined as ACS. This pressure limit is associated almost invariably with signs of organ dysfunction, although these may be subtle.

A consensus on definitions of issues related to increased IAP is needed in order to approach scientific accuracy in comparing different clinical reports and to plan for future clinical trials. These definitions should be comprehensive, detailed, simple, practical and acceptable to the majority of the scientific community working in this particular field. Until such a consensus is achieved, this chapter seeks to put forward the following definitions to be used as a basis for further work.

Intra-abdominal hypertension (IAH) is defined as a peak IAP value of $\geq$ 12 mm Hg recorded, at a minimum, as two standardized measurements obtained 1-6 hours apart.

Abdominal Compartment Syndrome (ACS) is defined as severe IAH ($\geq$ 20 mm Hg) associated with the new onset of single or multiple organ failure (as assessed by the daily SOFA score or an equivalent scoring system, with organ failure defined as SOFA organ system score of ≥ 3).

Primary (surgical) ACS is defined as a condition associated with injury or disease in the abdomino-pelvic region that requires early surgical or angioradiological intervention, or that develops following abdominal surgery (such as abdominal organ injuries that require surgical repair or damage control surgery, secondary peritonitis, bleeding pelvic fractures or other causes of massive retroperitoneal hematomas, liver transplantation). Patients that undergo an initial trial of nonoperative management for solid organ injuries who subsequently develop ACS are included in the Primary (surgical) category.

Secondary (medical) ACS refers to conditions that do not require early surgical or angioradiological intervention (such as sepsis and capillary leak, severe acute pancreatitis, major burns and other conditions requiring massive fluid resuscitation). Tertiary ACS refers solely to the condition where ACS develops following prophylactic or therapeutic surgical treatment of primary or secondary ACS (e.g., persistence of ACS after decompressive laparotomy or the development of a new ACS episode following the definitive closure of the abdominal wall after the previous utilization of temporary abdominal wall closure).

References

1. Malbrain ML. Different techniques to measure intra-abdominal pressure (IAP): Time for a critical reappraisal. Intensive Care Med 2004; 30(3):357-371.
2. Malbrain ML, Chiumello D, Pelosi P et al. Prevalence of intra-abdominal hypertension in critically ill patients: A multicentre epidemiological study. Intensive Care Med 2004; 30:822-829.
3. Malbrain ML. Is it wise not to think about intraabdominal hypertension in the ICU? Curr Opin Crit Care 2004; 10(2):132-145.
4. Kelman GR, Swapp GH, Smith I et al. Caridac output and arterial blood-gas tension during laparoscopy. Br J Anaesth 1972; 44(11):1155-1162.
5. Simon RJ, Friedlander MH, Ivatury RR et al. Hemorrhage lowers the threshold for intra-abdominal hypertension-induced pulmonary dysfunction. J Trauma 1997; 42(3):398-403.
6. Maxwell RA, Fabian TC, Croce MA et al Secondary abdominal compartment syndrome: An underappreciated manifestation of severe hemorrhagic shock. J Trauma 1999; 47(6):995-999.
7. Burrows R, Edington J, Robbs JV. A wolf in wolf's clothing—the abdominal compartment syndrome. S Afr Med J 1995; 85(1):46-48.
8. Mayberry JC, Welker KJ, Goldman RK et al. Mechanism of acute ascites formation after trauma resuscitation. Arch Surg 2003; 138(7):773-776.
9. Balogh Z, McKinley BA, Cocanour CS et al. Supranormal trauma resuscitation causes more cases of abdominal compartment syndrome. Arch Surg 2003; 138(6):637-642.
10. Burch JM, Moore EE, Moore FA et al. The abdominal compartment syndrome. Surg Clin North Am 1996; 76(4):833-842.
11. Dickens C. Martin chuzzlewit. Martin Chuzzlewit 1982.
12. Ivatury RR, Porter JM, Simon RJ et al. Intra-abdominal hypertension after life-threatening penetrating abdominal trauma: Prophylaxis, incidence, and clinical relevance to gastric mucosal pH and abdominal compartment syndrome. J Trauma 1998; 44(6):1016-1021.
13. Ivatury RR, Sugerman HJ. Abdominal compartment syndrome: A century later, isn't it time to pay attention? Crit Care Med 2000; 28:2137-2138.
14. Schein M, Ivatury R. Intra-abdominal hypertension and the abdominal compartment syndrome. Br J Surg 1998; 85(8):1027-1028.
15. Hong JJ, Cohn SM, Perez JM et al. Prospective study of the incidence and outcome of intra-abdominal hypertension and the abdominal compartment syndrome. Br J Surg 2002; 89(5):591-596.
16. Schein M, Wittmann DH, Aprahamian CC et al. The abdominal compartment syndrome: The physiological and clinical consequences of elevated intra-abdominal pressure. J Am Coll Surg 1995; 180(6):745-753.
17. Sugrue M, Buist MD, Lee A et al. Intra-abdominal pressure measurement using a modified nasogastric tube: Description and validation of a new technique. Intensive Care Med 1994; 20(8):588-590.
18. Sugrue M, Bauman A, Jones F et al. Clinical examination is an inaccurate predictor of intraabdominal pressure. World J Surg 2002; 26(12):1428-1431.

19. Sugrue M. Intra-abdominal pressure: Time for clinical practice guidelines? Intensive Care Med 2002; 28(4):389-391.
20. Sanchez NC, Tenofsky PL, Dort JM et al. What is normal intra-abdominal pressure? Am Surg 2001; 67(3):243-248.
21. Kirkpatrick AW, Brenneman FD, McLean RF et al. Is clinical examination an accurate indicator of raised intra-abdominal pressure in critically injured patients? Can J Surg 2000; 43(3):207-211.
22. Sugerman H, Windsor A, Bessos M et al. Intra-abdominal pressure, sagittal abdominal diameter and obesity comorbidity. J Intern Med 1997; 241(1):71-79.
23. Ridings PC, Bloomfield GL, Blocher CR et al. Cardiopulmonary effects of raised intra-abdominal pressure before and after intravascular volume expansion. J Trauma 1995; 39(6):1071-1075.
24. Saggi BH, Sugerman HJ, Ivatury RR et al. Abdominal compartment syndrome. J Trauma 1998; 45(3):597-609.
25. Diebel LN, Wilson RF, Dulchavsky SA et al. Effect of increased intra-abdominal pressure on hepatic arterial, portal venous, and hepatic microcirculatory blood flow. J Trauma 1992; 33(2):279-282.
26. Diebel LN, Dulchavsky SA, Wilson RF. Effect of increased intra-abdominal pressure on mesenteric arterial and intestinal mucosal blood flow. J Trauma 1992; 33(1):45-48.
27. Sugrue M, Jones F, Lee A et al. Intraabdominal pressure and gastric intramucosal pH: Is there an association? World J Surg 1996; 20(8):988-991.
28. Sugrue M, Buist MD, Hourihan F et al. Prospective study of intra-abdominal hypertension and renal function after laparotomy. Br J Surg 1995; 82(2):235-238.
29. Sugrue M, Jones F, Janjua KJ et al. Temporary abdominal closure: A prospective evaluation of its effects on renal and respiratory physiology. J Trauma 1998; 45(5):914-921.
30. Sugrue M, Jones F, Deane SA et al. Intra-abdominal hypertension is an independent cause of postoperative renal impairment. Arch Surg 1999; 134(10):1082-1085.
31. Malbrain ML. Abdominal pressure in the critically ill: Measurement and clinical relevance. Intensive Care Med 1999; 25(12):1453-1458.
32. Biancofiore G, Bindi ML, Romanelli AM et al. Intra-abdominal pressure monitoring in liver transplant recipients: A prospective study. Intensive Care Med 2003; 29(1):30-36.
33. Papavassiliou V, Anderton M, Loftus IM et al. The physiological effects of elevated intra-abdominal pressure following aneurysm repair. Eur J Vasc Endovasc Surg 2003; 26(3):293-298.
34. Cheatham ML, White MW, Sagraves SG et al. Abdominal perfusion pressure: A superior parameter in the assessment of intra-abdominal hypertension. J Trauma 2000; 49(4):621-626.
35. Tiwari A, Haq AI, Myint F et al. Acute compartment syndromes. Br J Surg 2002; 89(4):397-412.
36. Saggi BH, Sugerman HJ, Ivatury RR et al. Abdominal compartment syndrome. J Trauma 1998; 45(3):597-609.
37. Ghimenton F, Thomson SR, Muckart DJ et al. Abdominal content containment: Practicalities and outcome. Br J Surg 2000; 87(1):106-109.
38. Malbrain ML. Intra-abdominal pressure in the intensive care unit: Clinical tool or toy? In: Vincent JL, ed. Yearbook of Intensive Care and Emergency Medicine. Berlin: Springer-Verlag, 2001:547-585.
39. Ivatury RR, Diebel L, Porter JM et al. Intra-abdominal hypertension and the abdominal compartment syndrome. Surg Clin North Am 1997; 77(4):783-800.
40. Kron IL, Harman PK, Nolan SP. The measurement of intra-abdominal pressure as a criterion for abdominal reexploration. Ann Surg 1984; 199(1):28-30.
41. McNelis J, Soffer S, Marini CP et al. Abdominal compartment syndrome in the surgical intensive care unit. Am Surg 2002; 68(1):18-23.
42. McNelis J, Marini CP, Jurkiewicz A et al. Predictive factors associated with the development of abdominal compartment syndrome in the surgical intensive care unit. Arch Surg 2002; 137(2):133-136.
43. Yang EY, Marder SR, Hastings G et al. The abdominal compartment syndrome complicating nonoperative management of major blunt liver injuries: Recognition and treatment using multimodality therapy. J Trauma 2002; 52(5):982-986.
44. Meldrum DR, Moore FA, Moore EE et al. Prospective characterization and selective management of the abdominal compartment syndrome. Am J Surg 1997; 174(6):667-672.
45. Ivy ME, Atweh NA, Palmer J et al. Intra-abdominal hypertension and abdominal compartment syndrome in burn patients. J Trauma 2000; 49(3):387-391.
46. Raeburn CD, Moore EE, Biffl WL et al. The abdominal compartment syndrome is a morbid complication of postinjury damage control surgery. Am J Surg 2001; 182(6):542-546.
47. Loftus IM, Thompson MM. The abdominal compartment syndrome following aortic surgery. Eur J Vasc Endovasc Surg 2003; 25(2):97-109.

48. Balogh Z, McKinley BA, Holcomb JB et al. Both primary and secondary abdominal compartment syndrome can be predicted early and are harbingers of multiple organ failure. J Trauma 2003; 54(5):848-859.
49. Offner PJ, de Souza AL, Moore EE et al. Avoidance of abdominal compartment syndrome in damage-control laparotomy after trauma. Arch Surg 2001; 136(6):676-681.
50. Cheatham M. Intra-abdominal hypertension and abdominal compartment syndrome. New Horiz 1999; 7:96-115.
51. Malbrain ML. Abdominal perfusion pressure as a prognostic marker in intra-abdominal hypertension. In: Vincent JL, ed. Yearbook of Intensive Care and Emergency Medicine. Berlin: Springer-Verlag, 2002:792-814.
52. Eddy V, Nunn C, Morris Jr JA. Abdominal compartment syndrome. The Nashville experience. Surg Clin North Am 1997; 77(4):801-812.
53. Demetriades D. Abdominal compartment syndrome. Trauma 2000; 2:277-281.
54. Stassen NA, Lukan JK, Dixon MS et al. Abdominal compartment syndrome. Scand J Surg 2002; 91(1):104-108.
55. Vincent JL, Moreno R, Takala J et al. The SOFA (Sepsis-related Organ Failure Assessment) score to describe organ dysfunction/failure. Intensive Care Med 1996; 22:707-710.
56. Balogh Z, McKinley BA, Cocanour CS et al. Secondary abdominal compartment syndrome is an elusive early complication of traumatic shock resuscitation. Am J Surg 2002; 184(6):538-543.
57. Biffl WL, Moore EE, Burch JM et al. Secondary abdominal compartment syndrome is a highly lethal event. Am J Surg 2001; 182(6):645-648.
58. Gracias VH, Braslow B, Johnson J et al. Abdominal compartment syndrome in the open abdomen. Arch Surg 2002; 137(11):1298-1300.
59. Ertel W, Oberholzer A, Platz A et al. Incidence and clinical pattern of the abdominal compartment syndrome after "damage-control" laparotomy in 311 patients with severe abdominal and/or pelvic trauma. Crit Care Med 2000; 28(6):1747-1753.

Intra-Abdominal Pressure Measurement Techniques

Manu L. N. G. Malbrain* and Felicity Jones

Abstract

The diagnosis of intra-abdominal hypertension (IAH) or abdominal compartment syndrome (ACS) is heavily dependant on the reproducibility of the intra-abdominal pressure (IAP) measurement technique. This chapter will discuss the gold standard direct IAP measurement, followed by the value of clinical estimation of IAP by abdominal girth or by examiner's feel of the tenseness of the abdomen. Afterwards it will give an overview of various indirect IAP estimates, as bladder, gastric or rectal pressure. Finally it will raise some questions with regard to the role of the intrabladder pressure (IBP) as the gold standard for indirect IAP. The outline of this chapter was based on a recently written extensive overview on this subject.[1]

Introduction

Although there is a rising interest on intra-abdominal hypertension (IAH) and abdominal compartment syndrome (ACS) in the literature, there is still controversy about the ideal method for measuring IAP.[1] Direct IAP measurement is not always possible and over the years the indirect IAP estimation via the bladder evolved as the gold standard, however, considerable variability in the measurement technique has been noted, not only between individuals but also between institutions.

The purpose of this chapter is (1) to identify patients at risk that might benefit IAP monitoring; (2) to review the most commonly used direct and indirect techniques for IAP measurement; (3) to provide the reader with a full description and important advantages and disadvantages of each technique; (4) to describe some new or revised techniques and (5) to highlight the cost-effectiveness of each method.

What Is IAP?

If you want to measure something you first need to define what you're about to measure. The IAP is the steady state pressure concealed within the abdominal cavity. The abdomen and its contents can be considered as relatively noncompressive and primarily fluid in character. The pressure values therefore follow the hydrostatic laws of Pascal and the IAP can be measured in nearly every part of it. The degree of flexibility of the walls and the specific gravity of its contents will determine the pressure at a given point and a given position (prone, supine...). In real life things are complicated by the movable diaphragm, the shifting costal arch, the contractions of the abdominal wall, and the intestines that may be empty or filled with air, liquid or

*Corresponding Author: Manu L. N. G. Malbrain—Intensive Care Unit, ZiekenhuisNetwerk Antwerpen, Campus Stuivenberg, Lange Beeldekensstraat 267, B-2060 Antwerp, Belgium. Email: manu.malbrain@skynet.be

Abdominal Compartment Syndrome, edited by Rao R. Ivatury, Michael L. Cheatham, Manu L. N. G. Malbrain and Michael Sugrue. ©2006 Landes Bioscience.

Table 1. Indications for IAP monitoring

1) Postoperative patients (abdominal surgery)
2) Patients with open or blunt abdominal trauma
3) Mechanically ventilated ICU patients with other organ dysfunction as assessed by daily Sequential Organ Failure Assessment (SOFA) score.[15]
4) Patients with a distended abdomen and signs and symptoms consistent with abdominal compartment syndrome
    a. Oliguria
    b. Hypoxia
    c. Hypotension
    d. Unexplained acidosis
    e. Mesenteric ischemia
    f. Elevated ICP
5) Patients with abdominal packing after temporary abdominal closure for multiple trauma or liver transplantation
6) Patients with open abdomens, especially if they have an IV bag closure and are in the early postoperative period, may still develop abdominal compartment syndrome.
7) Patients who have not had an operation but have received large volumes of fluid resuscitation in the context of an underlying capillary leak problem (pancreatitis, septic shock, trauma, etc.)

fecal mass.[2] The IAP shifts with respiration as evidenced by an inspiratory increase (diaphragmatic contraction) and an expiratory decrease (relaxation). The invasive gold standard for IAP monitoring is via direct needle puncture; the noninvasive standard is an indirect method via the bladder.

Indications for IAP Monitoring

In a recent multicentre prevalence study in mixed ICU patients, univariate analysis identified body mass index (BMI), fluid resuscitation, polytransfusion, total sequential organ failure assessment (SOFA) score and SOFA respiratory, renal and coagulation subscores to be significantly associated with IAH.[3] The Odd's ratio for IAH was 3.3 (95%CI 1.2-9.2) for fluid resuscitation and 7.3 (95%CI 0.9-60.3) for polytransfusion. The only risk factor independently associated with IAH (on multivariate analysis) was the BMI, while massive fluid resuscitation (probably associated with a positive net fluid balance), renal and coagulation impairment were only at limit of significance. This is not surprising since BMI affects baseline IAP.[4-6]

In a study looking for predictive factors for ACS in a matched cohort of 22 ACS patients it was found that only 24 hour net fluid balance and peak airway pressure were independent predictors for the development of ACS.[7] This is a bit confusing since peak airway pressure was part of the initial ACS definitions and is merely a consequence of ACS rather than a cause.[8] Others also found a strong correlation between IAH and fluid balance or blood transfusions.[9-14]

However, the main message from the recent data suggests that neither a single factor nor a group of factors can predict with sufficient accuracy which patients are likely to develop IAH.[3] Keeping this in mind we would like to give the reader some suggestions for conditions where IAP monitoring could be of clinical benefit (Table 1) or etiologic factors and predisposing conditions where the clinician should have a high clinical index of suspicion for IAH (Tables 2 and 3).

An algorithm with regard to IAP monitoring and IAH treatment can be found in a recent review on the subject.[17]

Clinical Evaluation

In analogy with the paradigm "if you don't take a temperature you can't find a fever" (in Samuel Shem *The House of God*, Dell Publishing, ISBN: 0-440-13368-8) one can state that "if

Table 2. Etiologic factors for IAH

1) abdominal surgery
   a. laparoscopy,
   b. reduction of hernia, tight closure
   c. abdominal banding with postoperative Velcro belt to prevent incisional hernia)
2) massive fluid resuscitation defined as more than 5 litres of colloids or crystalloids in a 24 hour period
3) ileus whether paralytic, mechanical or pseudo-obstructive defined as abdominal distension or absence of bowel sounds or failure of enteral feeding; evidenced by gastric dilatation or massive gastroparesis with a gastric residual of more than 1000 mL in a 24 hour period
4) abdominal infection (pancreatitis, peritonitis, abscess,...)
5) pneumoperitoneum
6) haemoperitoneum either caused by an intra- or retroperitoneal bleeding

you don't measure IAP you cannot make a diagnosis of IAH or ACS". This is illustrated in Figures 1-6.

Besides these clinical clues pointing towards IAH or ACS not much additional information is given by the clinical examination and to have an idea of the exact magnitude of the IAP we need to measure it!

Abdominal Girth

A recent study looking at 132 paired IAP and abdominal perimeter measurements in mixed ICU patients found no correlation ($R^2 = 0.12$) (Figs. 7, 8).[18]

Clinically significant IAH may be present in the absence of abdominal distension as can be seen in case of an acute increase in IAP without sufficient time for abdominal wall compensation (e.g., localised rectus sheath haematoma).[19] Vice versa chronic abdominal distension with sufficient time for adaptation as seen with pregnancy, obesity, cirrhosis, or ovarian tumours is an example of increased abdominal perimeter that is not necessarily accompanied by an increase in IAP.

Table 3. Associated conditions

1) acidosis defined as an arterial pH below 7.2
2) hypothermia defined as a core temperature below 33°C
3) polytransfusion defined as the transfusion of more than 10 Units of packed red cells in a 24 hour period
4) coagulopathy defined as a platelet count below 55000/mm³ or an activated partial thromboplastin time (APTT) more than 2 times normal or a prothrombin time (PTT) below 50% or an international standardised ratio (INR) more than 1.5.
5) sepsis defined according to the American-European Consensus Conference definitions[16]
6) bacteraemia defined as the presence of bacteria in the bloodstream determined by blood cultures
7) liver dysfunction defined as decompensated or compensated cirrhosis or other liver failure with ascites (paraneoplastic, cardiac failure, portal vein thrombosis, ischemic hepatitis)
8) mechanical ventilation
9) use of positive end expiratory pressure (PEEP) or the presence of auto-PEEP
10) pneumonia as defined according to standard criteria

The combination of acidosis, hypothermia and coagulopathy is in the literature often referred to the deadly triad inevitableleading to abdominal compartment syndrome

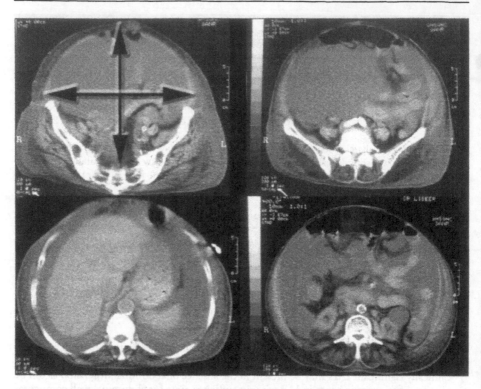

Figure 1. If the transversal and sagittal diameter of the abdomen are equal on an abdominal CT examination, the chance that IAP will be increased is substantial.

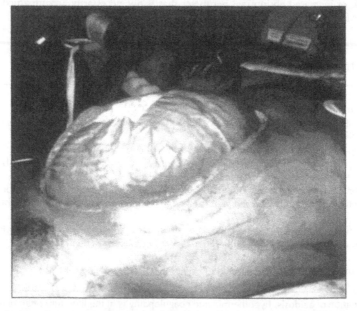

Figure 2. If the intestines protrude from the abdomen there is a great chance that IAP will be increased.

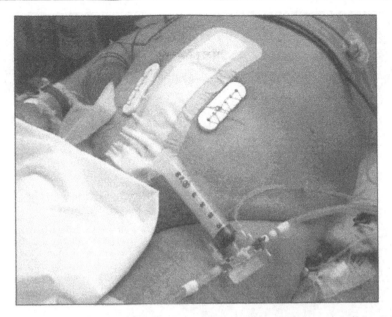

Figure 3. In case of tight abdominal closure with an iron wire or the so-called "ventre-au-fils" in Belgium, there is a great chance that IAP will be increased.

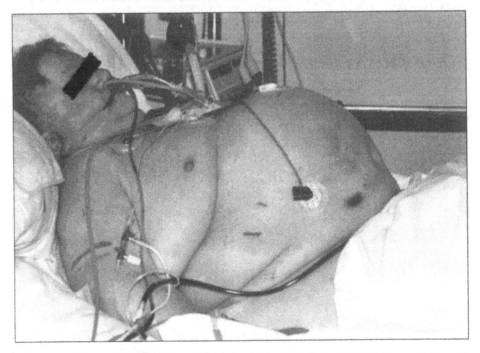

Figure 4. Any abdominal problem, like a faecal peritonitis in this patient with high body mass index, will further increase baseline IAP. We know that morbid obesity is associated with baseline IAH, therefore the risk for ACS is substantially higher in these patients.

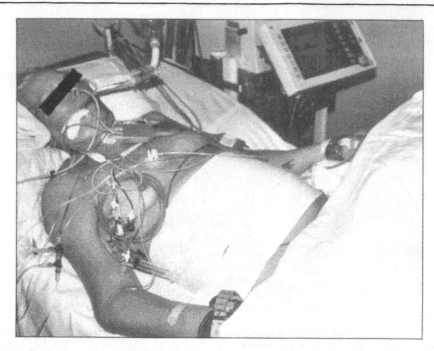

Figure 5. The use of an abdominal Velcro belt to prevent incisional hernias substantially increases the risk for IAH and ACS.

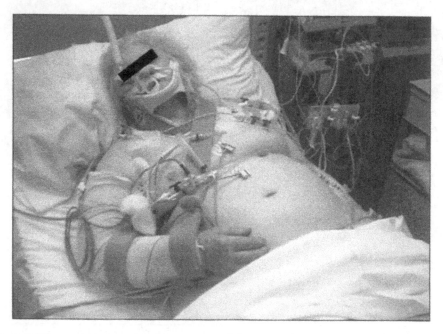

Figure 6. The most important group of patients at risk for IAH in the ICU are those with massive fluid resuscitation and polytransfusion as was the case with this trauma patient. We have all seen the so-called "Michelin" mannequins in our ICU.

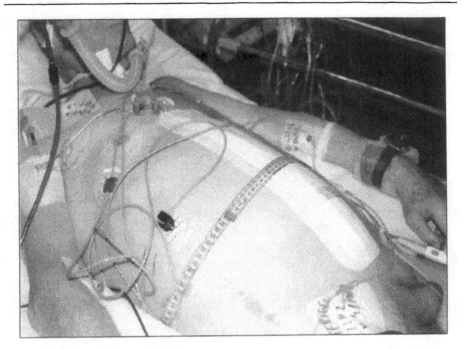

Figure 7. The use of the abdominal perimeter is not a good tool in assessing IAH or ACS.

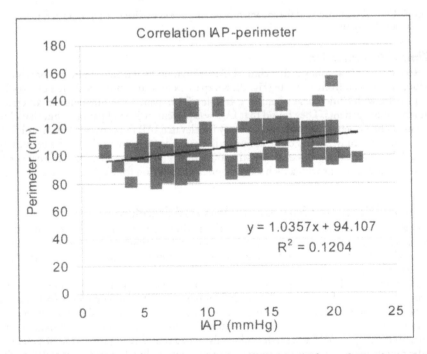

Figure 8. Absence of correlation between abdominal girth and IAP. Adapted from ref. 18 with permission, Van Mieghem N, Verbrugghe W, Daelemans R et al. Can abdominal perimeter be used as an accurate estimation of intra-abdominal pressure? Crit Care 2003; 7[Suppl 2]:P183.

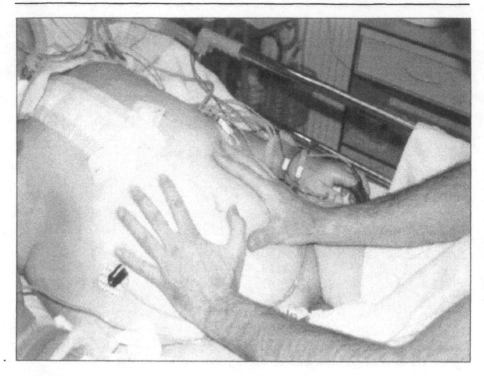

Figure 9. Clinical examination of the abdomen by putting one or two hands on it is far from accurate with a sensitivity of only 40%.

Clinical Examination

Recent studies have shown that clinical IAP estimation by putting one or two hands on the abdomen is also far from accurate (Fig. 9). Kirkpatrick and coauthors compared urinary bladder pressure with clinical assessment in a sample of 147 paired measurements in 42 critically injured trauma patients. They found that IAH was present in 50% of patients, but that clinical assessment had a poor sensitivity and accuracy for elevated IAP, respectively of 40% and 77%.[19]

In a similar study Sugrue came about to the same conclusions and found a sensitivity of 60%.[20] Finally, Castillo also concluded that 60% of clinically estimated IAPs are inaccurate.[21] Therefore in conditions with a high clinical index of suspicion for IAH, one needs to measure IAP.[18-22]

Invasive Direct IAP Measurement (Gold Standard)

Direct measurement by cannulation of the peritoneal cavity with a metal cannula or a wide-bore needle and attached to a saline-manometer or pressure transducer has been used historically and experimentally but has no advantages over the more accessible and simple indirect techniques.

The limitation of inflation pressure during laparoscopic surgery is an example of direct IAP measurement. Although in many animal and human studies laparoscopic insufflation pressure has been used as the so-called invasive gold standard to validate other indirect less invasive IAP measurement techniques, it is difficult to extrapolate these single observer comparisons in patients undergoing general anesthesia and paralysis to a mixed ICU population of patients not under muscle relaxation as well as subject to other confounding factors (nurse shifts, position, zero reference, ...). Direct IAP measurement via a laparoscopic insufflator is prone to errors by flow dynamics resulting in rapid increases in pressure during insufflation.[1,8] The Verres needle opening can be blocked by tissue or fluid leading to over- or underestimation of IAP and pressures can be influenced by muscle relaxation. Laparoscopy remains an artificial environment, this makes it even more difficult to validate indirect IAP measurement methods.

Finally recent data suggests to measure IAP directly during chronic ambulatory peritoneal dialysis (CAPD) via the PD or Tenckhoff catheter. High IAP has recently been identified as a risk factor for abdominal wall complications in patients on CAPD.[23] Besides IAP, advanced age, polycystic kidney disease and high BMI were also independent risk factors for these complications. The authors suggest that automated CAPD with low daytime fill volumes and pressures (below 14 mm Hg) should be considered in all patients at risk for hernias and/or leaks.[23] Others also noted a strong correlation between BMI and IAP in children on PD.[24,25] This correlation gives a better understanding of the interindividual variability of the above described unique relationship between IAP and intraperitoneal volume.

Noninvasive Indirect IAP Measurement (Gold Standard)

Before discussing the different indirect methods to measure IAP we will briefly describe a standard method for intravesicular pressure monitoring.

Equipment Required

Cardiac monitor with invasive pressure recording capabilities; 500 mL pressurized bag of isotonic sodium chloride solution, or D5W; Foley catheter; one standard pressure transducer setup with two three-way stopcocks; and 1 transducer connection cable; a 60 mL Luer lock syringe; a Kocher, or Kelly clamp; a 18-gauge needle or better an intravenous access catheter disinfection fluid.

Set-Up

Wash hands and follow universal antiseptic precautions connect the pressurized IV bag to the standard pressure transducer and flush the system. Place the two three-way stopcocks at the distal end of the pressure transducer. Connect the 60 mL syringe to one end of the distal stopcock. Connect the 18-gauge needle or intravenous access catheter to the end of the pressure tubing. Connect the transducer to the monitor via the special pressure module and ensure a normal waveform on the scope. Select a scale of 20 to 40 mm Hg.

Method of Measurement

If the patient is awake, explain the procedure. If the patient is sedated, ensure good sedation. Place the patient in a complete supine position.

Zero the pressure module at the symphysis pubis of the patient (mark for future reference) by turning the proximal stopcock on to the air and the transducer.

Clamp the Foley catheter distal to the sampling port.

Disinfect the culture aspiration port and insert aseptically the 18-gauge needle or intravenous access catheter already connected to the pressure tubing and flushed.

Draw 50 mL of IV solution into the syringe by turning the distal stopcock off to the patient and open to the syringe and pressurized bag. Turn the stopcock off to the pressurized bag and open to the syringe and patient and inject the 50 mL of IV solution into the bladder.

To verify correct measurement gentle compression of the abdomen should give instant variations on the IAP reading in form of oscillations (Fig. 10), if a damped signal is noted then purge any air seen between the clamp and the Foley catheter by releasing the clamp and allowing the IV solution to flow back past the clamp, then reclamp. In case of persistent damped signal perform a rapid flush test.

Turn the two three-way stopcocks to transmit the pressure from the bladder via the Foley catheter to the pressure transducer.

Allow the pressure to equilibrate and record the mean IAP at end-expiration on the scope of the monitor (Fig. 11).

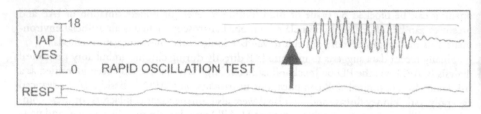

Figure 10. Rapid oscillation test: Confirmation of correct IAP measurement can be done by inspection of respiratory variations and by gently applying oscillations to the abdomen that should be immediately transmitted and seen on the monitor with a quick return to baseline. Reprinted with permission from ref. 1: Malbrain ML. Different techniques to measure intra-abdominal pressure (IAP): Time for a critical reappraisal. Intensive Care Med 2004; 30(3):357-371.

Once a measurement has been obtained remove the 18-gauge needle. If an intravenous access catheter was used, it should be left in place. Unclamp the drainage system between measurements.

Substract 50 mL of instilled IV solution from the patient's next hourly urine output.

Interpreting the Results

The IAP is expressed in mm Hg (for cm H_2O a conversion factor of 1.34 should be taken into account). Normal IAP is around 0 to 5 mm Hg. Sanchez found a mean IAP of 6.5 mm Hg in a randomly selected sample of 77 hospitalised patients.[4] Pressures up to 15 mm Hg may be normal after abdominal surgery; however pressures of 10-15 mm Hg may indicate early IAH. Not necessarily the absolute value but the trend of IAP over time together with the presence of organ dysfunction should alert the clinician to prevent ACS.

Different Indirect IAP Measuring Methods

This section will give an overview of the most common described and widely used indirect IAP monitoring methods. For each method a short description will be given with the technique properties (first publication, author, reference, fluid or air-filled, manipulation, difficulty, costs and time-consumption for initial set-up and next measurement, ...). Afterwards the properties (advantages and disadvantages) of the measurement itself will be listed (possibility for repeated measurements, continuous trend, automated, auto- or manual recalibration, standardisation, accuracy and reproducibility, validation with direct gold standard or other indirect methods, problems related to hydrostatic fluid columns, air bubbles, zero-reference, over- or underdamping, effect of body position, interference with urine output,...), as well as specific problems related to manometry (multiple menisci, filter blocking, standardisation). Finally the major risks (needle stick injury, urinary tract infection, sepsis) and contraindications (bladder, gastric or abdominal trauma) will be shortly addressed.

Before going any further it is important that the reader is familiar with some problems related to the use of a hydrostatic fluid column. Even with the consensus of using the symphysis

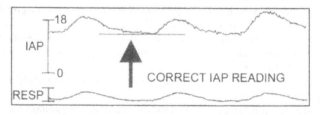

Figure 11. Intra-abdominal pressure waveform. Correct IAP reading at end-expiration.

pubis as zero reference, problems can arise when the same pressure transducer is used for IAP and CVP monitoring, with traditional zero-reference at the midaxillary line. Putting the patient upright with concomitant rise in the transducer may lead to underestimation of IAP, while putting the patient in the Trendelenburg position can lead to overestimation.

The fact that recalibration needs to be done before every measurement augments the risk for errors. We have all seen the "magic" drop or rise in CVP at changes of nurse shifts when pressure transducers are recalibrated, the same can happen with IAP. Variations in IAP from -6 to + 30 mm Hg have been noted previously.[26] Therefore it is important to perform IAP measurements in the complete supine position and to note the IAP value at end-expiration.

Should the patient be unable to remain supine, the position at which the first measurement is taken should be noted and subsequent measurements should be taken with the patient in the same position. Although the absolute value of the IAP readings may be inaccurate, subsequent measurements are reproducible and the trend in IAP can give valuable information. Furthermore, a fluid-filled system can produce artefacts that can further distort the IAP pressure waveform. Failure to recognise these recording system artefacts can lead to interpretation errors.[27] It can oscillate spontaneously, and these oscillations can distort the IAP pressure curve. The performance of a resonant system is defined by the resonant frequency (this is the inherent oscillatory frequency) and the damping factor (this is a measure of the tendency of the system to attenuate the pressure signal). Therefore, any fluid-filled system is prone to changes in body-position and over- or underdamping due to the presence of air-bubbles, a tubing that is too compliant or too long etc. A rapid flush test should therefore always be performed before IAP reading in order to obtain an idea of the dynamic response properties and to minimise these distortions and artefacts.[28] Off course these distortions are mostly seen in high frequency observations e.g., arterial pressure waveforms ranging from 50 to 150 bpm with systolic and diastolic excursions ranging from 40 to 200 mm Hg. Looking at an IAP tracing these high frequency variations are mostly absent, so that the dynamic response properties may play only a little role in these more static conditions. The only systolic to diastolic excursions seen in IAP tracings are those caused by respiration, however when IAP is measured in proximity to the heart (e.g., when measuring via the stomach) or the great vessels high frequency tracings can interfere with the IAP tracing. In some conditions IAP tracings can resemble dynamic tracings as seen with arterial pressure: just imagine a patient with secondary ARDS and ACS ventilated with a high respiratory rate of 30, low tidal volumes and high PEEP. Due to the diminished chest and abdominal wall compliance it is not impossible to observe excursions in IAP from 15 to 45 mm Hg mainly due to diminished chest wall compliance and the greater transmission from the intrathoracic pressure to the abdomen. An IAP going from 15 to 45 mm Hg at a rate of 30 might resemble an arterial tracing (of an extremely sick patient), hence exhibiting the same dynamic response properties. Confirmation of correct measurement can be done by inspection of respiratory variations and by gently applying oscillations to the abdomen that should be immediately transmitted and seen on the monitor with a quick return to baseline. In case of a damped signal the flush test should be repeated.

Bladder

The Original Open System Single Measurement Technique (Kron)[29,30]

Description

Traditionally the bladder has been used as the method of choice for measuring IAP and evolved as the so-called gold standard indirect method. The technique was originally described by Kron and disrupts for each IAP measurement what is normally a closed sterile system:[29,30] measuring IAP involves disconnecting the patient's Foley catheter and instilling 50 to 100 mL of saline via a 60 mL (non Luer-lock) syringe directly connected to the Foley catheter using a sterile field. After reconnection the urinary drainage bag is clamped distal to the culture aspiration port (Fig. 12).

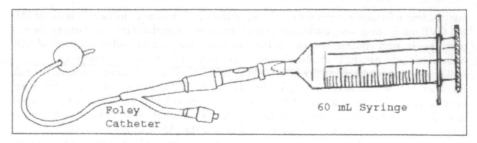

Figure 12. Kron technique: 60 mL syringe connected to Foley.

After reconnection the urinary drainage bag is clamped distal to the culture aspiration port. For each individual IAP measurement a 16 gauche needle is then used to Y-connect a manometer or pressure transducer using the symphysis pubis as reference line (Fig. 13).

Advantages and Disadvantages

This technique implicates a lot of time-consuming manipulations that disrupts a closed sterile system at each measurement. It has all the problems that come along with the hydrostatic convective fluid column.

Other disadvantages are: it is an intermittent technique that interferes with urine output without the possibility to obtain a continuous trend, it places the patient at increased risk of urinary tract infection or sepsis and subjects healthcare providers to the risk of needle stick injuries and exposure to blood and body fluids. In conclusion the Kron technique has at the present time no clinical implications.

The Closed System Single Measurement Technique (Iberti)[31,32]

Description

Iberti and colleagues reported the use of a closed system drain and transurethral bladder pressure monitoring method (Fig. 14).[31,32]

Using a sterile technique they infused an average of 250 mL of normal saline through the urinary catheter to purge catheter tubing and bladder. The bladder catheter is clamped and a 20 gauche needle is inserted through the culture aspiration port for each IAP measurement.

Figure 13. Using a 16 gauche needle the bladder is Y-connected to a pressure transducer

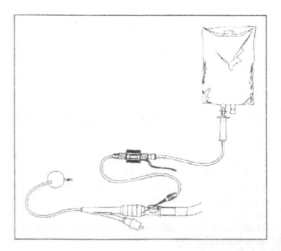

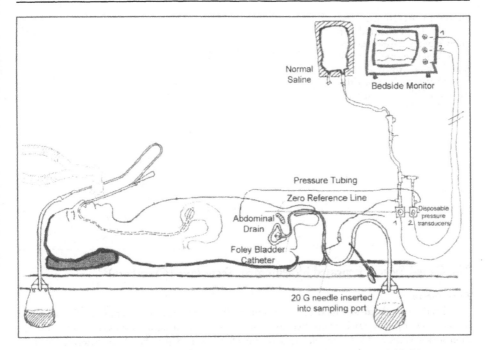

Figure 14. The original Iberti technique, comparing direct intraperitoneal pressures with bladder pressures.

The transducer is zeroed at the symphysis and mean IAP is read after a 2-min equilibration period (Fig. 15).

Advantages and Disadvantages

It has the same disadvantages related to the hydrostatic fluid column as the Kron technique, and since it is not needle free it also subjects health care workers to needle stick injuries.[31,32]

The advantage compared to the Kron technique is that it is simpler, less time-consuming, and with less manipulations. In conclusion, the Iberti technique has at the present time limited clinical implications (e.g., screening for IAH).

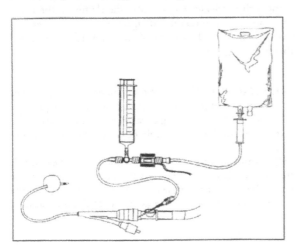

Figure 15. Practical implementation at the bedside of the Iberti technique.

Figure 16. Revision of the original Kron method for intravesicular pressure measurement. Reprinted with permission from the American College of Surgeons (Journal of the American College of Surgeons, 1998; 186:594-595).

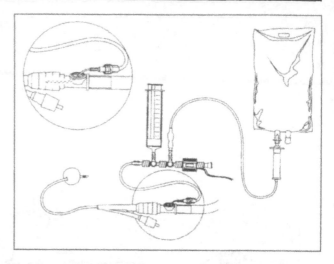

The Closed System Repeated Measurement Technique (Cheatham)[33]

Description

Cheatham and Safcsak reported a revision of the Kron's original technique.[33] A standard intravenous infusion set is connected to 1000 mL of normal saline, two stopcocks, a 60 mL Luer lock syringe and a disposable pressure transducer. An 18-gauche plastic intravenous infusion catheter is inserted into the culture aspiration port of the Foley catheter and the needle is removed.

The infusion catheter is attached to the first stopcock via arterial pressure tubing after being flushed with saline and "zeroed" at the level of the symphysis pubis (or the midaxillary line when the patient is in complete supine position), the Foley catheter is clamped immediately distal to the culture aspiration port. The stopcocks are turned "off" to the patient and pressure transducer and 50 mL of saline is aspirated from the intravenous bag. The first stopcocks is turned "on" to the patient and the 50 mL of saline are instilled into the bladder. The stopcocks are turned "off" to the syringe and the intravenous tubing. After equilibration the patient's IAP is then measured at end-expiration on the bedside monitor. To verify correct measurement gentle compression of the abdomen should give instant variations on the IAP reading in form of oscillations, if a damped signal is noted then release momentarily the clamp on the Foley catheter in order to ensure that all air is flushed and measure IAP again. After correct reading the clamp is removed, the bladder allowed to drain, and the volume of saline utilised is substracted from the patient's urine output for that hour (Fig. 16).

Advantages and Disadvantages

It has the same inconveniencies related to any fluid-filled system as described with the Kron and Iberti technique. It can pose problems after a couple of days because the culture aspiration port membrane can become leaky or the catheter kinky, leading to false IAP measurement. The fact that the infusion catheter needs to be replaced after a couple of days could increase the infection risk and needle-stick injuries.

This technique has minimal side effects and complications, e.g., without an increased risk for urinary tract infection.[34] It is safer and less invasive, takes less than one minute, is more efficient with repeated measurements possible and thus is more cost effective.[33] This technique is ideal for screening and monitoring for a short period of time (a couple of days) because of leakage.

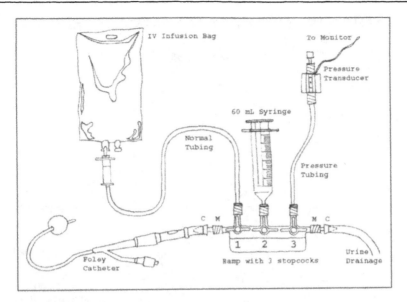

Figure 17. A closed needle-free method for repeated IAP measurement. Reprinted with permission from ref. 1: Malbrain ML. Different techniques to measure intra-abdominal pressure (IAP): Time for a critical reappraisal. Intensive Care Med 2004; 30(3):357-371.

The Revised Closed System Repeated Measurement Technique (Malbrain and Sugrue)[1,35]

Description

The technique by Cheatham was modified: A Foley catheter is sterile placed and the urinary drainage system connected. Using a sterile field and gloves, the drainage tubing is cut (with sterile scissors) 40 cm after the culture aspiration port after disinfection. A ramp with 3 stopcocks (Manifold set, Pvb Medizintechnik Gmbh, a SIMS Trademark, 85614 Kirchseeon, Germany, REF: 888-103-MA-11; or any other manifold set or even 3 stopcocks connected together will do the job) is connected to a conical connection piece (Conical Connector with female or male lock fitting, B Braun, Melsungen, Germany, REF: 4896629 or 4438450) at each side with a male/male adaptor (Male to Male connector piece, Vygon, Ecouen, France, REF: 893.00 or 874.10). The ramp is then inserted in the drainage tubing.[1] A standard intravenous (IV) infusion set is connected to a bag of 1000 mL of normal saline and attached to the first stopcock. A 60 mL syringe is connected to the second stopcock and the third stopcock is connected to a pressure transducer via rigid pressure tubing. The system is flushed with normal saline and the pressure transducer is zeroed at the symphysis pubis (or the midaxillary line when the patient is in complete supine position). The pressure transducer is fixed at the symphysis or the thigh. At rest the 3 stopcocks are turned "off" to the IV bag, the syringe and transducer giving an open way for urine to flow into the urometer or drainage bag, said otherwise the 3 stopcocks are turned "on" to the patient (Fig. 17).

To measure IAP, the urinary drainage tubing is clamped distal to the ramp-device and the third stopcock is turned "on" to the transducer and the patient and "off" to the drainage system. The third stopcock also acting as a clamp. The first stopcock is turned "off" to the patient and "on" to the IV infusion bag, the second stopcock is turned "on" to the IV bag and the 60 mL syringe. Hence 50 mL of normal saline can be aspirated from the IV bag into the syringe. The first stopcock is turned "on" to the patient and "off" to the IV bag and the 50 mL of normal saline is instilled in the bladder through the urinary catheter. The first and second

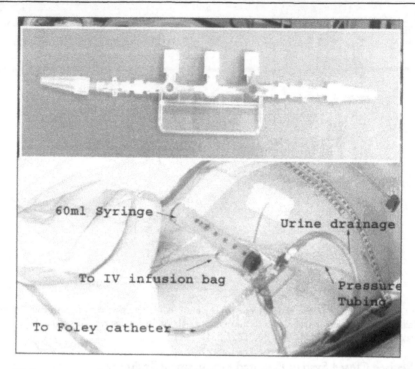

Figure 18. Patient mounted view of the device and close up of the manifold and conical connection pieces. Reprinted with permission from ref. 1: Malbrain ML. Different techniques to measure intra-abdominal pressure (IAP): Time for a critical reappraisal. Intensive Care Med 2004; 30(3):357-371.

stopcock are then turned "on" to the patient, and thus turned "off" to IV tubing and the syringe. The third stopcock already being turned "on" to the transducer and patient allows then immediate IAP reading on the monitor (Fig. 18).

In case of a damped signal the flush test should be repeated. After correct reading the third stopcock is turned "on" to the patient and "off" to the transducer, the clamp is removed, the bladder allowed to drain, and the volume of saline utilised is subtracted from the patient's urine output for that hour.

Sugrue described a similar modified closed system for repeated measurements via the bladder using a T-piece bladder pressure device.[35] Firstly, it is important to observe universal precautions and aseptic techniques at all times. The patient requires a urinary catheter. After removing the giving set from the transducer, prime the transducer with sterile saline solution, ensuring all air bubbles are removed. Connect to the side port on the bladder T-piece and turn the white tap "off".

Clean the urinary catheter and drainage tubing connection with alcohol. Insert the bladder T-piece pressure device between the catheter and drainage tubing. The drainage tubing needs to be clamped. This can be achieved by bending it over and inserting into the barrel of a 10 mL syringe.

50 mLs of sterile saline solution is then injected into the bladder. Connect the pressure monitoring cable to the transducer. Select the appropriate label on the monitor. The patient should be supine for the IAP measurements. The transducer is placed in line with the iliac crest, at the mid axillary line. This position should be marked with a reference line in order to reduce intra- and inter-observer variability. Next the pressure transducer needs to be zeroed. Turn the tap "off" to the patient and zero the transducer. "Open" the tap to the patient and

monitoring system in order to get the reading. When the reading has been recorded unclamp the 10 mL syringe from the drainage tubing and remember to deduct 50 mLs from the next hour's urine output measurement. We will now show a step by step explanation for this technique illustrated with photographs (Fig. 19, steps 1-8).

Advantages and Disadvantages

It has the same inconveniencies related to a fluid-filled system as described with the Kron, Iberti or Cheatham technique. This technique has the same advantages as the Cheatham technique with a required nursing time less than 2 minutes per measurement, a minimized risk of urinary tract infection and sepsis since it is a closed sterile system, the possibility of repeated measurements and reduced cost. Since it is a needle-free system it does not interfere with the culture aspiration port and the risk of injuries is absent. This technique can be used for screening or for monitoring for a longer period of time (2 to 3 weeks).

The Revised Closed System Continuous Measurement Technique (Malbrain and Sugrue)[1]

In an anuric patient continuous IAP recordings are possible via the bladder using a closed system connected to the Foley catheter after the culture aspiration port or directly to the Foley catheter using a conical connection piece connected to a standard pressure transducer via pressure tubing (Fig. 20).

After initial "calibration" of the system with 50 mL of saline and zeroing at the sypmhysis pubis the transducer is taped at the symphysis or thigh and a continuous IAP reading can be obtained. Daily calibration can be done in oliguric patients after voiding of rest diuresis.

Description

Size 18 F standard three-way catheters is inserted into the patient. The continuous IAP measurement was performed via the irrigation port of the three-way catheter, in which continuous sterile normal irrigation was maintained at 2 mL/h and connected through a two-way stopcock and normal saline filled tubing to a pressure transducer placed in-line with the iliac crest at the mid axillary line. The transducer is zeroed and the continuous IAP measurement is recorded on the bedside monitor.

Advantages and Disadvantages

This technique does not require any major change in the present practice apart from the use of three-way urinary catheters. The monitoring is continuous and does not interfere with the urinary flow through the drainage port of the catheter. This also reduces the potential risk for urinary tract infection. The continuous measurement of IAP also makes it possible to monitor the abdominal perfusion pressure (APP). This method has the advantage in that it includes the patient's hemodynamics to a certain extent. This method is also time saving for nursing and medical staff. One of the disadvantages of this method is the price difference between the two-way and three-way Foley catheters.

Intermittent IAP Measurement with the AbViser™ (Wolfe Tory)

Description

The AbViser™ is a sterile two-way valve that is placed between the Foley catheter and the urine collection bag (www.wolfetory.com). One side of the valve is connected to the drainage tubing and the other side to a 3-way stopcock and pressure transducer via arterial pressure tubing. The 3-way stopcock is then connected to an infusion syringe and saline infusion bag via a double check valve. Between IAP measurements urine can drain freely trough the AbViser™ valve in the "drain" position. When doing an IAP measurement the AbViser™ valve is turned into the "measure" position, 50 mL of sterile saline is aspirated from the saline infusion bag and injected into the bladder via the double check valve. Then the IAP can be read on the monitor screen (Fig. 21).

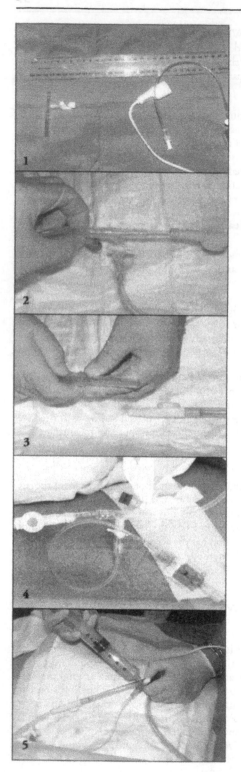

Figure 19. Step 1) Preparation showing a close up of the T-piece and pressure transducer used for IAP monitoring. Step 2) Close-up of the T-piece inserted in sterile conditions between the standard Foley catheter and the urinary drainage system. Step 3) A 2- or 3-way stopcock is connected to the free Luer-lock connection of the mounted T-piece, arterial pressure tubing is connected to the pressure transducer. Step 4) Close up of the mounted T-piece and pressure transducer. Step 5) Instillation of 50 mL of sterile saline solution with the urinary catheter clamped distal to the culture aspiration port. Figure continued on next page.

Figure 19. Continued. Step 6) IAP measurement showing T-piece, pressure transducer and urinary catheter. Step 7) The pressure transducer is positioned at the midaxillary line and the reference point is marked to reduce inter- and intra-observer variability. Step 8) The transducer is positioned and zeroed with the patient in the supine position and the actual reading is shown on the monitor. In this patient with temporary abdominal closure the IAP value was 9 mm Hg.

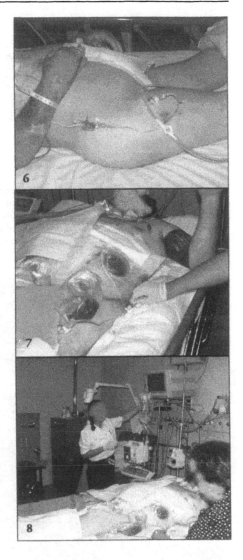

Advantages and Disadvantages

This technique allows IAP measurements to be performed in a more consistent, standardized way, basically it resembles the modified techniques allowing repeated measurements with a T-piece or ramp with stopcocks as presented above. It has however all other inconveniences that come with any fluid filled system. Although it is much more expensive, it does not allow IAP measurements to be performed that are more accurate and reproducible than the other intermittent techniques, since it is practically based on the same principles. Furthermore until now no clinical data is available.

Conclusion

In conclusion, if one wants to use IBP as estimate for IAP the Cheatham or revised techniques are preferred over the Kron or Iberti technique. The revised methods for IAP measurement via the bladder maintain the patient's Foley catheter as a closed system, limiting the risk of infection. Since these are needle-free system they also avoid the risks of needle stick injury

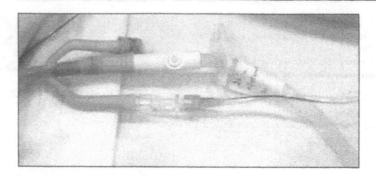

Figure 20. Close up view of a closed needle-free system for continuous intra-abdominal pressure measurement in an anuric patients, using a conic connection piece (Conical Connector with female or male lock fitting, B Braun, Melsungen, Germany, REF: 4896629 or 4438450) connected to a standard pressure transducer via pressure tubing. Reprinted with permission from ref. 1: Malbrain ML. Different techniques to measure intra-abdominal pressure (IAP): Time for a critical reappraisal. Intensive Care Med 2004; 30(3):357-371.

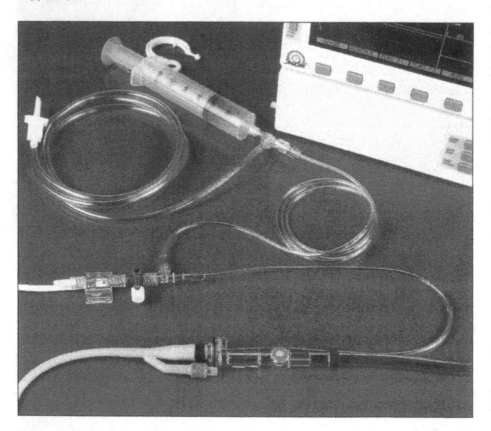

Figure 21. IAP measurement with the AbViser™.

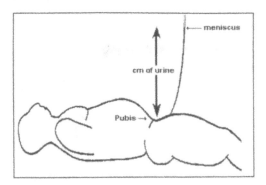

Figure 22. The classic manometer technique.

and overcome the problems of leakage and catheter knick in the method described by Cheatham. They are more cost-effective, and facilitate repeated measurements of IAP.

Manometry

The Classic Technique (Malbrain and Harrahill)[2,17,36]

Description

A quick idea of the IAP can also be obtained in a patient without a pressure transducer connected by using his own urine as transducing medium, first described by nurse Harrahill.[2,17,36] One clamps the Foley catheter just above the urine collection bag. The tubing is then held at a position of 30 to 40 centimetres above the symphysis pubis and the clamp is released (Figs. 22, 23).

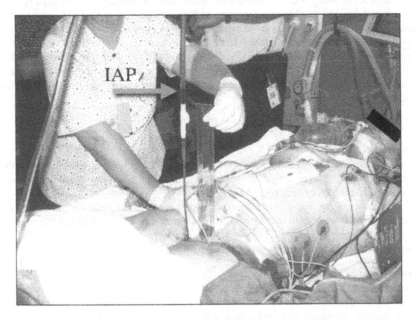

Figure 23. Patient set-up of the classic manometer technique.

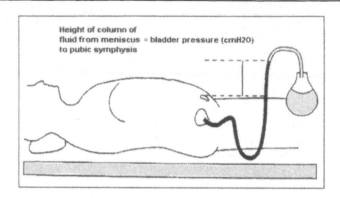

Figure 24. The U-tube manometer technique.

The IAP is indicated by the height (in cm) of the urine column from the pubic bone. The meniscus should show respiratory variations. This rapid estimation of IAP can only be done in case of sufficient urine output. In an oliguric patient 50 mL saline can be injected as priming.

Advantages and Disadvantages

It has all the inconveniencies that come along with a fluid-filled system as described before. However since it is needle-free it poses no risks for injuries. It allows repeated measurements, is very inexpensive and fast with minimal manipulation. Harrahill has described a technique, which is a simpler method of IAP measurement. It resembles central venous pressure manometry performed in the days before transducers. This technique has the advantage that it can be performed by any member of the health care team at the bedside, and also outside an intensive care location. The diagnosis of raised IAP can be made quickly and easily. This technique also has the potential of being available worldwide. It is a cheaper method and would be suitable for resource limited countries. However, care must be taken to ensure that the urinary catheter system used has an air inlet to avoid the generation of erroneously high pressures due to a closed system. If no air-inlet is available the drainage system needs to be disconnected probably increasing the risk of infection. When using this technique a conversion has to be made from cm H_2O to mm Hg and introduces the potential risk for error.[37] Since the volume reinstilled into the bladder is not constant raising questions on accuracy and reproducibility, it has limited clinical implications.

The U-Tube Technique (Lee)[38]

Description

In a recent animal study Lee and coauthors compared direct insufflated abdominal pressure with indirect bladder, gastric and inferior vena cava pressures.[38] IBP was measured by both the standard and U-tube technique (Fig. 24).

With the U-tube technique, the catheter tubing was raised approximately 60 cm above the animal to form a U-tube manometer, and IBP was measured as the height of the meniscus of urine from the pubic symphysis. The authors found a good correlation between the U-tube pressure and other direct and indirect techniques.

Advantages and Disadvantages

It has the same advantages and inconveniences as the classic "Harrahill" technique, as with the previous technique the clinical validation is poor. The major advantage of this technique is that the volume reinstilled into the bladder is more stable (but still not well defined) so it can be used as a quick screening method.

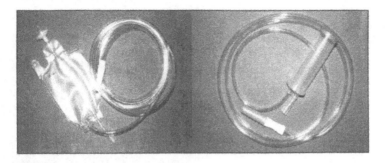

Figure 25. FoleyManometer first and second generation prototypes. ©Holtech Medical. Reprinted with permission from ref. 1: Malbrain ML. Different techniques to measure intra-abdominal pressure (IAP): Time for a critical reappraisal. Intensive Care Med 2004; 30(3):357-371.

The FoleyManometer Technique (Holtech)[1,39]

Description

In 2002 we tested a prototype (Holtech Medical, Copenhagen, Denmark, www.holtech-medical.com) for IAP measurement using the patients' own urine as pressure transmitting medium.[1,39] A 30 mL container fitted with a bio-filter for venting is inserted between the Foley catheter and the drainage bag (Figs. 25-33).

The container fills with urine during drainage; when the container is elevated, the clamp between the FoleyManometer and the biofilter is released and 30 mL urine flows back into the patient's bladder. The zero-reference mark is directly positioned at the symphysis pubis and IAP can be read from the position of the meniscus in the clear manometer tube between the container and the Foley catheter. This manometer tube has units in mm Hg (initially in cm H_2O) (Fig. 28).

We found a good correlation between the IAP obtained via the FoleyManometer and the "gold standard" in 119 paired measurements (R^2 = 0.71, p<0.0001). The analysis according to

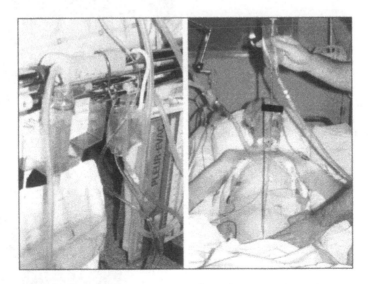

Figure 26. Urine drainage and IAP measurement with first FoleyManometer prototype.

Figure 27. Drainage with second Foley-Manometer prototype.

Figure 28. Close up view of the fourth generation FoleyManometer.

Figure 29. Patient mounted view of the FoleyManometer inserted between the Foley catheter and the urinary drainage bag.

Figure 30. Bed mounted view on the FoleyManometer, note the horizontal position of the drainage tubing, preventing intermittent blocking of urine flow and incomplete bladder emptying.

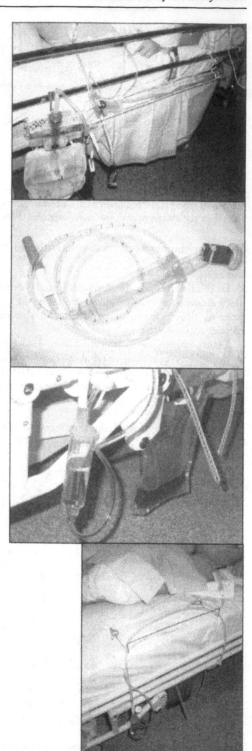

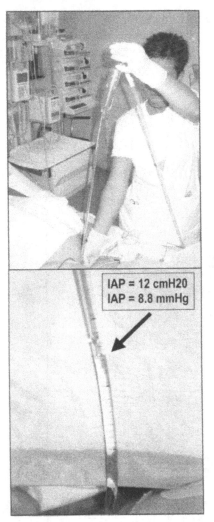

Figure 31. IAP measurement with FoleyManometer.

IAP = 12 cmH20
IAP = 8.8 mmHg

Figure 32. Close up view of the urine column and meniscus. In this patient the IAP was 12 cm H_2O. In the latest version of the FoleyManometer units are in mm Hg so that conversions from cm H_2O to mm Hg is no longer necessary.

Bland and Altman showed that both measurements were almost identical with a mean bias of 0.17±0.8 (SD) mm Hg (95% CI 0.03 to 0.3).

Advantages and Disadvantages

The FoleyManometer has no major disadvantages. It allows repeated measurements, is very cost-effective and fast with minimal manipulation, and the 0 mm Hg pressure reference at the symph. pubis is confirmed at each IAP determination. Product refinement and multicentric validation has been done so that this technique can be used in a clinical setting.

Conclusion

The manometry techniques give a rapid and cost-effective idea of the magnitude of IAP and may be as accurate as other direct and indirect techniques. They can easily be done 2-hourly together with and without interfering with urine output measurements. Moreover, the risk of infection and needle stick injury is absent. Multicentre validation is currently undertaken, making them ready for general clinical usage.

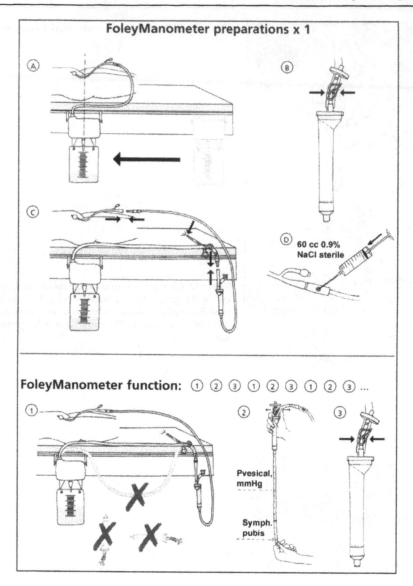

Figure 33. FoleyManometer instructions.

Stomach

The Classic Intermittent Technique (Collee)[40,41]

Background and Description

The IAP can also be measured by means of a nasogastric or gastrostomy tube and this method can be used when the patient has no Foley catheter in place, or when accurate bladder pressures are not possible due to the absence of free movement of the bladder wall.

Basically it is the same technique as described by Kron but applied to the stomach. A normal nasogastric tube is inserted into the stomach (PharmaPlast Duodenal Tube Levin, 14 F

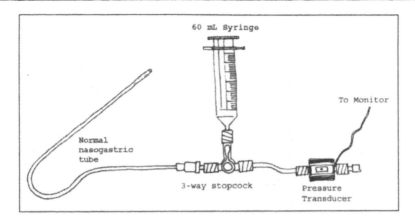

Figure 34. The classic Collee technique.

122 cm, Maersk Medical A/S, Denmark) (Fig. 34). The intragastric position of the tube is confirmed by aspiration of gastric juice, auscultation of air insufflation into the stomach, portable chest/upper abdomen X-ray, and confirmation of a rise in IAP following external epigastric pressure.

A three-way stopcock is connected to the nasogastric tube, one end is connected to a pressure transducer via arterial tubing and the other end is connected to a infusion bag of 1000 mL saline. All air is aspirated from the stomach and 100 mL of saline is injected. The transducer is zeroed at the midaxillary line with the patient in the supine position and IAP is read end-expiratory.

In case of head elevation, bladder trauma, peritoneal adhesions, pelvic haematomas or fractures, abdominal packing, a contracted or neurogenic bladder, IBP may overestimate IAP, and the procedure used for the bladder can then be applied via the stomach.[40,41]

Advantages and Disadvantages

Same inconveniences as with every fluid-filled system. Another disadvantage is that gastric pressures might interfere with the migrating motor complex or with nasogastric feeding. Furthermore all air needs to be aspirated from the stomach before measuring IAP, something that is difficult to verify.

The advantages are that it is cheap, does not interfere with urine output, and the risk for infection and needle-stick injuries is absent. This cost-effective technique is ideal for screening.

The Semi-Continuous Technique with a Tonometer (Sugrue)[42,43]

Background and Description

Sugrue and coauthors assessed the accuracy of measuring simultaneous IBP and IAP via the balloon of a gastric tonometer during laparoscopic cholecystectomy.[42] They found a good correlation between both methods. This technique allows to obtain a trend.

A pressure volume curve of the gastric tonometer balloon at 37°C was obtained to confirm that instillation of 3 mL of air allows the balloon to act as a pressure transducer. Each balloon was individually checked prior to insertion. A gastric tonometer with a balloon attached was inserted in all patients ("TRIP" TGS tonometry catheter, 16 F with stopcock 122 cm x 5.3 mm, Tonometrics Inc, Bethesda, USA and Datex-Engstrom Division, Instrumenatrium Corp, Helsinki, Finland). The intragastric position of the tonometer will be confirmed by aspiration of gastric juice, auscultation of air insufflation into the stomach, portable chest/upper abdomen X-ray, and confirmation of a rise in IAP following external epigastric pressure. To measure

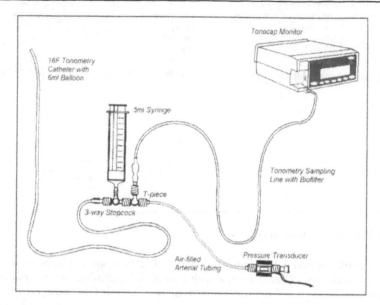

Figure 35. The tonometer technique.

regional pCO_2 and intramucosal pH the balloon is filled with 6 mL of air and is distended and measurements are obtained via a tonocap monitor (Fig. 35).

To read IAP, 3 mL of air is withdrawn of the tonometer balloon immediately after a pCO2 sampling cycle by turning the stopcock from "syringe off" to "T-piece off" and aspirating 3 mL of air from the balloon. Then the stopcock is turned back to "syringe off" and IAP is read end-expiratory after zeroing of the transducer at the mid-axillary line with the patient in the supine position. To measure prCO2 again the stopcock is turned to "T-piece off" again and the 3 mL of air are reinjected into the balloon. The stopcock is then turned to "syringe off" for measuring pCO2.

We recently validated these results and found good correlation between the classic gastric method, the tonometer method and IBP.[43] Simultaneous IAP via the tonometer (IAPtono) and PrCO2 measurement was also possible (Fig. 36).

Advantages and Disadvantages

Measurement via the tonometer balloon limits the risks and has major advantages over the standard intravesical method: no infection risk and no interference with estimation of urine output. Simultaneous measurement of IAP and $PrCO_2$ is possible, however only in an intermittent way. Since it is air-filled it has none of the disadvantages associated with fluid-filled systems: no problem with zero-reference, over- or underdamping or body position. A possible disadvantage is the effect on interpretation of IAP values by the Migrating Motor Complex. Recording the "diastolic" value of IAP at end-expiration can solve this problem (Fig. 36). Other problems are that a 5 mL glass syringe is needed and that no data is available on effects of enteral feedings on these IAP measurements. This technique could be used for study purposes and clinicians interested in simultaneous $PrCO_2$, pHi and IAP monitoring.

The Revised Semi-Continuous Technique (Malbrain)[1]

Description

An oesophageal balloon catheter is inserted into the stomach (Oesophageal balloon catheter set, adult size with PTFE coated stylet, Ackrad Laboratories, Inc, Cranford NJ, USA REF:

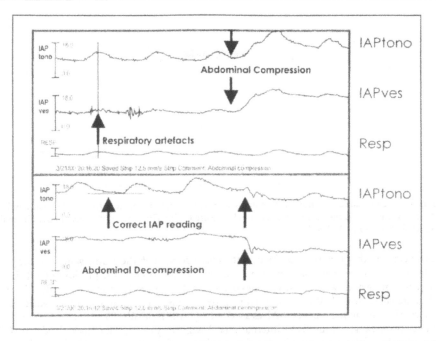

Figure 36. Simultaneous IAP tracings obtained from the stomach via a tonometer balloon (IAPtono) and via the bladder (IAPves). Respiratory and ECG artefacts are more pronounced in the stomach due to proximity of the heart and lungs. Correct IAP should be read at end-expiration.

47-9005, see at www.ackrad.com/products/c-balloon_catheter.cfm or Compliance catheter female or male, International Medical Products BV, Zuthpen, Netherlands, distributed by Allegiance, REF: 84310). The intragastric position of the tube is confirmed by portable chest/upper abdomen X-ray, and confirmation of a rise in IAP following external epigastric pressure or by a rapid oscillation test. When the balloon is correctly in the stomach, the whole respiratory IAP pressure wave will be positive and the positive pressure will increase upon inspiration with a functional diaphragm. If the balloon is too high in the thorax the pressure will flip from positive to negative on inspiration measuring oesophageal or pleural pressure instead of IAP. The preferred position for the procedure is to place the patient in a 30° Fowler position. The catheter is advanced to the 55-60 cm marks which places the balloon in the stomach (the International Medical Products catheter has no marks) (Figs. 37, 38).

 A standard 3-way stopcock is connected to the now "nasogastric" tube and one end is connected to a pressure transducer via arterial tubing. All air is evacuated from the balloon with a glass syringe and 1 mL of air reintroduced to the balloon. A glass syringe is recommended to minimize the risk of pulling a negative pressure inside the catheter prior to reintroducing the 1 mL air. The negative pressure will use up much of the volume and the balloon may not get enough air to get inflated. The balloon is connected via a "dry" system to the transducer, the transducer itself is NOT classically connected to a pressurized bag and NOT flushed with normal saline in order to avoid air/fluid interactions. The transducer is zeroed to atmosphere and IAP is read end-expiratory. By using this technique the cost of IAP is further reduced depending on the catheter used, moreover a semi-continuous trend can be obtained (Fig. 39).

Advantages and Disadvantages

 A disadvantage is that the air in the balloon gets resorbed after a couple of hours, so that "recalibration" of the balloon is necessary with a 2 to 5 mL glass syringe for continuous

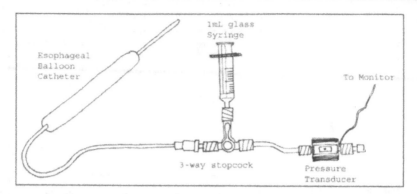

Figure 37. The revised semi-continuous technique. Reprinted with permission from ref. 1: Malbrain ML. Different techniques to measure intra-abdominal pressure (IAP): Time for a critical reappraisal. Intensive Care Med 2004; 30(3):357-371.

measurement, this might cause inaccurate measurement if the nurse waits too long for recalibration or if the reinstilled volume is not exactly the same as the previous one. It is less time-consuming and has all the advantages of an air-filled system (cfr tonometer). By using this technique the cost of IAP is further reduced depending on the catheter used. Moreover, a semi-continuous measurement of IAP as a trend over time is possible. The oesophageal balloon catheter price ranges from €15 (International Medical Systems, The Netherlands) to €55 (Ackrad, USA). This technique is ideal for monitoring for a longer period of time, however when using multiple tubes the risk of sinusitis or infection needs to be evaluated in the future.

The Continuous Fully-Automated Technique (Spiegelberg)

Description

IAP Measurement with the Air-Pouch System (www.spiegelberg.de). The IAP-Catheter is introduced like a nasogastric tube, it is equipped with an air pouch at the tip. The catheter has one lumen that connects the air-pouch with the IAP-Monitor and one lumen that takes the guide wire for introduction. The pressure transducer, the electronic hardware, and the device for filling the air-pouch are integrated in the IAP-Monitor. On the digital display the mean pressure and the amplitude of the pressure wave are shown. At the monitor output both the mean pressure and the pulsatile signal are available. Once every hour the IAP-Monitor opens the pressure transducer to atmospheric pressure for automatic zero adjustment. The air-pouch is then filled with the exact air volume required for accurate pressure transmission (about 0.1

Figure 38. Close-up view of the esophageal balloon catheter. Reprinted with permission from ref. 1: Malbrain ML. Different techniques to measure intra-abdominal pressure (IAP): Time for a critical reappraisal. Intensive Care Med 2004; 30(3):357-371.

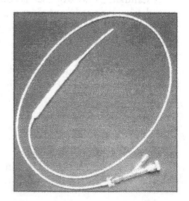

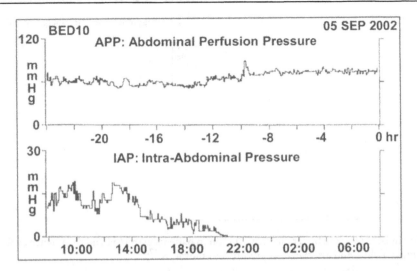

Figure 39. A trend of 24 hour IAP and APP recordings obtained with an esophageal balloon placed in the stomach (Ackrad). Note the resorption of air after a couple of hours with loss of IAP signal, confirming the need for recalibration. Reprinted with permission from ref. 1: Malbrain ML. Different techniques to measure intra-abdominal pressure (IAP): Time for a critical reappraisal. Intensive Care Med 2004; 30(3):357-371.

mL). Initial validation in ICU patients and laparoscopic surgery showed good correlation with the standard IBP method (Fig. 40).[44]

The IAP-Monitor can be used in a stand-alone fashion, with an ICU monitor, or with the Spiegelberg: APP-Monitor. The guide wire is coated with PTFE for easy introduction. The IAP-Catheter is supplied sterile for single use (Figs. 41-46).

Recently Schachtrupp and coauthors used the same technique to directly measure IAP in a porcine model and found a very good correlation between the air pouch system and direct insufflator pressure (R^2 0.99) with a mean bias of 0.5±2.5 mm Hg and small limits of agreement (-4.5 to 5.4 mm Hg) (Fig. 47).[45]

Advantages and Disadvantages

This technique has no major disadvantage except that validation in humans is still in its infant stage. The advantages are those related to other gastric and air-filled methods. In summary it is simple, fast, accurate, reproducible, and fully automated, so that a real continuous 24 hour trend can be obtained (Figs. 43, 45). This technique is not suited for screening but is best for continuous fully automated monitoring for a long period of time. Since it is less prone to errors and most cost-effective if in place for a longer period of time this technique has a lot of potential in becoming the future standard for multicentre research purposes. Recently another device (the CiMON, Fig. 48) became commercially available allowing simultaneous measurement of both intrathoracic and intragastric pressure (www.pulsion.com).

Conclusion

The revised methods via the stomach have the advantage to be free of interference caused by wrong transducer positions since they do not need the creation of a conductive fluid column as they use air as transmitting medium. The last described fully automated technique (Spiegelberg and CiMON) also gives a continuous tracing of IAP together with abdominal perfusion pressure (APP) in analogy with intracranial pressure and cerebral perfusion pressure, allowing both parameters to be monitored as a trend over time. The APP is calculated by substracting IAP from the mean arterial blood pressure. Recent data showed the importance of APP as a superior

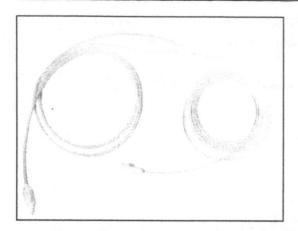

Figure 40. The Spiegelberg IAP catheter.

marker in IAH to titrate better our resuscitation in patients with IAH and ACS hence avoiding end-organ failure and associated morbidity and mortality.[17,46]

Rectal Pressure[47]

Description

Rectal pressures are used routinely as estimate for IAP during urodynamic studies to calculate the transmural detrusor muscle pressure as IBP minus IAP.[47,48] Rectal pressures can be obtained by means of an open rectal catheter with a continuous slow irrigation (1 mL/min), but special fluid-filled balloon catheters are used more routinely, although more expensive.

First the catheter has to be prepared by putting a 3-way stopcock and a 5 mL syringe with saline on the "FILL" end of the catheter. This step is easily accomplished with the catheter held so that the balloon is pointed upward. Using the syringe, water is infused into the catheter. This action forces the air inside to rise and then discharge out of the lumen extension labeled "PURGE ONLY". After all the air has been flushed through the balloon and catheter, the cap on the lumen extension labeled "PURGE ONLY" is closed. The balloon is collapsed by withdrawing the water from the catheter with the syringe.

The 3-way stopcock is then turned "OFF" to the catheter that is now ready for insertion. The balloon-tipped catheter is placed 10 to 15 cm in the rectum or vagina/uterus and taped in place. The stopcock is set "ON" to the syringe and approximately 1 to 1.5 mL of saline is infused into the balloon with the syringe (Fig. 49).

The 3-way stopcock is then turned "OFF" to the catheter and a standard Pressure Transducer is inserted on the last free connection of the stopcock with pressure tubing. The 3-way stopcock is turned "OFF" to the syringe and "ON" to the Pressure Transducer and "ON" to the catheter. The pressure waveforms are observed on the channels recording IAP and pdet (detrusor pressure). At the beginning of a cystic myogram (CMG), when the bladder is (presumably) empty, it is recommended that the starting pdet (detrusor) pressure is "0" mm Hg. To achieve this value, pabd (abdominal pressure) must equal IBP (pdet = IBP-IAP). To change the pdet value add water to lower it or remove water to raise it. This is done with the 3-way stopcock turned so that all three ports are open. After doing so the pressure changes can be observed while adjusting the water content of the balloon. After finishing the stopcock is turned "OFF" to the syringe as it was initially. Remember that adding water decreases pdet (increases IAP) and removing water increases pdet (decreases IAP) (Fig. 50).

Advantages and Disadvantages

The major problem with the open catheter is that residual faecal mass can block the catheter tip opening leading to overestimation of IAP. Other disadvantages of this technique are that it

Figure 41. The Spiegelberg IAP monitor.

Figure 42. Combination of the Spiegelberg IAP and APP monitor allowing continuous fully automated monitoring. IAP: intra-abdominal pressure; ABP: arterial blood pressure; APP: abdominal perfusion pressure.

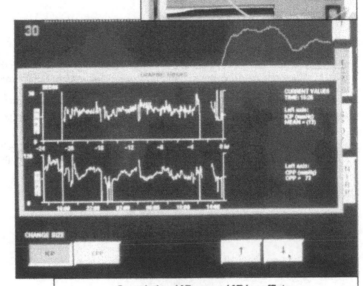

Figure 43. Connection of the IAP signal to the bedside monitor allowing to obtain a continuous IAP and APP trend for further reference.

Figure 44. Correlation between intragastric (IAPgas) pressure obtained with the Spiegelberg IAP catheter versus direct insufflation pressure during laparoscopy. Adapted from ref. 44, Malbrain ML. Validation of a novel fully auto-mated continuous method to measure intra-abdominal pressure. Intensive Care Med 2003; 29[Supplem 1]:S73.

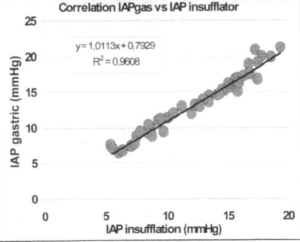

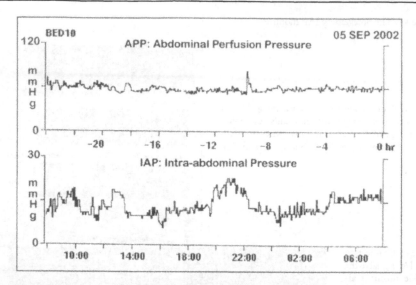

Figure 45. A continuous trend of 24 hour IAP and APP recordings obtained with the Spiegelberg balloon-tipped IAP catheter placed in the stomach. Note the absence of resorption of air due to automated recalibration every hour. Note also the effect of CAPD fluid inflow on IAP. Reprinted with permission from ref. 1: Malbrain ML. Different techniques to measure intra-abdominal pressure (IAP): Time for a critical reappraisal. Intensive Care Med 2004; 30(3):357-371. CAPD: chronic ambulatory peritoneal dialysis.

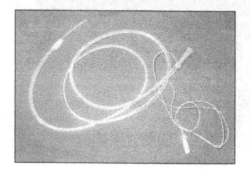

Figure 46. Options for the future, a catheter allowing gastric nutrition or emptying together with IAP monitoring.

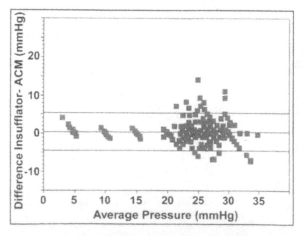

Figure 47. Low bias and reasonable limits of agreement between direct insufflator pressure and Spiegelberg balloon catheter (ACM). Adapted from ref. 45: Schachtrupp A, Tons C, Fackeldey V et al. Evaluation of two novel methods for the direct and continuous measurement of the intra-abdominal pressure in a porcine model. Intensive Care Med 2003; 29(9):1605-1608.

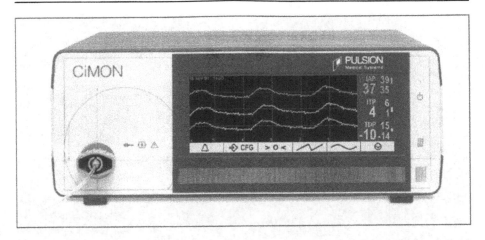

Figure 48. Close-up view of the CiMON monitor (compartmental intrathoracic intra-abdominal monitoring) displaying the intra-abdominal pressure (IAP); the intrathoracic pressures (ITP) and the transdiaphragmatic pressure (TDP).

is more difficult, implicates more manipulation, is intermittent, cannot be used in patients with lower gastro-intestinal bleeding or profound diarrhoea. There is also a great reluctance among nurses to use it. Since it is fluid-filled it has all the problems associated with an hydrostatic fluid column, but since it is needle free it decreases patient and healthcare worker infections or injuries. The fluid-filled balloon catheters are more expensive and, even though could theoretically stay in place for a longer period of time, interfere with gastro-intestinal transit and can cause erosions and even necrosis of the anal sphincter and rectal ampulla. Finally these techniques have not been validated in the ICU setting. This technique has no clinical implications in the ICU setting.

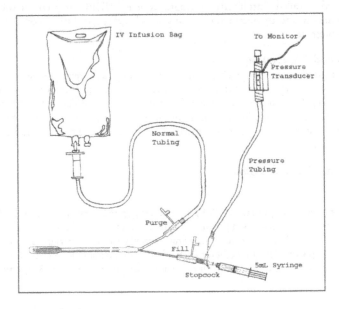

Figure 49. Uterine or rectal catheter set-up.

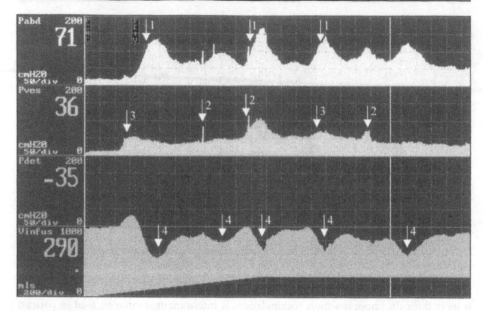

Figure 50. Example of urodynamic investigation. Big rectal contractions are seen at arrow 1. Arrow 2 indicates well cancelled artefacts. Arrow 3 indicates genuine unstable contractions while arrow 4 shows good cancellation.

Uterine Pressure

Description

Basically this technique is mostly done with the same catheters as for the rectal route. Uterine pressures are used routinely by gynaecologists during pregnancy and labour. Most classically a standard so-called "intrauterine pressure catheter" (IUPC) is used for this purpose.[49] Uterine pressures are mostly obtained by means of a closed special fluid-filled balloon catheter (as for rectal pressure).

Advantages and Disadvantages

The major disadvantages of this technique are the same as for rectal pressures, i.e., it is that more difficult, implicates more manipulation, is intermittent, and cannot be used in patients with gynaecological bleeding or infection. Since it is also fluid-filled it has all the problems associated with an hydrostatic fluid column, but is needle free. Finally these technique has not been validated in specific ICU patient populations. This technique has no clinical implications in the ICU setting.

Inferior Vena Cava Pressure

Description

The inferior vena cava pressure (IVCP) has been suggested as estimation for IAP. Basically it is the same techniques as described previously but applied to an IVC catheter. A normal central venous line is inserted into the inferior vena cava via the left or right femoral vein. The intra-abdominal position of the catheter is confirmed by portable lower abdomen X-ray, and confirmation of a rise in IAP following external abdominal pressure. A 3-way stopcock is connected to the distal lumen, one end is connected to a pressure transducer via arterial tubing and the other end is connected to a pressurized infusion bag of 1000 mL saline. The transducer is zeroed at the midaxillary line with the patient in the supine position and IAP is read end-expiratory as with CVP.

Advantages and Disadvantages

The major disadvantage of this technique is the risk of (possible catheter related) bloodstream infections and septic shock. The initial placement is more time-consuming. It has also the problems inherent to fluid-filled systems and poses potential injury to patient and healthcare workers. The major advantages are that a continuous trend can be obtained, it does not interfere with urine output, and that it could be used in bladder-trauma patients. Finally this technique has not been validated in specific ICU patient populations. In an animal study comparing different methods of indirect IAP measurement Lacey found a good correlation between bladder and inferior vena cava pressure with direct intraperitoneal IAP measurement, but not with gastric, femoral or rectal pressure.[48] Lee also found a good correlation in 30 patients during laparoscopy.[38] A recent study in man, comparing superior vena cava pressure (SVCP) with common iliac venous pressure (CIVP) in various conditions of IAP and PEEP showed that the difference between CIVP and SVCP was not affected by the IAP, which implies that CIVP does not reflect IAP correctly [32]. The most likely explanation is the differing anatomy and experimental model used to induce increased IAP in canine studies. In humans both CIVP and SVCP increase as IAP increases [32]. Recently Joynt also found a good correlation between SVCP and IVCP regardless of IAH.[50] This technique has limited implications in the ICU setting.

Microchip Transducer Tipped Catheters

Description

Different types of catheters, tipped with microchip transducers are nowadays available on the market. They can either be placed via the rectal, uterine, vesical or gastric route. These catheters can either have a 360° membrane pressor sensor in the organ (rectum, uterus, bladder, stomach) connected to an external transducer in a reusable cable or they can have a fibreoptic in vivo pressure transducer in the tip of the catheter itself (Figs. 51, 52).

These catheters provide true zero in-situ calibration. By disconnecting and checking for zero on the monitor, clinicians can instantly validate and check the zero status of the monitor and the transducer.[49] Recently Schachtrupp and coauthors found a good correlation between IAP calculated be a piezoresistive pressure measurement and direct insufflator pressure (R^2 0.92) with a difference of 1.6±4.8 mm Hg however the limits of agreement were large (-8 to 11.2 mm Hg).[45] This might have been due to an unknown measurement to measurement drift due to the fact that the device cannot be zeroed to the environment when placed intra-abdominally.

Advantages and Disadvantages

The major disadvantages of this technique is that it is very expensive with catheter-price ranging from €1000 to €1500. These catheters are said to be reusable a couple of times after cleaning with soap and water and gas sterilisation, but no data in ICU patients is available. These catheters are mostly used during urodynamic studies and labour for a limited period of time (hours); none of them have been tested in ICU patients for longer periods of time (days to weeks). The major advantages are that a continuous trend can be obtained, it is less time-consuming, it does not interfere with urine output. This technique has no clinical implications in the ICU setting.

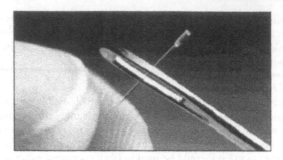

Figure 51. Microchip transducer. Close up view of an ultra-miniature microchip pressure transducer (Fiso technologies). These transducers provide high fidelity, robustness and exceptional measurement performance for in vivo catheter pressure applications.

Cost-Effectiveness

Costs

A cost estimation (in Euros) was performed for the different indirect IAP measurement techniques, based on cost of disposables and nursing time. Costs were scored based on initial set-up, the cost for the first measurement and the cost evolution in a hypothetical situation where IAP measurements were performed twice, six times or twelve times a day for 1 up to 4 weeks. Costs were compared based on costs of initial set-up, the first and next measurements as well as the costs based on the number of IAP measurements per day and duration of measurement period. The initial setup cost was the highest for the microchip transducer systems (above 1000€) and the lowest for the manometry techniques (around 1 to 20€). Costs for intermittent IAP monitoring techniques varied around 20 to 40€ whereas the setup cost for (semi-)continuous techniques was between 70 to 200€ (Fig. 53).

The cost for the first measurement was the highest for the oldest described techniques (Kron, Iberti, Collee) at around 2.5 to 3.5€. For the other techniques the cost per measurement varied between 0.5 and 1€ (Figs. 54, 55).

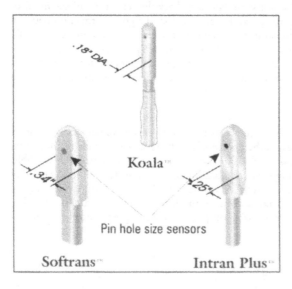

Figure 52. Comparison of different intrauterine pressure catheters.

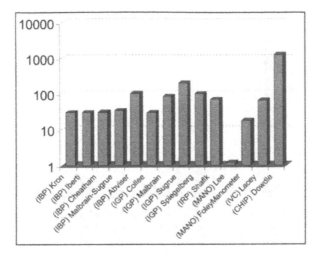

Figure 53. Comparison of the initial set-up costs (in €) for different indirect IAP monitoring techniques. IBP: intrabladder pressure; IGP: intragastric pressure; IRP: intrarectal pressure; MANO= manometry; IVC: inferior vena cava.

The total cost score was calculated as a percentage based on the rang order for the different cost comparisons (setup, first measurement, further monitoring). Manometry techniques were least expensive followed by automated (semi-)continuous techniques, the revised intermittent bladder techniques. Finally the time-consuming techniques were most expensive mainly due to the cost of nursing time (Fig. 56).

Effectiveness

Effectiveness analysis was based on available literature information on measurement properties, the technique itself, whether the system is fluid or air-filled, the associated risks and contraindications. Technique properties looked at the difficulty, manipulation and time consumption, as well as the possible interference with urine output or whether or not special material was needed. Measurement properties looked at the presence or absence of the possibility to do repeated measurements with the same equipment, to obtain a continuous trend, to have an automated measurement, the need for recalibration, the standardization, the accuracy and reproducibility, the validation with the gold standard. Other topics looked at was the problems possibly associated with air-bubbles, multiple menisci, filter blocking or intereference with the migrating motor complex. Specific problems related to fluid-filled systems were identified as the problems associated with zero-reference, over or underdamping and body position. Possible risks were needle stick injury, urinary tract infection or sepsis. Contraindications were bladder trauma, neurogenic bladder, haematuria, gastric or other abdominal trauma. The effectiveness score was calculated as the fraction of advantages over the total number of possible advantages or disadvantages. As suspected the fully automated method via the stomach had the best effectiveness score although clinical validation of this technique is still in its infant stage. The oldest reported intermittent techniques (Kron, Iberti, Collee) were least effective mainly because of manipulation and associated risks (Fig. 57).

Combined Cost and Effectiveness

The combined cost and effectiveness score was calculated as a percentage obtained by adding twice the effectiveness score to the cost score and dividing that sum by 3 (since we feel that the efficiency of a technique values more than the actual cost if it helps to better monitor our patients).

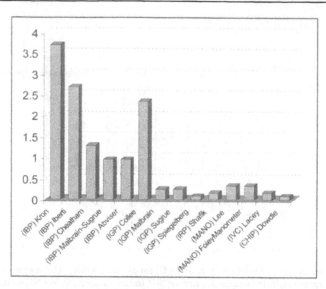

Figure 54. Comparison of cost per measurement for different indirect IAP monitoring techniques. The cost comparison has taken into account initial setup cost and cost per measurement in a hypothetical condition where IAP is monitored 12 times a day together with urine output for 1 up to 4 weeks. The costs were between 1 to 10€ per measurement. IBP: intrabladder pressure; IGP: intragastric pressure; IRP: intrarectal pressure; MANO: manometry; IVC: inferior vena cava.

Again the fully automated technique via the stomach had the best cost-effectiveness score followed by the manometry techniques (Fig. 58).

Reproducibility of IAP Measurement

As stated previously, the intravesical route evolved as the gold standard. However, considerable variability in the measurement technique has been noted and the common pitfalls are briefly addressed below.

1. Malpositioning of the pressure transducer with regard to the symphysis pubis after repositioning of the patient: This may lead to over- and underestimation of IAP and are commonly seen at changes of nurse shifts. Incorrect zeroing affects the accuracy of the IAP measurement

2. All fluid-filled systems connected to a pressure transducer have their own static and dynamic response properties that can create distortions or artefacts in the IAP pressure waveform, leading to signal over- or underdamping.[27,28] A common example is the loss of pressure in the pressurized bag.

3. IBP is the most used and validated technique, but with inadequate accuracy and reproducibility. The inaccuracy can come from the presence of air-bubbles in any fluid-filled system leading to over- or underestimation. If the measurement itself is inaccurate, this also implicates that it is not reproducible leading to intra- and inter-observer variability. However, when the pressure transducer position is consistently too high or too low with a fully compliant transducer system the IAP value obtained will be too low or too high respectively but may be reproducible, so that the trend over time can still provide valuable information. In order to get an idea of these reproducibility problems with bladder pressure we performed a multicentre snapshot study (4 IAP measurements each every 6 hours) on a given day.[3] The mean IAP was 10.2±2.7 mm Hg, (range 7.6±4 to 12.7±5.7). Analysis according to Bland and Altman showed a global bias of IAP within 24 hours (difference between minimum and maximum value) of 5.1±3.8 (SD) mm Hg (95%CI 4.3 to 5.9); the limits of agreement

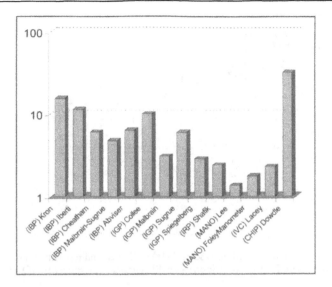

Figure 55. Comparison of cost per measurement for different indirect IAP monitoring techniques assuming 12 measurements per day were performed for up to 4 weeks and taken into account the initial setup costs.

were -2.5 to 12.7 mm Hg. The bias differed from centre to centre between 2.4 and 6.2 mm Hg with one outlier bias value as high as 11 mm Hg, raising questions on the reproducibility of the measurement technique used in that centre and making it difficult to compare literature data.[3]

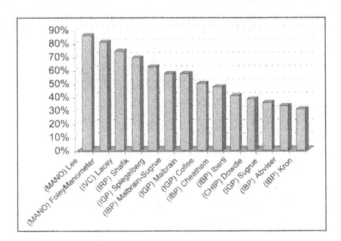

Figure 56. Comparison of total cost score (in %) of different indirect IAP monitoring techniques. The cost evaluation was based on the following estimates: transducer: 24.75€; 50 mL of saline: 0.3€; syringe: 0.36€; needle: 0.023€; Foley catheter: 0.53€; nasogastric tube: 0.53€; Esophageal catheter (Ackrad, IMS): 55€; Tonometer (Datex): 175€; IAP catheter (Spiegelberg): 100€; AbViser™ (WolfeTory): 75€; FoleyManometer (Holtech): 17.5€; Rectal/Uterine probe (Medtronic, Tyco, Kendall): 34.8€; Microchip transducer (Rehau): 1250€; conical connector: 2.2€; male-male connector: 0.4€; stopcock: 0.31€; sterile drapings: 1.36€; nursing costs: 25€ per hour. IBP: intrabladder pressure; IGP: intragastric pressure; IRP: intrarectal pressure; MANO: manometer; IVC: inferior vena cava.

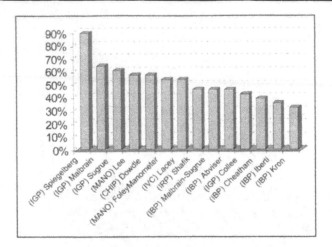

Figure 57. Comparison of total effectiveness score (in %) of different indirect IAP monitoring techniques. IBP: intrabladder pressure; IGP: intragastric pressure; IRP: intrarectal pressure; MANO: manometer; IVC: inferior vena cava.

4. An idea on how often IAP should be measured can be obtained from the analysis of the coefficient of variation (COVA) during one 24 hour period. In the above mentioned study the mean COVA (defined as the standard deviation divided by the mean IAP) was 25% which is comparable to daily fluctuations in other pressures like central venous pressure or pulmonary artery occlusion pressure. However, this coefficient ranged from 4 to 66% between centres. Since the literature provides no data on 24-hour continuous IAP-measurement in the ICU it is not possible to determine whether these variations or fluctuations in IAP during one study day were normal or related to the measurement technique used. Therefore IAP should be monitored as often as possible. This implicates that the prevalence or incidence of IAH and ACS is affected by the number of IAP measurements performed during the day.

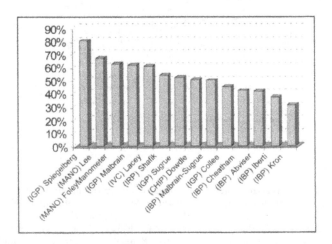

Figure 58. Comparison of total cost and effectiveness score (in %) of different indirect IAP monitoring techniques. IBP: intrabladder pressure; IGP: intragastric pressure; IRP: intrarectal pressure; MANO: manometer; IVC: inferior vena cava.

5. As IAP is a physiologic parameter as any other "body pressure" it probably substantially fluctuates during the day. Since the inception of IAP monitoring, measurements obtained every four to twelve hours have been assumed to accurately portray a patient's IAP state during the intervening time. It is now recognized however that these intermittent measurements are only "snapshots" that poorly illustrate the "moving picture" of the patient's response to injury and subsequent resuscitation. A recent study looking at the coefficient of variance during different 24 hour periods showed that the daily COVA (standard deviation divided by mean) was 18.7±7.4% for intragastric pressure with the Spiegelberg catheter (range 5-46%) and 17.5±12.4% for IBP (range 0-58%) (Figs. 59A,B, 60, 61).

6. Intrinsic bladder compliance is important. In case of a low bladder compliance a smaller amount of saline needs to be instilled into the bladder, otherwise there is a risk to increase intrinsic IBP and overestimate IAP. A common example of a decreased bladder compliance is IAH or ACS, therefore the higher the previous IAP measurement the smaller the amount needed for the next measurement (Fig. 62).

7. Baseline IAP and the volume instilled in the bladder are important. Gudmundsson found recently in an animal study that the IAP increase by instilling Ringer's solution into the abdominal cavity correlated well with intravesical pressures.[51] It was also found that IBP as an estimation for IAP is affected by the amount of fluid in the bladder that should not exceed 10-15 mL. If baseline IAP is lower than 8 mm Hg a 131 mL extra bladder volume is needed to increase IAP with 2 mm Hg, however if baseline IAP is 20 mm Hg only 39 mL extra bladder volume is needed for the same IAP increase.[52] We recently came to the same conclusions by analysing bladder pressure volume curves we found that IBP significantly increased depending on the volume instilled. The IBP rose from 4.2±3.2 mm Hg at baseline to 6.9±5 mm Hg with 50 mL and 23.7±16.1 at 300 mL (p<0.0001, ANOVA) (Fig. 63). If IBP is used as an estimate for IAP the volume instilled in the bladder should be between 50 and 100 mL, however in some patients with a low bladder compliance IBP can be raised at low bladder volumes. Ideally a bladder PV curve should be constructed for each individual patient before using IBP as an estimation for IAP. This study makes it difficult to compare the literature data. It raises not only questions with regard to the previously published definitions and IAP cut-offs but it also puts the IBP in question as the so-called gold standard. Ideally the bladder should be fully emptied before an IAP measurement, but how can you be really sure?

8. The bladder "gold standard" measurement techniques reported are not uniform, most authors recommend to inject 50 mL[1] others 0 mL,[32] 100 mL,[30] 200 mL[53] or even 250 mL[31] of saline in the bladder (Fig. 64).
   In fact, in the initial article from Iberti, data are presented from a canine model without stating the volume instilled in the bladder. The only statement was "*The bladder was continuously emptied between measurements.*"[32] In a next study Iberti presented human data stating "*using a sterile technique an average of 250 mL of normal saline was infused through the urinary catheter to gently fill the bladder and eliminate air in the drainage catheter*".[31]

9. Conflicting results are reported in the literature regarding the validation of IBP versus directly measured IAP during laparoscopy. In a recent study Yol compared bladder pressure with direct insufflation pressure during laparoscopic cholecystectomy in 40 patients and he found a very good correlation between the two measurements (R = 0.973, p<0.0001).[54] This was also shown by Fusco who compared direct laparoscopic insufflation pressure with bladder pressures measured with bladder volumes of 0, 50, 150 and 200 mL.[55] He found that there was a good correlation across the IAP range from 0 to 25 mm Hg between direct and indirect methods with all tested volumes. A bladder volume of 0 mL demonstrated the lowest bias, but when considering only elevated IAPs (25 mm Hg) a bladder volume of 50 mL revealed the lowest bias. He concluded that intravesicular pressure closely approximates IAP and that instilling 50 mL of saline improved the accuracy of the bladder pressure in measuring elevated IAPs. However Johna recently found that intravesicular pressure did not reflect actual intra-abdominal insufflation pressure (limited up to 15 mm Hg) during

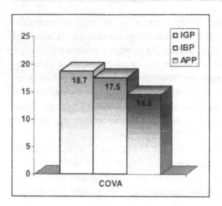

Figure 59A. Comparison of the COVA for IGP, IBP and APP during different 24 hour periods. COVA: coefficient of variation; IBP: intrabladder pressure; IGP: intragastric pressure; APP: abdominal perfusion pressure.

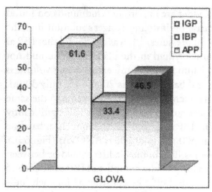

Figure 59B. Comparison of the GLOVA for IGP, IBP and APP during different 24 hour periods. The daily global bias (IAPmax minus IAPmin) was 5.6±2.6 mm Hg for IGP (range 1.2-16 mm Hg) and 3.1±2.3 for IBP (range 0-12 mm Hg); the daily global variance (GLOVA) defined as global bias divided by mean was 61.6±26.9% for IGP (19-162) and 33.4±23.9% for IBP (0-129); thus showing considerable variability that would be missed if a non-continuous technique was used to assess IAP. GLOVA: global variation; IBP: intrabladder pressure; IGP: intragastric pressure; APP: abdominal perfusion pressure.

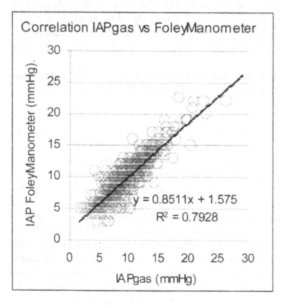

Figure 60. Good correlation between continuous gastric (IAPgas) and 2 hourly IBP via FoleyManometer.

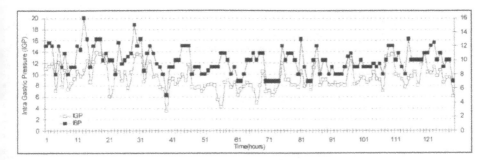

Figure 61. Daily fluctuations in IGP measured via the Spiegelberg (triangles) technique and IBP (squares) via a FoleyManometer. IGP: intragastric pressure; IBP: intrabladder pressure.

laparoscopy.[56,57] He concluded that further research is needed to identify possible variables that may play a role in the relationship between the urinary bladder and abdominal cavity pressures, providing better means for diagnosing ACS. Further reading shows that the methodology of this study was poor.

10. Although many articles have validated IBP against direct insufflation pressures it is difficult to extrapolate these single observer comparisons in patients undergoing general anesthesia and paralysis to a mixed ICU population of patients not under muscle relaxation as well as subject to other confounding factors (nurse shifts, position, zero reference, …). Direct IAP measurement via a laparoscopic insufflator is prone to errors by flow dynamics resulting in rapid increases in pressure during insufflation. The Verres needle opening can be blocked by tissue or fluid leading to over- or underestimation of IAP and pressures can be influenced by muscle relaxation. Laparoscopy remains an artificial environment, this makes it even more difficult to validate indirect IAP measurement methods.

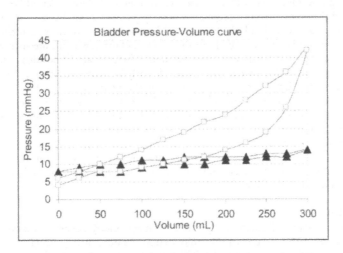

Figure 62. Bladder PV curve in a patient with a compliant bladder (closed triangles). Note that pressures are higher during insufflation than during deflation. Regardless of the amount of saline instilled in the bladder the pressures are comparable: 10 mm Hg at 50 mL, 11 mm Hg at 100 mL and 12 mm Hg at 200 mL. The open squares show a bladder PV curve in a septic patient with a poor bladder compliance. Note that pressures are higher during insufflation than deflation. Note the significant difference in IAP value with regard to the amount of saline instilled in the bladder: 10 mm Hg at 50 mL, 14 mm Hg at 100 mL and 24 mm Hg at 200 mL.

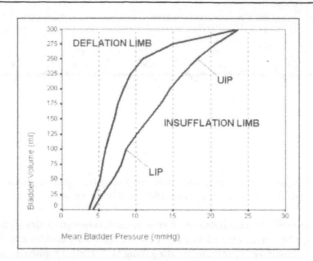

Figure 63. Plot of the "insufflation" and "deflation" PV curve as a curve fit of the means of 13 measurements in 6 mechanically ventilated patients. The Bladder PV curves were obtained by instilling sterile saline into the bladder with 25 mL increments. A lower inflection point (LIP) can be seen at a bladder volume of 50 mL to 100 mL and an upper inflection point (UIP) at a bladder volume of 250 mL. The difference in bladder pressure was 2.7±3.3 mm Hg between 0 or 50 mL volume, 1.7±1.2 mm Hg between 50 and 100 mL, 7.7±5.7 mm Hg between 50 and 200 mL and 16.8±13.4 mm Hg between 50 and 300 mL. Reprinted with permission from ref. 1, Malbrain ML. Different techniques to measure intra-abdominal pressure (IAP): Time for a critical reappraisal. Intensive Care Med 2004; 30(3):357-371.

11. Body position is important. Putting a patient in different body positions has significant effects on IAP (Fig. 65). This is in contradiction with the hypothesis that the abdominal compartment is primarily fluid in character and should follow the law of Pascal, since IAP would then remain constant regardless of body position as fluid is not compressible. The abdomen should in fact be looked at as a "fluidlike" compartment with different components that may influence IBP (the intrinsic weight of the organs, the presence of ascites, the air in the bowel, ...). Assessment of IAP should therefore always be done in the complete supine position. The upright position significantly increases IAP compared to the supine. The effects on IAP being more pronounced in obese patients (Fig. 65).[58]

Many of these drawbacks are not only true for the bladder but are also present when IAP is estimated via other routes. Not much has been studied on the effects of spontaneous breathing, mechanical ventilation, the presence of expiratory muscle activity, auto-PEEP, curarisation on IAP measurement via the different routes.

Definitions for IAH and ACS stand or fall with the correct measurement of IAP and its reproducibility. Recent literature data put the bladder pressure in question as the so-called gold standard for abdominal pressure.[1,8,51,55-59]

Conclusion

This chapter obtained from the analysis of advantages and disadvantages as well as the cost projection for each IAP measurement technique supports the concept that (1) there is no gold standard; (2) it is difficult to compare the different techniques; (3) cost-effectiveness is an issue; (4) manometry techniques are cost effective and can be used as estimation for IAP as a screening method to identify patients at risk; (5) IBP can be used as estimation for IAP for initial follow-up either with the Cheatham or revised bladder technique, the AbViser™ is an elegant technique however it is not cost-effective; (6) for (multicentre) study purposes, surgical

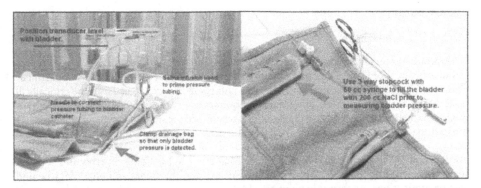

Figure 64. Instructions as found on the internet indicating that 200 mL of saline should be instilled before IAP measurement.

patients, trauma patients, patients at risk for IAH (Tables 1-3) it is preferable to switch to a continuous method for IAP monitoring via the stomach and focus therapy on optimising IAP and APP.

Acknowledgements

I'm indebted to my wife Ms. Bieke Depré for her patience, advice and technical assistance with the preparation of this manuscript, and to my 3 sons for providing a quiet writing environment.

Part of this work was presented at the 14th Annual Congress of the European Society of Intensive Care Medicine, Geneva, Switzerland, Sept. 30-Oct. 3, 2001; at the 22nd International Symposium on Intensive Care and Emergency Medicine, Brussels, Belgium, Mar. 19-22, 2002; at the 13th Symposium Intensivmedizin and Intensivpflege, Bremen, Germany, Feb. 19-21, 2003, on the 23rd International Symposium on Intensive Care and Emergency Medicine, Brussels, Belgium, Mar. 18-21, 2003, at the 16th Annual Congress of the European Society of Intensive Care Medicine, Amsterdam, The Netherlands, Oct. 5-8, 2003 and at the the 24th International Symposium on Intensive Care and Emergency Medicine, Brussels, Belgium, Mar. 29-Apr. 1, 2004.

Figure 65. Boxplot of mean IAP values in different body positions. The IAP was significantly higher in the anti-trendelenburg and upright position versus the supine ($p < 0.0001$, one-way Anova).

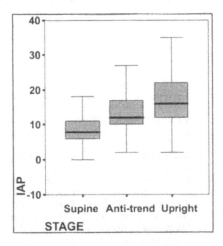

Commentary

Michael L. Cheatham

Accurate and timely assessment of intra-abdominal pressure (IAP) is crucial to the diagnosis and management of both intra-abdominal hypertension (IAH) and abdominal compartment syndrome (ACS), conditions that have been increasingly recognized to occur in the critically ill of all disciplines. There are few other disease processes where measurement of a single physiologic parameter can have such a dramatic impact on patient resuscitation and outcome. As elevated IAP may occur insidiously and clinical examination is notoriously inaccurate in detecting the presence of IAH, IAP measurements are an indispensable tool in the modern intensive care unit (ICU) setting. In this chapter, Dr. Malbrain and Ms. Jones superbly and comprehensively describe not only when to determine IAP, but also the theory and physiology behind such determinations. As they ably point out, IAP monitoring is an evolving technology that is still in its infancy with numerous measurement techniques having been described.

Since IAH and ACS cannot be diagnosed by physical examination alone and frequent IAP measurements are essential to effectively resuscitating such patients, choosing and implementing an IAP measurement technique in your ICU is paramount in improving patient morbidity and mortality. Obviously, the optimal measurement technique is the one that is safe, rapid, accurate, and cost-effective. While regional variations and financial considerations may guide your choice of technique to an extent, *the best technique is the one that you and nursing staff will use*. Cumbersome, onerous, and/or frustrating techniques will result in IAP measurements not being performed, thus defeating the purpose of IAP monitoring. As in most things, simple is best. Begin IAP monitoring with one of the traditional techniques described herein using equipment available in your ICU's supply room. As you gain experience and comfort with IAP monitoring and its application in the management of IAH and ACS, you may wish to try one or more of the other methods of IAP measurement to expand your ability to monitor these complex patients. In the near future, *continuous* IAP monitoring techniques will become widely available further improving our ability to diagnose and treat these life-threatening disease processes. In summary, choose an IAP measurement technique and use it frequently; your patient's life depends on it.

References

1. Malbrain ML. Different techniques to measure intra-abdominal pressure (IAP): Time for a critical reappraisal. Intensive Care Med 2004; 30(3):357-371.
2. Malbrain ML. Intra-abdominal pressure in the intensive care unit: Clinical tool or toy? In: Vincent JL, ed. Yearbook of Intensive Care and Emergency Medicine. Berlin: Springer-Verlag, 2001:547-585.
3. Malbrain ML, Chiumello D, Pelosi P et al. Prevalence of intra-abdominal hypertension in critically ill patients: A multicentre epidemiological study. Intensive Care Med 2004.
4. Sanchez NC, Tenofsky PL, Dort JM et al. What is normal intra-abdominal pressure? Am Surg 2001; 67(3):243-248.
5. Sugerman H, Windsor A, Bessos M et al. Intra-abdominal pressure, sagittal abdominal diameter and obesity comorbidity. J Intern Med 1997; 241(1):71-79.
6. Sugerman HJ. Effects of increased intra-abdominal pressure in severe obesity. Surg Clin North Am 2001; 81(5):1063-75, vi.
7. McNelis J, Marini CP, Jurkiewicz A et al. Predictive factors associated with the development of abdominal compartment syndrome in the surgical intensive care unit. Arch Surg 2002; 137(2):133-136.
8. Malbrain ML. Is it wise not to think about intraabdominal hypertension in the ICU? Curr Opin Crit Care 2004; 10(2):132-145.
9. Biancofiore G, Bindi ML, Romanelli AM et al. Postoperative intra-abdominal pressure and renal function after liver transplantation. Arch Surg 2003; 138(7):703-706.
10. Biancofiore G, Bindi ML, Romanelli AM et al. Intra-abdominal pressure monitoring in liver transplant recipients: A prospective study. Intensive Care Med 2003; 29(1):30-36.
11. Biancofiore G, Bindi L, Romanelli AM et al. Renal failure and abdominal hypertension after liver transplantation: Determination of critical intra-abdominal pressure. Liver Transpl 2002; 8(12):1175-1181.

12. Balogh Z, McKinley BA, Cocanour CS et al. Secondary abdominal compartment syndrome is an elusive early complication of traumatic shock resuscitation. Am J Surg 2002; 184(6):538-543.
13. Balogh Z, McKinley BA, Cocanour CS et al. Supranormal trauma resuscitation causes more cases of abdominal compartment syndrome. Arch Surg 2003; 138(6):637-642.
14. Mayberry JC, Welker KJ, Goldman RK et al. Mechanism of acute ascites formation after trauma resuscitation. Arch Surg 2003; 138(7):773-776.
15. Vincent JL, de Mendonca A, Cantraine F et al. Use of the SOFA score to assess the incidence of organ dysfunction/failure in intensive care units: Results of a multicenter, prospective study. Working group on "sepsis-related problems" of the European Society of Intensive Care Medicine. Crit Care Med 1998; 26(11):1793-1800.
16. Bone RC, Sibbald WJ, Sprung CL. The ACCP-SCCM consensus conference on sepsis and organ failure. Chest 1992; 101(6):1481-1483.
17. Malbrain ML. Abdominal perfusion pressure as a prognostic marker in intra-abdominal hypertension. In: Vincent JL, ed. Yearbook of Intensive Care and Emergency Medicine. Berlin: Springer-Verlag, 2002:792-814.
18. Van Mieghem N, Verbrugghe W, Daelemans R et al. Can abdominal perimeter be used as an accurate estimation of intra-abdominal pressure? Crit Care 2003; 7[Suppl 2]:P183.
19. Kirkpatrick AW, Brenneman FD, McLean RF et al. Is clinical examination an accurate indicator of raised intra-abdominal pressure in critically injured patients? Can J Surg 2000; 43(3):207-211.
20. Sugrue M, Bauman A, Jones F et al. Clinical examination is an inaccurate predictor of intraabdominal pressure. World J Surg 2002; 26(12):1428-1431.
21. Castillo M, Lis RJ, Ulrich H et al. Clinical estimate compared to intra-abdominal pressure measurement. Crit Care Med 1998; 26[Suppl 1]:78A.
22. Platell CF, Hall J, Clarke G et al. Intra-abdominal pressure and renal function after surgery to the abdominal aorta. Aust NZJ Surg 1990; 60(3):213-216.
23. Del Peso G, Bajo MA, Costero O et al. Risk factors for abdominal wall complications in peritoneal dialysis patients. Perit Dial Int 2003; 23(3):249-254.
24. Fischbach M, Terzic J, Gaugler C et al. Impact of increased intraperitoneal fill volume on tolerance and dialysis effectiveness in children. Adv Perit Dial 1998; 14:258-264.
25. Fischbach M, Terzic J, Provot E et al. Intraperitoneal pressure in children: Fill-volume related and impacted by body mass index. Perit Dial Int 2003; 23(4):391-394.
26. Pouliart N, Huyghens L. An observational study on intraabdominal pressure in 125 critically ill patients. Crit Care 2002; 6[Suppl 1]:S3.
27. Darovic GO, Vanriper S, Vanriper J. Dynamic response properties in hemodynamic monitoring. In: Darovic GO, ed. Hemodynamic monitoring. Philadelpjia: WB Saunders, 1995:149-175.
28. Kleinman B, Powell S, Kumar P et al. The fast flush test measures the dynamic response of the entire blood pressure monitoring system. Anesthesiology 1992; 77:1215-1210.
29. Kron IL. A simple technique to accurately determine intra-abdominal pressure. Crit Care Med 1989; 17(7):714-715.
30. Kron IL, Harman PK, Nolan SP. The measurement of intra-abdominal pressure as a criterion for abdominal reexploration. Ann Surg 1984; 199(1):28-30.
31. Iberti TJ, Lieber CE, Benjamin E. Determination of intra-abdominal pressure using a transurethral bladder catheter: Clinical validation of the technique. Anesthesiology 1989; 70(1):47-50.
32. Iberti TJ, Kelly KM, Gentili DR et al. A simple technique to accurately determine intra-abdominal pressure. Crit Care Med 1987; 15(12):1140-1142.
33. Cheatham ML, Safcsak K. Intraabdominal pressure: A revised method for measurement. J Am Coll Surg 1998; 186(5):594-595.
34. Sagraves SG, Cheatham ML, Johnson JL et al. Intravesicular pressure monitoring does not increase the risk of urinary tract or systemic infection. Crit Care Med 1007; 27:A48.
35. Sugrue M, Buist MD, Hourihan F et al. Prospective study of intra-abdominal hypertension and renal function after laparotomy. Br J Surg 1995; 82(2):235-238.
36. Harrahill M. Intra-abdominal pressure monitoring. J Emerg Nurs 1998; 24(5):465-466.
37. Crispin C, Jones W, Daffurn K. How consistently do RN's perform the procedure of collecting specimens for the measurement of gastric pHi and CO2. Intensive Crit Care Nurs 1995; 11:123-125.
38. Lee SL, Anderson JT, Kraut EJ et al. A simplified approach to the diagnosis of elevated intra-abdominal pressure. J Trauma 2002; 52(6):1169-1172.
39. Malbrain MLNG, Leonard M, Delmarcelle D. A novel technique of intra-abdominal pressure measurement: Validation of two prototypes. Crit Care 2002; 6[Suppl 1]:S2-S3.
40. Collee GG. Intra-abdominal pressure can be measured by measuring the pressure within the stomach. Intensive Care Med 1996; 22(3):269.

41. Collee GG, Lomax DM, Ferguson C et al. Bedside measurement of intra-abdominal pressure (IAP) via an indwelling naso-gastric tube: Clinical validation of the technique. Intensive Care Med 1993; 19(8):478-480.
42. Sugrue M, Buist MD, Lee A et al. Intra-abdominal pressure measurement using a modified nasogastric tube: Description and validation of a new technique. Intensive Care Med 1994; 20(8):588-590.
43. Debaveye Y, Bertiaux S, Malbrain MLNG. Simultaneous measurement of intra-abdominal pressure and regional CO_2 via a gastric tonometer. Intensive Care Med 2000; 26[Suppl 3]:S324.
44. Malbrain ML. Validation of a novel fully automated continuous method to measure intra-abdominal pressure. Intensive Care Med 2003; 29[Supplem 1]:S73.
45. Schachtrupp A, Tons C, Fackeldey V et al. Evaluation of two novel methods for the direct and continuous measurement of the intra-abdominal pressure in a porcine model. Intensive Care Med 2003; 29(9):1605-1608.
46. Cheatham ML, White MW, Sagraves SG et al. Abdominal perfusion pressure: A superior parameter in the assessment of intra-abdominal hypertension. J Trauma 2000; 49(4):621-626.
47. Shafik A, El Sharkawy A, Sharaf WM. Direct measurement of intra-abdominal pressure in various conditions. Eur J Surg 1997; 163(12):883-887.
48. Lacey SR, Bruce J, Brooks SP et al. The relative merits of various methods of indirect measurement of intraabdominal pressure as a guide to closure of abdominal wall defects. J Pediatr Surg 1987; 22(12):1207-1211.
49. Dowdle M. Evaluating a new intrauterine pressure catheter. J Reprod Med 1997; 42:506-513.
50. Joynt GM, Gomersall CD, Buckley TA et al. Comparison of intrathoracic and intra-abdominal measurements of central venous pressure. Lancet 1996; 347(9009):1155-1157.
51. Gudmundsson FF, Viste A, Gislason H et al. Comparison of different methods for measuring intra-abdominal pressure. Intensive Care Med 2002; 28(4):509-514.
52. Malbrain ML, Deeren D, Darquennes K et al. Estimating the optimal bladder volume for intra-abdominal pressure measurement by bladder PV curves. Intensive Care Med 2003; 29[Supplem 1]:S147.
53. Morgan B. Abdominal compartment pressure monitoring. 2001. www.lhsc.on.ca/critcare/icu/education/abdcompt.html, (Ref Type: Electronic Citation).
54. Yol S, Kartal A, Tavli S et al. Is urinary bladder pressure a sensitive indicator of intra-abdominal pressure? Endoscopy 1998; 30(9):778-780.
55. Fusco MA, Martin RS, Chang MC. Estimation of intra-abdominal pressure by bladder pressure measurement: Validity and methodology. J Trauma 2001; 50(2):297-302.
56. Johna S. Can we use the bladder to estimate intra-abdominal pressure? J Trauma 2001; 51(6):1218.
57. Johna S, Taylor E, Brown C et al. Abdominal compartment syndrome: Does intra-cystic pressure reflect actual intra-abdominal pressure? A prospective study in surgical patients. Crit Care (Lond) 1999; 3(6):135-138.
58. Malbrain MLNG, Van Mieghem N, Verbrugghe W et al. Effects of different body positions on intra-abdominal pressure and dynamic respiratory compliance. Crit Care 2003; 7(Suppl 2):P179.
59. Malbrain ML. Abdominal pressure in the critically ill: Measurement and clinical relevance. Intensive Care Med 1999; 25(12):1453-1458.

CHAPTER 4

Abdominal Perfusion Pressure

Michael L. Cheatham* and Manu L. N. G. Malbrain

Summary

E levated intra-abdominal pressure (IAP) is commonly encountered in the critically ill and is associated with significant morbidity and mortality. The "critical IAP" that causes end-organ dysfunction varies from patient to patient as a result of each individual's physiology and preexisting comorbidities. As a result, a single threshold value of IAP cannot be globally applied to the decision making of all critically ill patients. Calculation of "abdominal perfusion pressure" (APP), defined as mean arterial pressure (MAP) minus IAP, assesses not only the severity of IAP present, but also the adequacy of the patient's abdominal blood flow. APP is superior to both IAP and global resuscitation endpoints such as arterial pH, base deficit, and arterial lactate in its ability to both predict patient outcome and serve as a useful parameter for guiding the resuscitation and management of the patient with IAH or ACS.

Introduction

Although initially recognized almost 150 years ago, the pathophysiologic implications of elevated intra-abdominal pressure (IAP) have essentially been rediscovered only within the past decade.[1-15] Elevated IAP or "intra-abdominal hypertension" (IAH) is now commonly identified in the critically ill and acknowledged as a cause of significant morbidity and mortality.[7,16-18] IAH has been recognized as a continuum of pathophysiologic changes beginning with regional blood flow disturbances and ultimately culminating in the IAP-induced end-organ failure now known as the "abdominal compartment syndrome" (ACS). A recent multi-center epidemiologic study found IAH (defined as IAP $\geq$ 12 mm Hg) to be present in 51% of critically ill medical and surgical intensive care unit (ICU) patients and ACS (defined as IAP $\geq$ 20 mm Hg with one or more organ failures) to be present in 8%.[16] The prevalence of elevated IAP in patients developing organ failure suggests that IAH may well play a major role in the development of multiple system organ failure (MSOF), a major cause of ICU mortality.

Further complicating the diagnosis and management of these critically ill patients is the fact that IAH is difficult to detect by physical examination alone.[19] Elevated IAP can easily go undetected resulting in significant end-organ dysfunction and failure. Assessment of IAP, most commonly using the surrogate measurement of intravesicular or "bladder" pressure, has been identified as being essential to the accurate diagnosis and treatment of patients with IAH or ACS.[20-23] Given the prevalence of IAH in the ICU, the difficulty of diagnosis, and the significant associated morbidity and mortality, the importance of measuring IAP, identifying the presence of end-organ malperfusion, and restoring both systemic and regional organ perfusion in patients with IAH or ACS cannot be overemphasized.

*Corresponding Author: Michael L. Cheatham—Department of Surgical Education, Orlando Regional Medical Center, 86 West Underwood Street, Mailpoint #100, Orlando, Florida, 32806 U.S.A. Email: michael.cheatham@orhs.org

Abdominal Compartment Syndrome, edited by Rao R. Ivatury, Michael L. Cheatham, Manu L. N. G. Malbrain and Michael Sugrue. ©2006 Landes Bioscience.

As our recognition of the detrimental effects of elevated IAP has evolved, so has our understanding of what constitutes IAH. Whereas early studies suggested that an IAP of 30 to 40 mm Hg was acceptable, we now recognize that even small elevations in IAP to 10 to 15 mm Hg can have a tremendous impact on end-organ perfusion and patient outcome. As a result, the definition of IAH has required continual adjustment over the years and the "critical IAP" that mandates intervention has been revised downward.[5,7,18,22] The patient populations at risk for developing IAH and ACS have also been expanded. Originally considered a disease of the traumatically injured, IAH and ACS have now been recognized to occur in virtually all patient populations. The acuity of onset and presence of preexisting comorbid conditions have been identified to exert a significant impact upon the presentation and management of elevated IAP. Age, obesity, prior pregnancy, mechanism of injury, pulmonary, cardiac, or renal dysfunction or failure, need for mechanical ventilation, and development of systemic inflammatory response syndrome (SIRS) have all been recognized to modulate the development and appropriate treatment of these critically ill patients.

During the early evolution of our understanding of IAH and ACS, attempts were made to identify a single threshold value of IAP that could be simply and globally applied to the decision making of all critically ill patients with IAH. The "critical IAP" that mandates intervention may well be considered the "holy grail" of IAH and ACS research. Such a quest, however, oversimplifies what is actually a highly complex and variable physiologic process. While IAP is a major determinant of patient outcome during critical illness, the IAP that defines IAH and ACS clearly varies from patient to patient and even within the same patient as their disease process evolves. Within the clinically acceptable ranges, IAP is a specific, but nonsensitive predictor of illness and resuscitation adequacy. Improving the diagnostic sensitivity of IAP would require allowing patients to maintain higher IAP values. Such levels of IAH, however, are well-known to cause end-organ malperfusion and such a practice would not be clinically acceptable given our current understanding of the morbidity and mortality of IAH. Thus, although a useful screening tool, IAP lacks sufficient sensitivity to be useful as a resuscitation endpoint. Given the marked variation in IAP values that may be witnessed in the critically ill, it is unlikely that a single threshold value of IAP will ever be universally applicable to all critically ill patients.

An alternative approach to improving the sensitivity of IAP is to incorporate it in the assessment of abdominal perfusion as a resuscitation endpoint. Analogous to the widely accepted and utilized concept of cerebral perfusion pressure (CPP), calculated as mean arterial pressure (MAP) minus intracranial pressure (ICP), the abdominal perfusion pressure (APP), calculated as MAP minus IAP, has been proposed as a more accurate marker of critical illness and endpoint for resuscitation.[18,24] This chapter evaluates the scientific support for and potential clinical application of APP in the management of patients with IAH and ACS.

APP = MAP - IAP

Physiology

Elevated IAP causes significant impairment of cardiac, pulmonary, renal, gastrointestinal, and hepatic function with each organ demonstrating its own unique vulnerability. The IAP that may induce malperfusion in one organ system may have little effect on another. This differential response, coupled with the augmented susceptibility to IAP seen with hypovolemia and comorbid disease, complicates the management of these critically ill patients and use of IAP as a resuscitative endpoint. It also emphasizes the importance of assessing perfusion pressure as opposed to compartment pressure alone. The detrimental effect of IAP on individual organ systems is discussed in detail in other chapters; the following summarizes the salient points as they pertain to the validity of IAP and APP as resuscitative endpoints.

Cardiovascular

As originally described over 80 years ago by Emerson, rising IAP increases intrathoracic pressure through cephalad deviation of the diaphragm.[25] Increased intrathoracic pressure significantly reduces venous return and cardiac output and compresses both the aorta and pulmonary parenchyma raising systemic vascular resistance.[13,26-30] Such reductions have been demonstrated to occur at an IAP of only 10 mm Hg.[26,30] Hypovolemic patients, those with marginal cardiac contractility, and those requiring positive pressure ventilation appear to sustain reductions in cardiac output at lower levels of IAP than do normovolemic patients.[27,28]

Pulmonary

Increases in intrathoracic pressure, through cephalad elevation of the diaphragm, also result in extrinsic compression of the pulmonary parenchyma with development of alveolar atelectasis, decreased diffusion of oxygen and carbon dioxide across the pulmonary capillary membrane, and increased intrapulmonary shunt fraction and alveolar dead space.[26,29,30] This dysfunction begins at an IAP of 15 mm Hg and is accentuated by the presence of hypovolemia.[30] In combination, these effects lead to the arterial hypoxemia and hypercarbia that characterize ACS.[7,26,30]

Renal

Elevated IAP significantly decreases renal artery blood flow and compresses the renal vein leading to renal dysfunction and failure.[31,32] Oliguria develops at an IAP of 15 mm Hg and anuria at 30 mm Hg in the presence of normovolemia and at lower levels of IAP in the patient with hypovolemia.[32] Renal perfusion pressure (RPP) and renal filtration gradient (FG) have been proposed as key factors in the development of IAP-induced renal failure.[33] The FG is the mechanical force across the glomerulus and equals the difference between the glomerular filtration pressure (GFP) and the proximal tubular pressure (PTP).

$$FG = GFP - PTP$$

In the presence of IAH, PTP may be assumed to equal IAP and GFP can be estimated as MAP - IAP. The FG can then be calculated by the formula:

$$FG = MAP - 2 \times IAP$$

Thus, changes in IAP have a greater impact upon renal function and urine production than will changes in MAP. It should not be surprising, therefore, that decreased renal function, as evidenced by development of oliguria, is one of the first visible signs of IAH.

Gastrointestinal

The gut appears to be particularly sensitive to IAH with virtually all intra-abdominal and retroperitoneal organs demonstrating decreased blood flow in the presence of elevated IAP.[34] Reductions in mesenteric blood flow may appear with an IAP of only 10 mm Hg.[35] Celiac artery blood flow is reduced by up to 43% and superior mesenteric artery blood flow by as much as 69% in the presence of an IAP of 40 mm Hg.[35,36] The negative effects of IAP on mesenteric perfusion are augmented by the presence of hypovolemia or hemorrhage.[11,28,35] Bowel ischemia and inadequate perfusion initiates a vicious cycle of worsening perfusion, increased capillary leak, decreased intramucosal pH, and systemic metabolic acidosis.[11,14,28] An IAP of 20 mm Hg diminishes intestinal mucosal perfusion and has been speculated as a possible mechanism for subsequent development of bacterial translocation, sepsis, and MSOF.[11,14,28,36-38] Bacterial translocation to mesenteric lymph nodes has been demonstrated to occur in the presence of hemorrhage with an IAP of only 10 mm Hg.[38]

Table 1. The cranial and abdominal compartments

	Cranium	Abdomen
Organ(s)	Brain	Liver, spleen, kidneys, stomach, intestines
Fluid(s)	Cerebrospinal fluid	Ascites, air, feces
Enclosure	Skull	Abdominal cage
Lesions	Tumor, hematoma	Blood, edema, ascites, air, tumor
Pressure	ICP	IAP
Perfusion	CPP = MAP - ICP	APP = MAP - IAP

ICP: intracranial pressure; IAP: intra-abdominal pressure, CPP: cerebral perfusion pressure MAP: mean arterial pressure, APP: abdominal perfusion pressure

Hepatic

Hepatic artery blood flow is directly affected by IAP-induced decreases in cardiac output while hepatic vein and portal vein blood flow are reduced by extrinsic compression.[28] These changes have been documented with IAP elevations of only 10 mm Hg and in the presence of both normal cardiac output and mean arterial blood pressure.[28]

Theory

The perfusion pressure of any anatomic compartment is dependent upon three factors
  1. the arterial inflow pressure
  2. the venous outflow pressure
  3. the compliance or ability of the compartment to expand in response to increases in volume

Perhaps the most clinically accepted example of perfusion pressure is that of the traumatized brain. As the brain is enclosed within a bony "box" and generally cannot expand past the confines of the skull, cranial compliance becomes a minor issue and CPP may be calculated as arterial inflow (MAP) minus venous outflow (ICP). MAP is determined by intravascular volume, cardiac contractility, and systemic vascular resistance while ICP is dependent upon the respective volumes of the brain, cerebrospinal fluid, intracranial blood, and any space-occupying lesion such as hematoma or tumor. By the Monro-Kellie Doctrine, any increase in the volume of one or more of these four constituents of the cranium will result in an increase in ICP. Optimization of CPP is a matter of either increasing MAP (through use of appropriate resuscitation fluids or vasopressors) or decreasing ICP (through administration of diuretics, drainage of cerebrospinal fluid, or evacuation of space-occupying lesions). Interestingly, recent studies have resurrected the concept of hemicraniectomy (removal of a large portion of the skull) to increase cranial compliance much as abdominal decompression increases compliance of the abdominal cavity.[39]

Abdominal perfusion may be considered analogous to that of the brain (Table 1). The abdomen contains a variety of solid organs (liver, kidneys, spleen, etc.) and fluids (blood, urine, enteric contents, etc.) of limited compliance, but also air-filled structures (stomach, small intestine, colon) of marked distensibility as well as potential spaces (peritoneum, retroperitoneum) that may expand significantly in response to injury or illness. The abdomen may also contain pathologic space-occupying lesions such as blood, air, ascites, or tumor. As with the Monro-Kellie Doctrine and the brain, an increase in the volume of the abdominal cavity contents will result in an increase in IAP. Although not enclosed in a rigid shell as is the brain, the abdomen is far from being freely compliant and expandable. Portions of the abdomen, such as the spine, pelvis, and costal arch, are fairly rigid while others, such as the diaphragm and especially the abdominal wall, are compliant only up to a point determined by a variety of factors. Age, obesity, abdominal wall musculature, prior pregnancy, and previous abdominal surgery may each alter abdominal wall compliance. Pain and third-space edema also impair compliance and augment the effects of elevated IAP. All of these factors may significantly impact upon the patient's ability to tolerate IAH as well as the adequacy of abdominal perfusion.

Clinical Studies

Cheatham et al first proposed the concept of APP as a predictor of survival in patients with IAH or ACS in 2000.[18] A retrospective study was performed evaluating all patients admitted to a surgical / trauma ICU with evidence of IAH (defined as IAP ≥ 15 mm Hg). IAP monitoring was instituted whenever one or more signs of IAH-induced organ dysfunction were present including abdominal distention, oliguria refractory to volume administration, hypercarbia, hypoxemia refractory to increasing inspired oxygen fractions and positive end-expiratory pressure, elevated peak inspiratory pressures, or refractory hypotension. IAP monitoring was also instituted in the absence of the above conditions if there was sufficient physician concern for the presence of elevated IAP. Intravesicular pressure was measured, using the technique previously described by Cheatham and Safcsak, every 4 hours until either IAP normalized or measurements were stable and the patient was no longer felt to be at risk for IAH-induced organ dysfunction.[21] Commonly advocated resuscitation endpoints including arterial pH, base deficit, arterial lactate, and hourly urinary output were recorded. Open abdominal decompression and temporary abdominal closure were performed for symptomatic IAH or development of ACS (defined as IAP ≥ 25 mm Hg in the presence of one or more signs of IAH-induced organ dysfunction). Multivariate logistic regression analysis was performed to identify those physiologic variables and resuscitation endpoints significantly associated with patient survival.

During the 25-month study period, 144 patients (68% trauma, 14% general surgery, 14% vascular surgery, 2% colorectal surgery, and 2% obstetrics / gynecology) underwent 2298 IAP measurements during resuscitation for IAH or ACS. The mean IAP for all patients was 22 ± 8 mm Hg (range 2-94) despite relatively liberal application of abdominal decompression. On average, patients underwent IAP monitoring for 3 ± 2 days (range 1-11 days) during which time 16 ± 14 measurements (range 1-61) were performed. Overall mortality was 53%. Significantly fewer patients developed ACS in the second half of the study (64% vs. 43%; p = 0.01) and mortality was significantly decreased (44% vs. 28%; p = 0.049) as a result of physician education and increased acceptance of abdominal decompression.

Multiple logistic regression analysis demonstrated that IAP, MAP, APP, arterial lactate, arterial pH, base deficit, and hourly urinary output were all significantly associated with patient survival from IAH with hourly urinary output and APP being most predictive (Table 2). Further analysis utilizing the worst values measured during the patient's resuscitation identified that lowest APP was significantly superior to the other resuscitation endpoints in its ability to predict patient survival from IAH (Table 3).

Receiver operator characteristic (ROC) curves were generated for IAP and APP in order to identify the threshold values of each endpoint that were most predictive of patient outcome. ROC curves graph the sensitivity of a diagnostic test (true positive proportion) versus 1 minus specificity (false positive proportion) and provide an improved measure of the overall discriminatory power of a test as they assess all possible threshold values. A test that always predicts survival has an area under the ROC curve of 1.0 and a test that predicts survival no more often than would be done by chance has an area under the ROC curve of 0.5. The point on the ROC curve closest to the upper left corner is generally considered to optimize the sensitivity and specificity of the test. In this study, the area under the ROC curve was 0.726 for APP and 0.748 for IAP (Fig. 1; IAP has been plotted against mortality instead of survival as in the original study).[18] Although the areas under the ROC curves for APP and IAP are not statistically different, the curves demonstrate that the sensitivity and specificity of APP are both superior to that of IAP for the clinically useful decision thresholds (Table 4). Maintenance of an APP of at least 50 mm Hg appears to maximize both the sensitivity (76%) and specificity (57%) of APP as a predictor of patient survival. The commonly utilized MAP resuscitation endpoint of 70 mm Hg achieved a sensitivity of only 57% and specificity of 61%. While an IAP threshold of 30

Table 2. Resuscitative endpoints in IAH / ACS stratified by survival[18]

	Survivors	Significance of Multiple Non-Survivors	Logistic Regression
Hourly urine output (mL/h)	113 ± 112	79 ± 111	< 0.0001
APP (mm Hg)	69 ± 17	61 ± 18	0.0001
Arterial l actate (mmol/L)	2.9 ± 1.5	4.5 ± 2.5	0.0002
MAP (mm Hg)	88 ± 15	85 ± 15	0.0004
Arterial pH	7.34 ± 0.08	7.27 ± 0.10	0.03
Base deficit	3.6 ± 4.8	7.5 ± 5.6	0.04
IAP (mm Hg)	20 ± 6	24 ± 8	0.05

MAP: mean arterial pressure; IAP: intra-abdominal pressure; APP: abdominal perfusion pressure

mm Hg achieved a sensitivity of 70% and specificity of 72%, this endpoint exceeds what is now recognized as being clinically acceptable and its application would place the patient at risk for significant end-organ malperfusion. Within the currently advocated ranges of 10 to 25 mm Hg, IAP was specific, but not sensitive for predicting patient outcome. APP appears to be a clinically superior resuscitation endpoint and predictor of patient survival during treatment of IAH and ACS as it addresses not only the severity of IAH, but also the adequacy of end-organ perfusion.

To evaluate the clinical validity of various resuscitation endpoints, including APP, in patients with IAH, Malbrain et al prospectively evaluated 8 patients treated by their surgeon with an abdominal binder to reduce postoperative dehiscence and incisional hernia.[24] IAP was noted to increase significantly upon application of the abdominal binder with significant decreases in cardiac output and increases in central venous and pulmonary artery occlusion pressure appearing within 30 to 45 minutes. With elevations in IAP, visceral perfusion worsened as evidenced by decreased APP, FG, arterial pH, intramucosal pH (pHi), and arterial-gastric mucosal carbon dioxide difference (CO_2-gap). The addition of positive end-expiratory pressure (PEEP) of 15 cm H_2O in combination with elevated IAP had even greater deleterious effects on APP and visceral perfusion. These effects were reversible with release of the abdominal binder (mimicking abdominal decompression) and a decrease in PEEP. This study confirmed the significant deleterious impact of both elevated IAP and intrathoracic pressure on APP and

Table 3. Outcome variables in IAH / ACS stratified by Survival[18]

	Survivors	Significance of Multiple Non-Survivors	Logistic Regression
Lowest APP (mm Hg)	52 ± 17	39 ± 18	0.002
Lowest MAP (mm Hg)	74 ± 14	69 ± 14	0.05
Highest IAP (mm Hg)	29 ± 12	38 ± 14	0.21
Highest arterial lactate (mmol/L)	5.4 ± 2.2	8.0 ± 3.7	0.38
Highest base deficit	9.0 ± 7.0	13.1 ± 6.9	0.44
Lowest arterial pH	7.24 ± 0.10	7.14 ± 0.13	0.66
Lowest hourly urinary output (mL/h)	47 ± 48	44 ± 68	0.85

MAP: mean arterial pressure; IAP: intra-abdominal pressure; APP: abdominal perfusion pressure

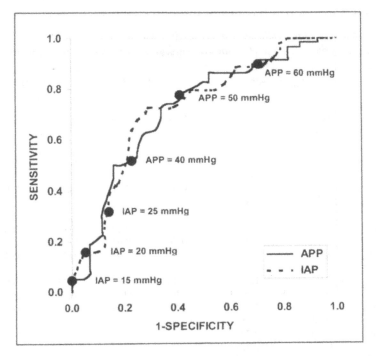

Figure 1. Receiver operator characteristic (ROC) curves for IAP and APP with clinically useful decision points. IAP: intra-abdominal pressure; APP: abdominal perfusion pressure

visceral perfusion. Abdominal decompression increased APP and decreased IAP restoring abdominal perfusion. Regional markers of perfusion adequacy such as APP, FG, pHi, and CO_2-gap were noted to change more rapidly than global indices such as MAP, arterial pH, base deficit, and calculated bicarbonate (HCO_3^-) suggesting that the regional markers are superior resuscitation endpoints.

Malbrain et al subsequently prospectively studied the development of IAH in 405 patients admitted to a mixed ICU over a 12-month period.[24] IAP was routinely assessed in all patients and the maximal IAP and lowest APP values measured within the first 72 hours were recorded. IAH (defined as IAP > 12 mm Hg) was associated with a significantly higher ICU mortality (65% vs. 8%; p<0.0001) and hospital mortality (69% vs. 18%; p<0.0001). APP was significantly lower among nonsurvivors (61 ± 23 vs. 76 ± 23 mm Hg; p<0.0001) as was MAP (72 ± 22 vs. 83 ± 22 mm Hg; p<0.0001) while IAP was significantly higher (11± 5 vs. 7± 4 mm Hg; p<0.0001). An APP of 60 mm Hg was identified as having a sensitivity of 55% and specificity of 76% for predicting survival while a MAP of 70 mm Hg had a sensitivity of 58% and specificity of 64% (Table 4). An IAP of 9 mm Hg had the best sensitivity (65%) and specificity (72%) for predicting patient outcome, however, this IAP threshold may be unrealistic for a more critically ill patient population. As only 18% of the patients in this study had evidence of IAH and only 2% had ACS (defined as IAP > 20 mm Hg), the findings of this study suggest that APP may have application as a resuscitative endpoint in not only the patient with IAH or ACS, but in the broader ICU patient population as well.

The Critically Ill and Abdominal Hypertension (CIAH) study group subsequently performed a prospective international multi-center trial in which 257 patients were screened for IAH (defined as IAP ≥ 12 mm Hg).[24] Patient demographics included 47% medical, 28% elective surgery, 17% emergency surgery, and 9% trauma. Overall hospital mortality was 35% (48% medical, 11% elective surgery, 44% emergency surgery, 18% trauma). Conditions

Table 4. *Sensitivity and specificity for predicting patient survival from intra-abdominal hypertension and abdominal compartment syndrome*

	Cheatham[18]		Malbrain[24]		CIAH[24]	
	SENS	SPEC	SENS	SPEC	SENS	SPEC
APP						
40 mm Hg	0.53	0.78	0.14	0.98	0.14	0.95
50 mm Hg	0.76	0.55	0.39	0.91	0.53	0.85
60 mm Hg	0.92	0.25	0.55	0.76	0.79	0.62
70 mm Hg	0.97	0.18				
MAP						
60 mm Hg	0.23	0.87	0.34	0.87	0.40	0.87
70 mm Hg	0.57	0.60	0.58	0.64	0.69	0.61
80 mm Hg	0.83	0.21	0.74	0.42	0.85	0.41
IAP						
12 mm Hg	0.05	1.0	0.44	0.93	0.75	0.59
15 mm Hg	0.05	1.0	0.23	0.97	0.47	0.75
20 mm Hg	0.16	0.85	0.06	0.99	0.17	0.92
25 mm Hg	0.32	0.86				
30 mm Hg	0.70	0.72				
35 mm Hg	0.80	0.47				
40 mm Hg	0.89	0.32				

MAP: mean arterial pressure; IAP: intra-abdominal pressure; APP: abdominal perfusion pressure

associated with the development of IAH included acidosis, hypothermia, polytransfusion, volume resuscitation, coagulopathy, sepsis, abdominal surgery, ileus, and hepatic dysfunction. IAP was significantly correlated with the development of both organ failure and mortality. APP was significantly lower among nonsurvivors (54 ± 16 vs. 69 ± 23 mm Hg; $p<0.0001$) as was MAP (68 ± 15 vs. 81 ± 23 mm Hg; $p<0.0001$) while IAP was significantly higher (15 ± 6 vs. 12 ± 5 mm Hg; $p<0.0001$). Within the subgroup of 145 patients with IAH, APP continued to be significantly lower among nonsurvivors (52 ± 14 vs. 65 ± 23 mm Hg; $p<0.001$) while IAP was no longer different (17 ± 5 vs. 17 ± 4 mm Hg) confirming the superiority of APP as a predictor of clinical outcome. The area under the ROC curve was 0.732 for APP and 0.678 for IAP. An APP of 60 mm Hg was associated with the best sensitivity (79%) and specificity (62%) while a MAP of 70 mm Hg had a sensitivity of 69% and specificity of 61%. An IAP threshold of 12 mm Hg had a sensitivity of 75% and specificity of 59%. Based upon these findings, the CIAH study group recommended maintaining an APP above 60 mm Hg and IAP below 12 mm Hg to optimize patient outcome.

The CIRFAH (Critically Ill Renal Failure and Abdominal Hypertension) study prospectively evaluated IAP and APP as predictors of outcome in 60 patients with acute renal failure (ARF) (defined as serum creatinine > 2 mg/dL). Over a 12-month period, patients admitted with or who developed ARF during their ICU stay were screened for IAH (defined as IAP $\geq$ 12 mm Hg) using intravesicular pressure measurements. IAP was recorded twice daily together with the highest and lowest APP, fluid balance, and Sequential Organ Failure Assessment (SOFA) score. There were 78% medical and 22% surgical patients. The renal Stuivenberg Hospital Acute Renal Failure (SHARF-II) score was 67 ± 22 on admission and 76 ± 24 after 48 hours. The SOFA score on day 1 was 9.4 ± 3.5 with 1.7 ± 1.1 organ failures. The IAP on day 1 was 12 ± 5 mm Hg while APP was 55 ± 18 mm Hg. Maximal IAP after 48 hours (IAPmax) was $14 \pm$

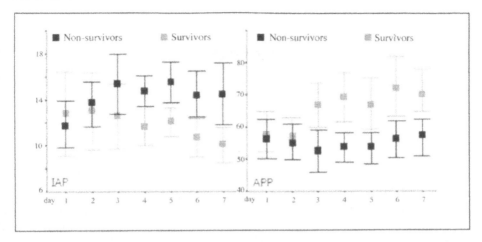

Figure 2. Evolution of IAP and APP in patients with acute renal failure stratified in survivors and non survivors. IAP: intra-abdominal pressure; APP: abdominal perfusion pressure.

6 mm Hg. IAH > 12 mm Hg within 48 hours of study inclusion was present in 63% patients. Outcome data on 46 patients showed a 28-day mortality of 63%. Outcome did not differ by presence of IAH although nonsurvivors had a significantly higher IAP and lower APP by day 3 (Fig. 2). There was also a trend towards a more positive daily fluid balance and cumulative net fluid balance in nonsurvivors. Thus, the incidence of IAH is high in patients with ARF and is associated with significant mortality that is underestimated by classic severity scores. The SHARF-II score better predicts outcome in these patients. The persistence of IAH and a low APP by day 3 was able to discriminate between survivors and nonsurvivors. Close monitoring of IAP and APP therefore seems warranted in patients with ARF.

Clinical Application

Early goal-directed therapy to restore end-organ perfusion and oxygen delivery at the cellular level is essential to reducing patient morbidity and mortality. As critical IAP varies from patient to patient, IAP alone cannot serve as a therapeutic goal in this patient population. APP, calculated as MAP minus IAP, assesses not only the severity of a patient's IAP, but also evaluates the adequacy of their abdominal perfusion. APP provides an easily calculated measure that has been demonstrated to exceed the clinical prediction of IAP alone in early clinical trials. APP improves upon the sensitivity of IAP while maintaining diagnostic specificity and appears to correlate well with visceral perfusion. Only time and further experience will determine whether APP, or another as yet undetermined resuscitation endpoint, proves its usefulness in large scale clinical trials.

Merging the results of the above studies with the available clinical literature on the pathophysiologic implications of IAH and management of ACS, an appropriate management algorithm for the patient with elevated IAP can be proposed (Fig. 3). First, serial IAP measurements should be performed liberally in the critically ill due to the high incidence of IAH in this patient population and its significant associated morbidity and mortality. Due to the inadequacy of physical examination in detecting elevated IAP, serial IAP measurements are currently the only method available by which to accurately diagnose IAH and direct appropriate treatment. Intravesicular pressure monitoring can be performed rapidly and inexpensively without specialized monitoring equipment using materials available in any ICU.[21]

Second, immediate abdominal decompression should be performed in any patient who demonstrates significant elevations in IAP or evidence of ACS. In surgical patients, this is best achieved by either creating or reopening their laparotomy incision and applying a temporary

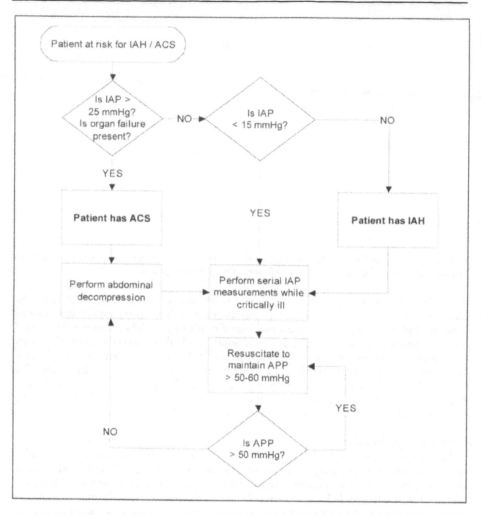

Figure 3. IAH / ACS resuscitation algorithm. IAH: intra-abdominal hypertension; ACS: abdominal compartment syndrome; IAP: intra-abdominal pressure; APP: abdominal perfusion pressure.

abdominal closure. This can be performed either in the operating room or at the patient's bedside in the ICU based upon their current hemodynamic stability.[7,18] Such a procedure should not be feared or delayed as rapid decompression following the diagnosis of ACS dramatically improves patient organ function and survival.[18] A recent long-term outcome study demonstrated no significant residual physical or mental health deficits and a return to normal functionality within 12 months among patients requiring emergent abdominal decompression.[17] In medical patients whose IAH is secondary to accumulation of ascites or resuscitation fluid, paracentesis should be considered as a viable alternative to open abdominal decompression. Leaving the paracentesis catheter in place until the patient's condition stabilizes allows ongoing drainage of peritoneal fluid, continued reduction in IAP, and a reduced incidence of tertiary or recurrent ACS. Patients whose IAH is secondary to retroperitoneal hemorrhage, visceral edema, or ileus will be best served by open abdominal decompression as paracentesis will not be effective in reducing the severity of IAH or restoring organ perfusion.

Third, in patients with IAH not requiring immediate decompression, APP should be maintained above 50-60 mm Hg through a combination of volume resuscitation and vasoactive medications. Intravascular volume status should be optimized, with vasopressors being reserved for those patients who continue to demonstrate an inadequate APP despite appropriate volume resuscitation. The use of volumetric measurements of intravascular volume status such as the right ventricular end-diastolic volume index (RVEDVI) or intrathoracic blood volume (ITBV) should be considered.[40,41] Traditional intracardiac filling pressure measurements such as pulmonary artery occlusion pressure (PAOP) and central venous pressure (CVP) are inaccurate in the presence of IAH and elevated intrathoracic pressure and reliance upon these parameters may lead to underresuscitation.[40]

Fourth, inability to maintain an APP of at least 50 mm Hg is an indication for open abdominal decompression and temporary abdominal closure until such time as the patient's clinical status improves. Once the patient's abdomen is open, IAP should continue to be monitored as IAH and ACS, contrary to popular belief, can recur despite an open abdomen. In the patient with an open abdomen, inability to maintain an adequate APP is an indication to decompress the patient's abdomen further through either a larger laparotomy or placement of a looser, more compliant temporary abdominal closure.

Fifth, attempts to close the patient's abdomen following decompression should be directed by the patient's IAP and APP. APP can be utilized to guide not only the difficult question of when to perform a decompressive laparotomy, but also the frequently more complicated decision of when and how to perform abdominal closure. Persistent elevations in IAP with marginal APP calculations should lead to a surgical decision for split-thickness skin grafting of the exposed viscera as opposed to attempts to close the patient's abdominal wall under tension. The latter will undoubtedly result in recurrent IAH, decreased visceral perfusion, and increased risk of incisional dehiscence.

In summary, the simple calculation of APP is superior to IAP alone as a resuscitative endpoint in patients with IAH or ACS as it assesses both the severity of IAH and the adequacy of end-organ perfusion. APP is also superior to global markers of resuscitation adequacy such as arterial pH, base deficit, and arterial lactate in its ability to quickly identify inadequate visceral perfusion. An APP of 50 mm Hg or greater should be considered a resuscitation endpoint in the patient with elevated IAP.

Commentary

Rao R. Ivatury

Cheatham and Malbrain, my distinguished co-editors and champions of the cause of intra-abdominal hypertension and the abdominal compartment syndrome, describe in this chapter the concept of abdominal perfusion pressure, which in the future may revolutionize our interpretation of the intra-abdominal pressure. The authors provide strong evidence why the effective "transmural pressure" (perfusion pressure, compartment perfusion pressure, transabdominal pressure, trans-thoracic pressure and such), modeled along the lines of compartment pressure of the extremities, will be important to be considered.

The authors provide persuasive arguments why APP is superior to random measurements of IAP alone. They also provide five sequential therapeutic steps to correct an abnormally low APP. If one believes in the concept of transmural or perfusion pressure, as in the thorax or in the kidney that drives perfusion, it makes sound physiologic sense to follow such perfusion pressures (APP) and not absolute readings of pressure (IAP) alone.

Obviously, the proof of a physiologic concept, however sound, will rest in the clinical validation of the concept. Future studies, based on sound outcome measures, will determine the merit of this concept. Till then, it makes perfect sense to mark APP our most desirable goal.

References

1. Schein M, Ivatury R. Intra-abdominal hypertension and abdominal compartment syndrome. Br J Surg 1998; 85:1027-1028.
2. Bendahan J, Coetzee CJ, Papagianopoulos C et al. Abdominal compartment syndrome. J Trauma 1995; 38:152-153.
3. Schein M, Wittmann DH, Aprahamian CC et al. The abdominal compartment syndrome: The physiological and clinical consequences of elevated intra-abdominal pressure. J Am Coll Surg 1995; 180:745-753.
4. Ivatury RR, Diebel L, Porter JM et al. Intra-abdominal hypertension and the abdominal compartment syndrome. Surg Clin N Am 1997; 77:783-800.
5. Burch JM, Moore EE, Moore FA et al. The abdominal compartment syndrome. Surg Clin N Am 1996; 76:833-842.
6. Saggi BH, Sugerman HJ, Ivatury RR et al. Abdominal compartment syndrome. J Trauma 1998; 45:597-609.
7. Cheatham ML. Intra-abdominal hypertension and abdominal compartment syndrome. New Horizons 1999; 7:96-115.
8. Mayberry JC, Mullins RJ, Crass RA et al. Prevention of abdominal compartment syndrome by absorbable mesh prosthesis closure. Arch Surg 1997; 132:957-962.
9. Rotondo MF, Schwab W, McGonigal MD et al. Damage control: An approach for improved survival in penetrating abdominal trauma. J Trauma 1993; 35:375-382.
10. Chang MC, Miller PR, D'Agostino R et al. Effects of abdominal decompression on cardiopulmonary function and visceral perfusion in patients with intra-abdominal hypertension. J Trauma 1998; 44:440-445.
11. Ivatury RR, Porter JM, Simon RJ et al. Intra-abdominal hypertension after life-threatening penetrating abdominal trauma: Prophylaxis, incidence, and clinical relevance to gastric mucosal pH and abdominal compartment syndrome. J Trauma 1998; 44:1016-1023.
12. Sugrue M, Buist MD, Hourihan F et al. Prospective study of intra-abdominal hypertension and renal function after laparotomy. Br J Surg 1995; 82:235-238.
13. Cullen DJ, Coyle JP, Teplick R et al. Cardiovascular, pulmonary, and renal effects of massively increased intra-abdominal pressure in critically ill patients. Crit Care Med 1989; 17:118-121.
14. Sugrue M, Jones F, Jangua KJ et al. Temporary abdominal closure: A prospective evaluation of its effects on renal and respiratory physiology. J Trauma 1998; 45:914-921.
15. Meldrum DR, Moore FA, Moore EE et al. Prospective characterization and selective management of the abdominal compartment syndrome. Am J Surg 1997; 174:667-673.
16. Malbrain MLNG, Chiumello D, Pelosi P et al. Prevalence of intra-abdominal hypertension in critically ill patients: A multicentre epidemiological study. Intensive Care Med 2004; 30:822-829.
17. Cheatham ML, Safcsak K, Llerena LE et al. Long-term physical, mental, and functional consequences of abdominal decompression. J Trauma 2004; 56:237-242.
18. Cheatham ML, White MW, Sagraves SG et al. Abdominal perfusion pressure: A superior parameter in the assessment of intra-abdominal hypertension. J Trauma 2000; 49:621-627.
19. Boulanger BR, Rapanos T, McLean RF et al. Intra-abdominal pressure in critically injured adults: Clinical assessment vs. bladder pressure. J Trauma 1997; 43:393.
20. Kron IL, Harman PK, Nolan SP. The measurement of intra-abdominal pressure as a criterion for abdominal reexploration. Ann Surg 1984; 199:28-30.
21. Cheatham ML, Safcsak K. Intraabdominal pressure: A revised method for measurement. J Am Coll Surg 1998; 186;594-595.
22. Malbrain MLNG. Abdominal pressure in the critically ill: Measurement and clinical relevance. Intensive Care Med 1999; 1453-1458.
23. Malbrain MLNG. Different techniques to measure intra-abdominal pressure (IAP): Time for a critical reappraisal. Intensive Care Med 2004; 30:357-371.
24. Malbrain MLNG. Abdominal perfusion pressure as prognostic marker in intra-abdominal hypertension. In: Vincent JL ed. Yearbook of Intensive Care and Emergency Medicine. New York: Springer, Berlin Heidelberg, 2002:792-814.
25. Coombs HC. The mechanism of the regulation of intra-abdominal pressure. Am J Physiol 1922; 61:159-170.
26. Richardson JD, Trinkle JK. Hemodynamic and respiratory alterations with increased intra-abdominal pressure. J Surg Res 1976; 20:401.
27. Kashtan J, Green JF, Parson EQ et al. Hemodynamic effects of increased abdominal pressure. J Surg Res 1981; 30:249-255.

28. Diebel LN, Wilson RF, Dulchavsky SA et al. Effect of increased intra-abdominal pressure on hepatic arterial, portal venous, and hepatic microcirculatory blood flow. J Trauma 1992; 33:279-282.
29. Ridings PC, Bloomfield GL, Blocher CR et al. Cardiopulmonary effects of raised intra-abdominal pressure before and after intravascular volume expansion. J Trauma 1995; 39:1071-1075.
30. Simon RJ, Friedlander MH, Ivatury RR et al. Hemorrhage lowers the threshold for intra-abdominal hypertension-induced pulmonary dysfunction. J Trauma 1997; 42:398-405.
31. Bradley SE, Bradley GP. The effect of increased intra-abdominal pressure on renal function in man. J Clin Invest 1947; 26:1010-1015.
32. Harman PK, Kron IL, McLachlan HD et al. Elevated intra-abdominal pressure and renal function. Ann Surg 1982; 196:594-597.
33. Ulyatt DB. Elevated intra-abdominal pressure. Australian Anaes 1992; 108-114.
34. Caldwell CV, Ricotta JJ. Changes in visceral blood flow with elevated intraabdominal pressure. J Surg Res 1987; 43:14-20.
35. Friedlander MH, Simon RJ, Ivatury R et al. Effect of hemorrhage on superior mesenteric artery flow during increased intra-abdominal pressures. J Trauma 1998; 45:433-439.
36. Diebel LN, Myers T, Dulchavsky S. Effects of increasing airway pressure and PEEP on the assessment of cardiac preload. J Trauma 1997; 42:585-591.
37. Diebel LN, Dulchavsky SA, Brown WJ. Splanchnic ischemia and bacterial translocation in the abdominal compartment syndrome. J Trauma 1997; 43:852-855.
38. Gargiulo NJ, Simon RJ, Leon W et al. Hemorrhage exacerabtes bacterial translocation at low levels of intra-abdominal pressure. Arch Surg 1998; 133:1351-1355.
39. Smith ER, Carter BS, Ogilvy CS. Proposed use of prophylactic decompressive craniectomy in poor-grade aneurismal subarachnoid hemorrhage patients presenting with associated large sylvian hematomas. Neurosurgery 2002; 51:117-124.
40. Cheatham ML, Safcsak K, Block EFJ et al. Preload assessment in patients with an open abdomen. J Trauma 1999; 46:16-22.
41. Rotondo MF, Cheatham ML, Moore FA et al. Abdominal compartment syndrome. Contemp Surg 2003; 59:260-270.

CHAPTER 5

Prevalence and Incidence of Intra-Abdominal Hypertension

Dries H. Deeren and Manu L. N. G. Malbrain*

Definitions

*P*revalence and *prevalence rate* are respectively the number and the proportion of persons in a given population (an ICU for example) who have a particular disease (for example IAH) at a specified point in time or over a specified period of time. These terms are distinct from *incidence*, which refers to the number of new cases of a disease during a given period in a specified population. The *incidence rate* is the number of new cases that occur in this population per unit of person-time at risk (expressed as person-years). The *cumulative incidence* is defined as the proportion of persons at risk who develop new cases of the disease during a specified period of observation (for example during their ICU stay).

To Compare Cats and Dogs

We are chiefly interested in the incidences of IAH and ACS in specific subgroups of patients. However, even within the subgroups of trauma, burn, postoperative and medical patients, a number of problematic issues may hamper the comparison between results obtained by different researchers. The study populations may differ in surgical techniques (e.g., primary fascial closure, mesh closure, abdominal packing), resuscitation practices (e.g., the amount of IV fluids administered), trauma severity, percentage total body surface area (TBSA) burned, and emergency versus elective surgery. Using various techniques, IAP may be measured routinely in every patient, or only when clinically indicated. ACS can be defined purely clinically, or based on IAP measurements. Likewise, IAH can be defined according to several cut-off values.

We determined the coefficient of variation (COVA, defined as the standard deviation divided by the mean value) of two techniques for IAP measurement in 15 sedated and ventilated patients: through a bladder Foley manometer (Holtech Medical, Copenhagen, Denmark) and with a fully-automated continuous technique using a balloon-tipped gastric catheter connected to an IAP monitor (Spiegelberg, Hamburg, Germany).[1] Measuring the IAP every two hours, the COVA's were 17.1% and 18.7%, respectively. These variations may be even more pronounced in nonsedated patients. Thus, intermittent measurements are only snapshots and prevalence and incidence of IAH are affected by the frequency of IAP measurements.

*Corresponding Author: Manu L. N. G. Malbrain—Intensive Care Unit, ZiekenhuisNetwerk Antwerpen, Campus Stuivenberg, Lange Beeldekensstraat 267, B-2060 Antwerp, Belgium. Email: manu.malbrain@skynet.be

Abdominal Compartment Syndrome, edited by Rao R. Ivatury, Michael L. Cheatham, Manu L. N. G. Malbrain and Michael Sugrue. ©2006 Landes Bioscience.

Prevalence Rate

In a multicentre, prospective one-day prevalence study, IAP was measured through a Foley bladder catheter according to the modified Krohn technique[2] every six hours during 24 hours in all patients hospitalised for more than 24 hours in 13 ICU's.[3] IAH was defined as a maximal IAP of 12 mm Hg or more, and ACS as a maximal IAP of 20 mm Hg or more associated with at least one organ failure. The mean IAP was 9.8 mm Hg (± 4.7 standard deviation). In medical ICU patients, the prevalence rate of IAH was 54.4% (31/57), whereas this proportion was higher (65% or 26/40) in surgical ICU patients. Overall the prevalence rate of IAH was 57/97 or 58.8%, and 8.2% of all patients classified as ACS. The mean COVA was 25%, but ranged from 11.8% to 46.3% among different centres.[4]

Cumulative Incidence

Mixed ICU Population

An international multicentre group recently performed an important, prospective epidemiologic study of IAH in a mixed ICU population.[5] Two hundred sixty five consecutive patients (mean APACHE II score 17.4) who stayed more than 24 hours in one of the 14 participating ICU's were followed until death, hospital discharge, or for a maximum of 28 days. Medical patients accounted for 46.8%, whereas elective surgery, emergency surgery, and trauma patients accounted for 27.9%, 16.6% and 8.7% of all study patients, respectively. Twice daily, IAP was measured in stable conditions. IAH was said to be present when the mean value of the two daily IAP measurements was higher than 12 mm Hg. ACS was diagnosed when an IAP above 20 mm Hg was associated with at least one organ failure.

On admission, 32.1% had IAH and 4.2% had ACS, although only one patient underwent decompressive laparotomy. Importantly, unlike the occurrence of IAH at day one, the occurrence of IAH during ICU stay was an independent predictor of mortality. Independent predictors of IAH at day one were liver dysfunction, abdominal surgery, fluid resuscitation with more than 3500 mL during the 24 hours before inclusion, and ileus.

Burn Patients

Data on the cumulative incidence of IAH and ACS in burn patients are depicted in Table 1. As expected, the occurrence of IAH was significantly correlated to burn size and the amount of infused fluids.[6-8] Interestingly, in Hobson's study, patients with ACS fell into one of two groups: ACS developing within the first 24 hours after admission (related to the initial fluid resuscitation), and ACS developing weeks to months later (related to sepsis).[9]

Postoperative Patients

Data on the cumulative incidence of IAH in postoperative patients from studies that exclusively investigated these patients are depicted in Table 2. Only the study of Biancofiore aimed to obtain the incidence of IAH.[10] The other studies investigated the association of postoperative IAH with renal insufficiency[11,12] and gastric intramucosal pH.[13] Emergency surgery, compared with elective surgery, carried a higher risk for postoperative IAH.[11]

Trauma Patients

Data on the cumulative incidence of IAH and ACS in trauma patients from studies that exclusively investigated these patients are depicted in Table 3.

ACS is a well recognised complication of damage control surgery.[14] The concept of damage control surgery is most commonly employed for complex liver injuries. Haemostasis is achieved by placing packs around the liver, allowing for further correction of coagulopathy.[15] The packs are later removed ("reconstruction") which requires return to the operating room.[16] Differences in the proportion of patients who receive abdominal packs may account for the difference in IAH incidence between groups of comparable patients. In Ertel's study[17] only 20% required abdominal packing, whereas 67% did in Meldrum's.[14]

Table 1. *Cumulative incidence of IAH and ACS in burn patients*

Reference	Design	Patients	n	Age*	TBSA Burned*	Period	Outcome	Criteria	Cumulative Incidence
Greenhalgh, 1994[6]	P	bladder catheter for monitoring urine output	30	5.5 y	56%	burn unit stay	IAH	≥ 30 mm Hg†	36.7%
Ivy, 2000[7]	P	> 20% TBSA burned	10	48 y	46%	burn resuscitation	IAH ACS	≥ 25 mm Hg† IAH plus pulmonary/ renal complications	70% 20%
Latenser, 2002[8]	R	≥ 40% TBSA burned	13	23 y	57%	NR	IAH ACS	≥ 25 mm Hg† ≥ 30 mm Hg† despite peritoneal catheter decompression at 25 mm Hg	69% 31%
Hobson, 2002[9]	R	all burn ICU admissions	1014	NR	NR	hospital stay	ACS	clinical, supported by IAP	1%

P: prospective; R: retrospective; n: number of patients included; TBSA: total body surface area; y: years; NR: not reported; *: mean or median value; †: in at least one measurement

Table 2. *Cumulative incidence of IAH in postoperative patients*

Reference	Design	Patients	n	Age*	APACHE II*	Period	Outcome	Criteria	Cumulative Incidence
Sugrue, 1995[11]	P	major abdominal surgery	88 (57 emergency surgery)	63 y	13.5	ICU stay	IAH	≥ 20 mm Hg†	33%
Sugrue, 1996[13]	P	major abdominal surgery	73 (44 emergency surgery)	62 y	16	ICU stay	IAH	≥ 20 mm Hg†	38.4%
Sugrue, 1999[12]	P	major abdominal surgery	263 (174 emergency surgery)	61 y	14.6	ICU stay	IAH	≥ 18 mm Hg†	40.7%
Biancofiore, 2003[10]	P	liver transplantation	108	54 y	NR	≥ 72 h	IAH	≥ 25 mm Hg‡	31.5%

P: prospective; n: number of patients included; y: years; NR: not reported; h: hours; *: mean or median value; †: in at least one measurement; ‡: in at least two consecutive measurements with an interval of six hours

Table 3. Cumulative incidence of IAH and ACS in trauma patients

Reference	Design	Patients	n	Age*	ISS*	Period	Outcome	Criteria	Cumulative Incidence
Morris, 1993[16]	R	damage control laparotomy	107	32.1 y‡	32.5‡	hospital stay	ACS	Clinical	15%
Meldrum, 1997[14]	P	ISS > 15 and laparotomy	145	39 y§	26§	ICU stay	ACS	>20 mm Hg† plus pulmonary/renal/cardiovascular complications	14%
Ivatury, 1998[18]	P	life threatening penetrating abdominal trauma	70	28.2 y	21.8	ICU stay	IAH	persistently %Q >25 cm H_2O	32.9%
Maxwell, 1999[22]	R	trauma without abdominal injury	1216	36 y§	25§	ICU stay	ACS	decompressive laparotomy performed	0.5%
Kirkpatrick, 2000[24]	P	blunt trauma and mechanical ventilation	42	44 y	36	first 5 days	IAH	>10 mm Hg†	50%
Ertel, 2000[17]	R plus P	damage control laparotomy	311	37.7 y	29.5	hospital stay	ACS	clinical	5.5%
Offner, 2001[19]	R	damage control laparotomy	52	33 y	28	hospital stay	ACS	>20 cm H_2O†%Q plus pulmonary/renal/cardiovascular complications	33%
Raeburn, 2002[15]	P	damage control laparotomy	77	35 y	29	ICU stay	ACS	>20 mm Hg† plus pulmonary/renal complications	36%
Hong, 2002[20]	P	all trauma patients	706	42 y	18	ICU stay	IAH ACS	Persistently > 20 mm Hg IAH plus MOF	2% 1%
Balogh, 2003[23]	R	major trauma	156	35.2 y	27.5	first 24 h	IAH ACS	>20 mm Hg† >25 mm Hg† plus pulmonary/renal complications	32% 12.8%
Balogh, 2003[21]	P	major torso trauma	188	39 y	28	first 24 h	ACS	>25 mm Hg† plus pulmonary/renal/cardiovascular complications	14%

P: prospective; n: number of patients included; ISS: injury severity score; y: years; h: hours; *: mean or median value; †: in at least one measurement; ‡: only data available from the group that survived until unpacking; §: only data available from the ACS group; %Q: 1 mm Hg = 1.3595 cm H_2O

Consistently, studies demonstrated that primary fascial closure is associated with a higher risk of IAH.[18-21] However, even with attempted prophylaxis by the use of a wide, loosely applied, prosthetic mesh for temporary abdominal closure, 25% of patients with penetrating abdominal trauma developed IAH.[18]

In a large review of the charts of trauma patients without abdominal injury, ACS occurred very early in the time course of fluid resuscitation.[22] The average time from admission to decompressive laparotomy was 18 hours in all patients, 3.1 hours in survivors, and 25 hours in nonsurvivors. Supranormal resuscitation, compared with normal resuscitation (to an oxygen delivery index of respectively > 600 mL/min/m^2 and > 500 mL/min/m^2), was associated with an increased incidence of IAH and ACS.[23]

Medical Patients

Unlike on prevalence, data on the incidence of IAH and ACS in medical patients have not been published yet. Medical reason for admission did however account for 46.8% of all patients in the above mentioned mixed ICU study.[5]

Conclusion

The incidence of IAH is difficult to determine and dependent on the measurement method of IAP, the definition of IAH and the study population. However, in burn, postoperative, trauma and medical ICU patients, its incidence is high and it can occur extremely early during the time course of resuscitation even when there was no primary abdominal injury. In a mixed ICU population, 32.1% developed IAH and 4.2% ACS on the first ICU day.

Commentary

Rao R. Ivatury

Deeren and Malbrain have discussed nicely the problems in establishing the true prevalence of IAH and ACS in our patients. They have underscored that it varies with the type of patient population, the method of measuring IAP and the definition of IAH and ACS. Undoubtedly, another important factor is the admission of the existence of these complications. The summarized data in the Tables should convince the clinicians that the cumulative incidence is high in all surgical patients. The more one looks for these adverse events, the more they will be found.

References

1. Malbrain ML, Deeren D, De Potter T et al. Coefficient of variation (COVA) during continuous intra-abdominal pressure (IAP) and abdominal perfusion pressure (APP) monitoring. Crit Care 2004; 8(Suppl 1):P171.
2. Cheatham ML, Safcsak K. Intraabdominal pressure: A revised method for measurement. J Am Coll Surg 1998; 186(5):594-595.
3. Malbrain ML, Chiumello D, Pelosi P et al. Prevalence of intra-abdominal hypertension in critically ill patients: A multicentre epidemiological study. Intensive Care Med 2004; 30(5):822-829.
4. Malbrain ML. Is it wise not to think about intraabdominal hypertension in the ICU? Curr Opin Crit Care 2004; 10(2):132-145.
5. Malbrain ML, Chiumello D, Pelosi P et al. Incidence and prognosis of intraabdominal hypertension in a mixed population of critically ill patients: A multiple-center epidemiological study. Crit Care Med 2005; 33(2):315-322.
6. Greenhalgh DG, Warden GD. The importance of intra-abdominal pressure measurements in burned children. J Trauma 1994; 36(5):685-690.
7. Ivy ME, Atweh NA, Palmer J et al. Intra-abdominal hypertension and abdominal compartment syndrome in burn patients. J Trauma 2000; 49(3):387-391.
8. Latenser BA, Kowal-Vern A, Kimball D et al. A pilot study comparing percutaneous decompression with decompressive laparotomy for acute abdominal compartment syndrome in thermal injury. J Burn Care Rehabil 2002; 23(3):190-195.
9. Hobson KG, Young KM, Ciraulo A et al. Release of abdominal compartment syndrome improves survival in patients with burn injury. J Trauma 2002; 53(6):1129-1133, (discussion 1133-1124).

10. Biancofiore G, Bindi ML, Romanelli AM et al. Intra-abdominal pressure monitoring in liver transplant recipients: A prospective study. Intensive Care Med 2003; 29(1):30-36.
11. Sugrue M, Buist MD, Hourihan F et al. Prospective study of intra-abdominal hypertension and renal function after laparotomy. Br J Surg 1995; 82(2):235-238.
12. Sugrue M, Jones F, Deane SA et al. Intra-abdominal hypertension is an independent cause of postoperative renal impairment. Arch Surg 1999; 134(10):1082-1085.
13. Sugrue M, Jones F, Lee A et al. Intraabdominal pressure and gastric intramucosal pH: Is there an association? World J Surg 1996; 20(8):988-991.
14. Meldrum DR, Moore FA, Moore EE et al. Prospective characterization and selective management of the abdominal compartment syndrome. Am J Surg 1997; 174(6):667-672, (discussion 672-663).
15. Raeburn CD, Moore EE, Biffl WL et al. The abdominal compartment syndrome is a morbid complication of postinjury damage control surgery. Am J Surg 2001; 182(6):542-546.
16. Morris Jr JA, Eddy VA, Blinman TA et al. The staged celiotomy for trauma. Issues in unpacking and reconstruction. Ann Surg 1993; 217(5):576-584, (discussion 584-576).
17. Ertel W, Oberholzer A, Platz A et al. Incidence and clinical pattern of the abdominal compartment syndrome after "damage-control" laparotomy in 311 patients with severe abdominal and/or pelvic trauma. Crit Care Med 2000; 28(6):1747-1753.
18. Ivatury RR, Porter JM, Simon RJ et al. Intra-abdominal hypertension after life-threatening penetrating abdominal trauma: Prophylaxis, incidence, and clinical relevance to gastric mucosal pH and abdominal compartment syndrome. J Trauma 1998; 44(6):1016-1021, (discussion 1021-1013).
19. Offner PJ, de Souza AL, Moore EE et al. Avoidance of abdominal compartment syndrome in damage-control laparotomy after trauma. Arch Surg 2001; 136(6):676-681.
20. Hong JJ, Cohn SM, Perez JM et al. Prospective study of the incidence and outcome of intra-abdominal hypertension and the abdominal compartment syndrome. Br J Surg 2002; 89(5):591-596.
21. Balogh Z, McKinley BA, Holcomb JB et al. Both primary and secondary abdominal compartment syndrome can be predicted early and are harbingers of multiple organ failure. J Trauma 2003; 54(5):848-859, (discussion 859-861).
22. Maxwell RA, Fabian TC, Croce MA et al. Secondary abdominal compartment syndrome: An underappreciated manifestation of severe hemorrhagic shock. J Trauma 1999; 47(6):995-999.
23. Balogh Z, McKinley BA, Cocanour CS et al. Supranormal trauma resuscitation causes more cases of abdominal compartment syndrome. Arch Surg 2003; 138(6):637-642, (discussion 642-633).
24. Kirkpatrick AW, Brenneman FD, McLean RF et al. Is clinical examination an accurate indicator of raised intra-abdominal pressure in critically injured patients? Can J Surg 2000; 43(3):207-211.

CHAPTER 6

Intra-Abdominal Hypertension and the Cardiovascular System

Michael L. Cheatham* and Manu L. N. G. Malbrain

Abstract

Cardiovascular dysfunction and failure are commonly encountered in the patient with intra-abdominal hypertension (IAH) or abdominal compartment syndrome (ACS). Accurate assessment and optimization of preload, contractility, and afterload is essential to restoring end-organ perfusion and function in such patients. Our understanding of traditional hemodynamic monitoring techniques and parameters, however, must be reevaluated in the patient with IAH/ACS as pressure-based estimates of intravascular volume such as pulmonary artery occlusion pressure (PAOP) and central venous pressure (CVP) have known limitations in patients with elevated intra-abdominal pressure (IAP). If such limitations are not recognized, misinterpretation of the patient's minute-to-minute cardiac status may result in the institution of inappropriate and potentially detrimental therapy. Volumetric monitoring techniques have been proven to be superior to PAOP and CVP in ensuring appropriate resuscitation of the patient with IAH/ACS. Application of an aggressive, goal-directed resuscitation strategy improves cardiac function, reverses end-organ failure, and minimizes IAH-related patient morbidity and mortality.

Introduction

Preload, contractility, afterload, and oxygen transport are commonly abnormal in the critically ill as a result of hemorrhage, "third space" fluid losses, preexisting cardiopulmonary disease, the inflammatory and coagulation cascades, and direct cellular and organ injury. The subsequent development of sepsis, progressive shock, or acute respiratory failure requiring mechanical ventilation can similarly cause worsening cardiac dysfunction. Inadequate resuscitation of such abnormalities and failure to restore adequate cellular oxygen delivery through improved organ blood flow has been demonstrated to result in ischemia, anaerobic metabolism, and development of multiple organ dysfunction syndrome (MODS) with its high attendant mortality rate. Appropriate resuscitation of such patients commonly requires the use of advanced hemodynamic monitoring, such as is afforded by insertion of a pulmonary artery catheter (PAC) or other hemodynamic monitoring technology, due to the poor predictive ability of bedside clinical assessment to identify abnormalities in intravascular volume status, cardiac output, systemic vascular resistance, and oxygen transport balance.

Multiple trials have confirmed that critically ill patients whose cardiopulmonary function is optimized to ensure that end organ perfusion and oxygen transport balance are restored demonstrate significantly decreased organ failure, improved survival, and increased functional

*Corresponding Author: Michael L. Cheatham—Department of Surgical Education, Orlando Regional Medical Center, 86 West Underwood Street, Mailpoint #100, Orlando, Florida 32806, U.S.A. Email: michael.cheatham@orhs.org

Abdominal Compartment Syndrome, edited by Rao R. Ivatury, Michael L. Cheatham, Manu L. N. G. Malbrain and Michael Sugrue. ©2006 Landes Bioscience.

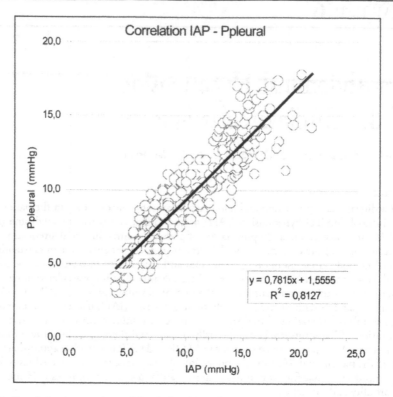

Figure 1. Correlation between intra-abdominal and intrathoracic pressure. A statistically significant correlation (p<0.001) correlation has been demonstrated between intra-abdominal pressure (IAP) and intrathoracic pressure (Ppleural) in patients with elevated IAP.

status.[1-7] The advantages of this "early goal-directed therapy" are especially applicable in the patient with elevated intra-abdominal pressure (IAP) who may manifest significant derangements in cardiac, pulmonary, renal, hepatic, gastrointestinal, and cerebral perfusion and function. The organ dysfunctions that characterize intra-abdominal hypertension (IAH) and abdominal compartment syndrome (ACS) may be related to either direct compressive effects, such as occur with IAP-induced pulmonary dysfunction, or more commonly inadequate end-organ perfusion as a result of pressure-mediated decreases in cardiac function. Thus, although IAH may appropriately be judged to exert deleterious effects on virtually every organ system within the human body, the cardiovascular system may arguably be considered the "cornerstone" of organ dysfunction and failure in patients with IAH/ACS and the primary target for early goal-directed resuscitation.

This chapter addresses the current understanding of the cardiovascular effects of IAH and validity of the commonly utilized hemodynamic monitoring parameters and techniques in the patient with elevated IAP. Based upon these observations, a rational evidence-based medicine approach to hemodynamic resuscitation in the patient with IAH/ACS is proposed.

Physiology

As originally described over 80 years ago by Emerson, elevated IAP increases intrathoracic pressure through cephalad deviation of the diaphragm (Fig. 1).[8] The resulting compression of the pulmonary parenchyma and displacement of the heart from its normal anatomic position can cause significant detrimental cardiovascular effects. Cardiac function may be distilled into three essential components: preload, contractility, and afterload. Elevated IAP negatively impacts upon all three of these interrelated components and restoration of each to adequate levels is essential to improving systemic perfusion, oxygen transport, and ultimately patient outcome.

Preload

Adequate intravascular volume or preload to the right heart is essential to the resuscitation of any critically ill patient who demonstrates malperfusion at either the global or regional level. Loss of intravascular fluid volume may be either absolute, as occurs with hemorrhage, or relative, where mechanical obstruction to blood flow impedes venous return to the heart. In patients with IAH/ACS, elevated intra-thoracic pressure decreases blood flow through the inferior vena cava (IVC) and limits blood return from below the diaphragm in a pressure-dependent manner.[9-11] IVC pressure increases significantly in the presence of IAH and has been demonstrated to parallel changes in IAP.[12,13] IVC pressure has even been advocated as an alternative to intravesicular pressure monitoring.[12,13] Cephalad deviation of the diaphragm can also induce mechanical narrowing of the IVC as it passes through the diaphragmatic crura, further reducing blood return to the heart.[14,15] Richardson et al and Diamant et al have separately demonstrated in animal models that an IAP of only 10 mm Hg can significantly reduce IVC blood flow and cardiac preload.[11,16] These effects have been widely witnessed in humans with the introduction of laparoscopic surgery where abdominal insufflation to similar IAP levels commonly causes decreased venous return and hypotension that are responsive to volume loading. As demonstrated by Bloomfield et al, reduced intra-thoracic pressure through median sternotomy or other method improves venous return and normalizes cardiac function.[17,18]

Reduced preload to the heart will have the immediate effect of decreasing cardiac output through decreased stroke volume. This can be especially pronounced in patients with concurrent hypovolemia. Kashtan et al demonstrated that increasing IAP to 40 mm Hg decreased cardiac output by 53% in hypovolemic and by 17% in normovolemic dogs.[10] Similar findings have been demonstrated in humans where hypovolemic patients appear to sustain reductions in cardiac output at lower levels of IAP than do normovolemic patients.[10,19] Hypervolemic patients actually demonstrate increased venous return in the presence of mild-to-moderate increases in IAP, suggesting that volume resuscitation may have a protective effect.[10,20-23] Maintenance of adequate intravascular volume will, to a point, overcome both the pressurerelated and anatomic restrictions to IVC blood flow, restoring systemic perfusion.[24] Ultimately, however, abdominal decompression and treatment of the patient's IAH will be necessary to restore normal IVC blood flow and structural integrity. Patients with underlying respiratory disease and limited myocardial reserve may well benefit from earlier decompression as a result of their inability to tolerate the pathophysiologic changes outlined above.[5,20-23]

Elevated IAP also causes pooling of blood in the pelvis and lower extremities, further limiting venous return to the heart. Femoral vein pressures are markedly increased and venous blood flow and pulsatility dramatically reduced.[25,26] The resulting increases in venous hydrostatic pressure may promote the formation of both genital and lower extremity edema. These changes also place the patient with IAH at risk for deep venous thrombosis.[26-28] Abdominal decompression, by acutely reducing IAP, restores femoral venous blood flow and rapidly augments intravascular volume status, but has anecdotally been witnessed to result in pulmonary embolism and cardiac arrest as the impediment to IVC blood flow is relieved.[28]

Contractility

Diaphragmatic elevation and increased intra-thoracic pressure can have marked effects on cardiac contractility. Traditionally, the right ventricle has been considered to be solely a conduit for delivering blood to the lungs and left ventricle. Compression of the pulmonary parenchyma increases pulmonary artery pressure and pulmonary vascular resistance (PVR) while simultaneously reducing left ventricular preload. As right ventricular afterload increases, the right side of the heart must play a more active role in maintaining cardiac output.[29] In response to worsening right ventricular afterload, the thin-walled right ventricle will dilate with a decrease in right ventricular ejection fraction (RVEF) and increases in ventricular wall tension and myocardial oxygen demand. This increased oxygen requirement, coinciding with an increase in right ventricular work requirement, places the myocardium at risk for subendocardial ischemia and further reductions in right ventricular contractility. Further, as the right ventricle enlarges within a closed pericardium, the interventricular septum may bulge into the left ventricular chamber impeding left ventricular function with decreases in cardiac output.[29,30] Reduced left ventricular output may contribute to the development of systemic hypotension, worsening right coronary artery blood flow. Right ventricular dysfunction can become severe in the presence of marked IAH leading to significant reductions in left ventricular contractility as a result of "ventricular interdependence". Initially responsive to fluid and perhaps inotropic support at lower levels of IAH, the reduced biventricular contractility of advanced IAH and ACS can only be treated by decompressive laparotomy with its resultant reductions in IAP and intra-thoracic pressure.

Afterload

Elevated intra-thoracic pressure and IAP can cause increased systemic vascular resistance (SVR) through direct compressive effects on the aorta and systemic vasculature and increased PVR through compression of the pulmonary parenchyma.[15,17,31,32] More commonly, however, increased SVR occurs as compensation for the reduced venous return and falling stroke volume outlined above. As a result of this physiologic compensation, mean arterial pressure (MAP) typically remains stable in the early stages of IAH/ACS despite reductions in venous return and cardiac output. These increases in afterload may be poorly tolerated by patients with marginal cardiac contractility or inadequate intravascular volume.[5,20-23] This may especially be the case in patients who develop pulmonary hypertension as a result of intra-thoracic pressure-induced compression of the pulmonary parenchyma and alveolar overdistention due to increased PEEP requirements. The resulting increase in PVR may lead to increased strain on the right ventricle and right heart failure as discussed above. Preload augmentation through volume administration appears to ameliorate, at least partially, the injurious effects of IAH-induced increases in afterload.[9-11,17,31,32]

Hemodynamic Monitoring

The past decade has seen much debate as to the safety, efficacy, and accuracy of hemodynamic monitoring in the critically ill. With increasing emphasis upon evidence-based justification for medical interventions and concern over the potential complications of invasive monitoring, the benefit of pulmonary artery catheterization, long considered the standard for hemodynamic monitoring in many intensive care units (ICU), has been called into doubt.[33,34] Such concern over the limitations of pressurebased monitoring techniques is not new, however. Soon after introduction of the PAC in the 1970s, numerous factors were found to influence the accuracy of PAC-derived data. Proper catheter positioning, pressure transducer calibration, and pressure waveform interpretation were identified as being essential to obtaining clinically reliable patient data. Further, the initial assumption that pressure-based physiologic measurements could be used as substitutes for volumetric physiologic variables was recognized as being flawed in the critically ill.

Simultaneously, there has been an increasing appreciation of the fact that traditional monitoring techniques provide only a limited "snapshot" of a critically ill patient's cardiopulmonary physiology when a continuous "moving picture" is what is truly needed. As a result, there has been an evolving trend towards the development of less invasive hemodynamic monitoring techniques that provide continuously updated physiologic data. Such monitoring technologies better illustrate the constantly changing hemodynamics of the critically ill patient, provide early warning of patient deterioration, and promote more effective resuscitation. Although a variety of hemodynamic monitoring technologies have been proposed, only those methods that have been adequately evaluated in the patient with IAH/ACS will be addressed.

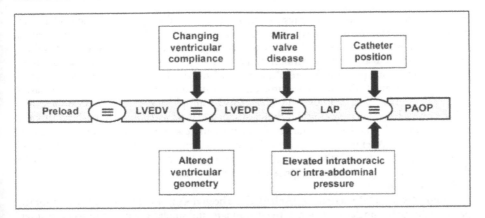

Figure 2. "The PAOP assumption": why intracardiac filling pressures do not accurately estimate preload status. LVEDV= left ventricular end-diastolic volume; LVEDP= left ventricular end-diastolic pressure; LAP= left atrial pressure; PAOP= pulmonary artery occlusion pressure. (Adapted from Cheatham ML. Right ventricular end-diastolic volume measurements in the resuscitation of trauma victims. Int J Intens Care 2000; 7:165-176.)

Inaccuracy of Intracardiac Filling Pressures

By the Frank-Starling principle, ventricular preload is defined as myocardial muscle fiber length at end-diastole (Fig. 2). Ideally, the appropriate clinical correlate would be left ventricular end-diastolic volume (LVEDV), but this physiologic parameter is not easily measured on a serial basis. If we assume that a patient's ventricular compliance remains constant, changes in ventricular volume should be reflected by changes in ventricular pressure through the following relationships:

$$\text{Compliance} = \Delta \text{ Volume} / \Delta \text{ Pressure and } \Delta \text{ Volume} = \Delta \text{ Pressure}$$

Based upon the assumption of stable ventricular compliance, pressurebased parameters such as left ventricular end-diastolic pressure (LVEDP), left atrial pressure (LAP), and pulmonary artery occlusion pressure (PAOP), obtained via a PAC, have long been utilized clinically as surrogate estimates of intravascular volume. Although likely valid in normal healthy individuals, the multiple assumptions necessary to utilize PAOP and central venous pressure (CVP) as estimates of left and right ventricular preload status respectively, are not necessarily true in the critically ill patient with IAH. These patients commonly demonstrate significant aberrations in cardiac function that can interfere with the accuracy of PAOP and CVP measurements as estimates of intravascular volume status. Each of these assumptions will be addressed.

First, ventricular compliance is now widely recognized to be constantly changing in the critically ill, resulting in a variable relationship between pressure and volume.[29,35] As a result, changes in intracardiac pressure no longer directly reflect changes in intravascular volume, reducing the accuracy of intracardiac filling pressure measurements such as PAOP and CVP as estimates of preload status. Altered ventricular geometry and disparate ventricular function as a result of acute respiratory failure-induced changes in right ventricular afterload and increased intra-thoracic pressure may also explain, in part, the poor correlation noted between right and left sided filling pressures in the critically ill.[35] The presence of IAH by itself has been shown to cause a flattening and rightward shift of the Frank-Starling ventricular compliance curve.[16]

Second, elevated intra-thoracic pressure as a result of mechanical ventilation strategies intended to treat acute-lung injury associated hypoxemia, such as positive end-expiratory pressure (PEEP) or inverse-ratio ventilation, has been demonstrated to increase PAOP and CVP measurements by an amount that is difficult to predict, further confounding the validity of intracardiac filling pressure measurements.[24,36] Paradoxically, PAOP and CVP typically increase with rising intra-thoracic pressure despite the reduced venous return and cardiac output that is characteristic of elevated intra-thoracic pressure.[5,10,15,16,20,21,23,30,32] This apparent deviation from Starling's Law of the Heart is due to the fact that both PAOP and CVP are measured relative to atmospheric pressure and are actually the sum of both intravascular pressure and intrapleural pressure. As a result, in the patient with IAH and elevated intra-thoracic pressure, PAOP and CVP tend to be erroneously elevated and no longer reflective of intravascular volume status (Fig. 2: PAOP and CVP plots). Multiple studies have demonstrated that reliance on such measurements to guide fluid resuscitation in patients with elevated intra-thoracic pressure may lead to under-resuscitation and inappropriate administration of diuretic medications.[24,32,35,37,38] Attempts to correct for this measurement error through the calculation of transmural pressure (i.e., PAOP minus pleural pressure) has identified that transmural PAOP decreases with rising intra-thoracic pressure, correctly reflecting the decreased venous return and cardiac preload.[39]

Third, elevated IAP as a result of IAH/ACS has been documented to increase intra-thoracic pressure through cephalad elevation of the diaphragm and similarly complicate accurate interpretation of PAOP and CVP measurements.[5,24,30,32,37] Such alterations in PAOP and CVP have been demonstrated with an IAP of only 10 mm Hg.[36] Despite significantly elevated PAOP and CVP values, fluid resuscitation commonly results in improved cardiac output and organ perfusion in patients with IAH.[5,24,32] Failure to appropriately fluid resuscitate based upon erroneously elevated PAOP and CVP measurements may result in under-resuscitation and worsened end-organ perfusion and failure as these elevated intracardiac filling pressures are artificial and do not reflect the patient's true preload status.

Fourth, mitral valve disease can confound the use of PAOP as an estimate of intravascular volume status. Mitral valve stenosis and regurgitation both interfere with the relationship between LAP and PAOP making the latter less indicative of left ventricular preload. The presence of mitral valve disease typically results in elevated PAOP values that may lead to under-resuscitation.

Fifth, accurate PAOP measurements are dependent upon proper placement of the PAC. Compression of the pulmonary parenchyma as a result of elevated IAP can markedly alter the normal progression of alveolar distention and pulmonary capillary pressures as illustrated in West's lung zones I, II, and III.[40] Loss of thoracic cavity volume as a result of cephalad deviation of the diaphragm decreases the size of all three lung zones, but especially the apical zone I and basilar zone III. The use of positive pressure ventilation and PEEP to restore alveolar volume, oxygenation, and ventilation increases the relative size of West's zone I through alveolar distention at the expense of zones II and III as manifested clinically by increased pulmonary artery pressures and decreased blood flow. IAH-induced cardiac and pulmonary dysfunction can alter the normal pulmonary artery waveforms making proper placement of the PAC tip in West's lung zone II difficult. Inadvertent placement of the tip in the apical zone I commonly results in PAOP measurements that more appropriately reflect alveolar pressure than pulmonary capillary pressure. This may result in measurement of erroneously high PAOP values that can lead to under-resuscitation. Verification of catheter tip location using lateral chest radiographs or blood sampling has been described, but such practices are time consuming, costly, and rarely performed. Further, several studies demonstrate that fewer than one-half of physicians inserting a PAC are able to recognize a normal PAOP tracing during catheter placement.[41,42] These issues suggest that many PACs may be improperly positioned in the patient with IAH/ACS, potentially providing erroneous information regarding the patient's preload status.

Hemodynamic monitoring can only improve patient care and outcome when clinicians thoroughly understand both the appropriate utilization as well as the potential measurement errors associated with the use of physiologic parameters. Due to the physiologic complexity of patients with IAH/ACS, intracardiac filling pressure measurements such as PAOP and CVP should be critically considered with a thorough appreciation of the potential errors and pitfalls of such measurements. Resuscitation to arbitrary, absolute PAOP or CVP values should be avoided as such a practice can lead to inappropriate therapeutic decisions, under-resuscitation, and organ failure.

Use of Transmural Cardiac Filling Pressures

Assuming proper placement of a PAC and the absence of other confounding factors, transmural PAOP and CVP may be appropriately represented by the following equations:

$$PAOP_{tm} = PAOP_{ee} - P_{pl}$$

$$CVP_{tm} = CVP_{ee} - P_{pl}$$

where tm = transmural, ee = end-expiratory, and P_{pl} = pleural pressure.

In patients with elevated intra-thoracic pressure due to acute respiratory failure and PEEP, some authors have advocated the practice of measuring PAOP during disconnection of the patient's airway (the so-called "pop-off" PAOP) in an attempt to account for the increase in P_{pl}.[43] Such a practice would not be valid in the patient with elevated IAP, as this does not reduce the contribution of IAP to the patient's P_{pl}. Safcsak et al calculated PAOP$_{tm}$ by substituting esophageal pressure for P_{pl}, but did not find this to improve upon the ability of PAOP$_{ee}$ alone to predict preload recruitable increases in cardiac output.[44] Malbrain et al recently evaluated four equations for determining PAOP$_{tm}$ in 5 patients with IAH and PEEP levels from 0 to 20 cm H_2O.[45]

$$PAOP_{tm} = PAOP_{ee} - P_{pl} \tag{1}$$

$$PAOP_{ta} = PAOP_{ee} - IAP \tag{2}$$

where ta = transabdominal

$$PAOP_{tm} = PAOP_{ee} - it \times PEEP \tag{3}$$

where it = index of transmission calculated as:

$$\Delta PAOP/DP_{alv} = (PAOP_{ei} - PAOP_{ee})/(P_{plat} - PEEP)$$

$$PAOP_{tm} = 0.8 \times PAOP_{ee} - 0.1 \times PEEP - 0.6 \times IAP + 0.02 \times C_{dyn} + 2.5 \tag{4}$$

Confirming the findings of previous authors, a significant correlation was found between IAP and P_{pl} with approximately 80% of the IAP being transmitted to the intrathoracic compartment ($P_{pl} = 0.8 \times IAP + 1.6$ (R^2=0.8, p<0.0001). Malbrain concluded that in patients with IAH, the simple calculation of subtracting half the IAP from PAOP$_{ee}$ or CVP$_{ee}$ may provide a rapid bedside estimate of transmural filling pressure.

Volumetric Pulmonary Artery Catheters

In the 1980s, a new generation of PAC was introduced allowing calculation of both right ventricular ejection fraction (RVEF) and right ventricular end-diastolic volume index (RVEDVI).[5,7,24,37,46-48] RVEF, reflecting the patient's right ventricular contractility and afterload, is utilized to calculate the RVEDVI using the following equation (where SVI = stroke volume index):

$$RVEDVI = SVI / RVEF$$

Independent of the effects of changing ventricular compliance and increased intra-thoracic pressure or IAP, RVEDVI provides clinicians with a valuable volumetric estimate of preload status. RVEDVI has been shown in multiple studies to be an accurate indicator of preload recruitable increases in cardiac index (CI) in a variety of patient populations and disease processes including hemorrhagic, cardiogenic, neurogenic, and septic shock, acute lung injury, and pulmonary hypertension.[4,5,7,21,24,37,44,46-50] These studies have consistently identified a significant correlation between RVEDVI and CI and a lack of correlation between PAOP or CVP and CI during preload assessment of patients undergoing resuscitation. Based upon the

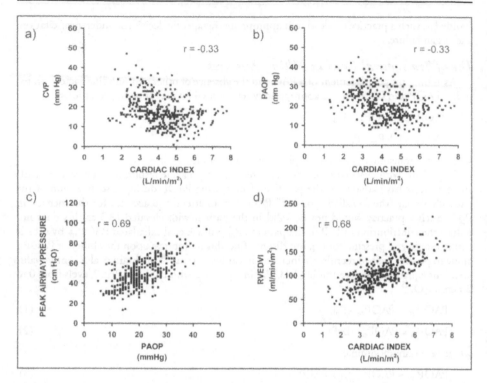

Figure 3. PAOP, CVP, and RVEDVI as estimates of intravascular preload. A and B demonstrate the poor and inverse correlation between CI and PAOP or CVP respectively in the presence of elevated intra-abdominal pressure, apparently contradicting Starling's Law of the heart. Figure 3C illustrates the strong correlation between peak airway pressure and end-expiratory PAOP, demonstrating the impact of elevated intrathoracic and intra-abdominal pressure on PAOP and CVP measurements and the potential validity of transmural PAOP calculations. Figure 3D depicts the strong correlation between CI and end-diastolic volume (RVEDVI) as described by Starling. (CI= cardiac index; PAOP= pulmonary artery occlusion pressure; CVP= central venous pressure; RVEDVI= right ventricular end-diastolic volume index.)

volumetric information provided by RVEDVI, Diebel et al demonstrated that PAOP measurements provide potentially misleading information regarding preload status in 52% of critically ill patients.[37]

The value of RVEDVI over traditional intracardiac filling pressures is especially notable in patients with elevated intra-thoracic pressure or IAP where PAOP and CVP are at greatest risk for providing erroneous information regarding preload status. Diebel et al and Cheatham et al assessed the impact of airway pressure and PEEP on preload assessment in surgical and trauma patients with ALI.[24,46] At levels of PEEP as high as 50 cm H_2O, CI consistently maintained a highly significant correlation with RVEDVI, while PAOP and CVP were frequently found to exhibit inverse correlations with CI, directly challenging the Frank-Starling principle. PAOP and CVP values as high as 60 mm Hg were documented despite the presence of clinical and echocardiographic evidence of intravascular volume depletion.[24] Cheatham et al and Chang et al independently compared PAOP, CVP and RVEDVI as estimates of preload status in patients with elevated IAP before and after abdominal decompression.[5,48] In both studies, CI was noted to correlate significantly with RVEDVI and inversely with both PAOP and CVP (Fig. 3).

Mathematical coupling, the interdependence of two variables when one is used to calculate the other, has been proposed to account for the significant correlation between CI and RVEDVI.[51] Since RVEDVI is calculated using SVI, CI and RVEDVI are, by definition,

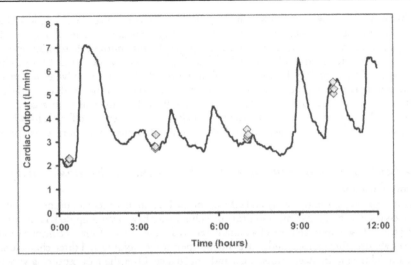

Figure 4. Continuous vs. intermittent cardiac output during patient resuscitation. Continuous cardiac output (solid line) vs. intermittent cardiac output measurements (diamonds) during the initial resuscitation of a critically ill patient. Continuous cardiac output technology provides significant insight into the dynamic nature of the critically ill patient previously unavailable with conventional cardiac output techniques. (Adapted from Cheatham ML. Right ventricular end-diastolic volume measurements in the resuscitation of trauma victims. Int J Crit Care 2000; 7:165-176.)

mathematically coupled variables. Chang, Durham, and Nelson have separately addressed the potential impact of mathematical coupling on the reliability of RVEDVI as a measurement of preload adequacy.[47,48,52] Chang independently measured cardiac output via indirect calorimetry and demonstrated a significant correlation between mathematically uncoupled CI and thermodilution RVEDVI.[48] Durham used mathematical modeling to correct for the shared measurement error introduced by mathematical coupling and found CI to remain significantly correlated with RVEDVI.[47] Nelson compared CI with RVEDVI measurements determined using two different thermodilution technologies and further confirmed the significant correlation between mathematically uncoupled CI and RVEDVI.[52]

In the late 1990s, a new generation of volumetric or "continuous cardiac output" (CCO) PACs was introduced that provides continuously updated measurements of CO, RVEF, and RVEDVI. CCO technology has several advantages over the traditional intermittent thermodilution PAC. First, many of the factors that may alter the accuracy of intermittent thermodilution measurements (such as injectate volume and temperature, injection technique, and injectate timing with regards to ventilation) do not play a role in the determination of CCO measurements. Second, by obviating the need for tedious thermal injectate boluses, measurement of cardiac output is possible without the potential volume load incurred by serial thermodilution measurements. Third, and most importantly, CCO monitoring provides a minute-by-minute assessment of patient response to therapeutic interventions, potentially allowing more rapid and effective resuscitation compared to traditional intermittent thermodilution techniques. CCO measurements have been shown to be equal in accuracy to intermittent thermodilution, indocyanine green dye dilution, radionuclide ventriculography, biplane angiography, and 2-D echocardiography.[24,49,53] These traditional techniques can be difficult and sometimes impossible to perform in the critically ill patient and, with the exception of echocardiography, cannot be used serially to guide therapy as can CCO determinations. CCO measurements are relatively inexpensive, safe, accurate, and reproducible, allowing serial determinations of hemodynamic function to guide therapeutic interventions.[4,5,7,24,49,53]

Clinical experience with this new technology has further confirmed the dynamic and constantly changing cardiopulmonary state exhibited by these critically ill patients (Fig. 4). These hemodynamic changes may either be missed completely by intermittent thermodilution measurements or not identified until potentially devastating events have occurred. CCO technology is capable of identifying these potentially untoward changes in hemodynamic function, allowing appropriate interventions to be made at an earlier point in time. With the addition of mixed or central venous oximetry, CCO technology provides clinicians with a continually updated, on-line assessment of oxygen transport balance and systemic perfusion by which to guide patient resuscitation, reduce organ dysfunction and failure, and improve patient outcome.

Pulse Contour Analysis and Volumetric Assessment via Transpulmonary Thermodilution

With the concern over the safety and efficacy of the PAC in management of the critically ill, several less invasive hemodynamic monitoring technologies have been proposed. Estimation of stroke volume based upon analysis of a patient's arterial pressure waveform or contour was first proposed almost 100 years ago and is dependent upon stroke volume and three characteristics of the arterial tree: resistance, compliance, and impedance. Over the past century, a variety of mathematical models have been suggested to improve the accuracy of stroke volume determination from the shape of the arterial pressure waveform. This technique, known as arterial pulse contour analysis, has been proposed as a less invasive alternative to the PAC for hemodynamic monitoring as it requires only an arterial pressure catheter and a central venous catheter, both of which are commonly present in most critically ill patients.[49,54,55]

Pulse contour analysis utilizes a bedside computer and dedicated thermistor-tipped arterial catheter to continuously analyze the patient's heart rate and arterial pressure waveform. By calculating the change in pressure over time from end-diastole to end-systole, and making several assumptions regarding the elastic and mechanical properties of the arterial tree, continuous beat-to-beat SV can be estimated and CO calculated. Due to the unique mechanical characteristics of each patient's arterial tree, initial calibration of the monitoring system using the transpulmonary thermodilution technique and the Stewart Hamilton equation greatly improves the accuracy of the stroke volumes subsequently calculated.[49] Given the constantly changing physiology and ventricular compliance of the critically ill patient, recalibration should be performed at least every 8 hours and whenever patients demonstrate significant changes in their physiology.[55]

Pulse contour analysis has several advantages over traditional intermittent thermodilution PAC monitoring. First, it is less invasive requiring only arterial and central venous catheters. Second, it provides an assessment of left as opposed to right ventricular CO. Third, as with CCO technology, it is independent of respiratory cycle variation and concerns over timing of thermodilution bolus injections. The combined transpulmonary thermodilution also provides for *off-line* calculation of global ejection fraction (GEF), an estimate of ventricular contractility, and several intravascular volume measurements including global end-diastolic volume index (GEDVI), intrathoracic blood volume index (ITBVI), and extravascular lung water (EVLW), as surrogate predictors of cardiac preload and capillary leak.[49] This technology can also be used to measure stroke volume variation (SVV), the variation in beat-to-beat stroke volume during a single respiratory cycle, as well as pulse pressure variation (PPV) which has been suggested to be a valuable predictor of hypovolemia and potential response to fluid administration.[56-58]

As with RVEDVI, GEDVI and ITBVI both appear to be superior to PAOP and CVP in predicting preload status, especially in patients with elevated intra-thoracic pressure or IAP where transmission of these pressures to the pulmonary capillaries can erroneously increase measured PAOP and CVP values.[49,59-62] Brienza et al demonstrated that CO correlated better with GEDVI and ITBVI than with PAOP in the presence of elevated intra-thoracic pressure.[62] Malbrain et al demonstrated in patients with IAH and PEEP that elevated intra-thoracic

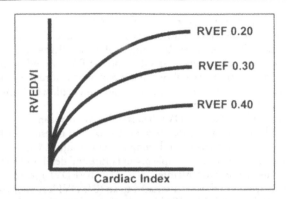

Figure 5. Ventricular function curves by RVEF. RVEDVI must be interpreted in conjunction with the patient's RVEF (RVEF= right ventricular ejection fraction; RVEDVI=right ventricular end-diastolic volume index). (Adapted from Cheatham ML. Right ventricular end-diastolic volume measurements in the resuscitation of trauma victims. Int J Crit Care 2000; 7:165-176.)

pressure and IAP resulted in significant increases in PAOP and CVP with decreases in GEDVI and ITBVI.[59] Based upon these studies and others, it is clear that IAH significantly depletes intravascular volume and that these changes in preload status are appropriately detected by volumetric measurements of intravascular volume such as RVEDVI, GEDVI, or ITBVI, but not by pressure-based measurements such as PAOP and CVP.

"Optimal" Volumes

The initial studies describing the use of volumetric preload measurements described "optimal" RVEDVI values of approximately 130-140 mL/m^2, and optimal GEDVI values of 640-800 mL/m^2, above which patients were felt to no longer respond to further volume administration with increases in CI.[37,47,49,61] As clinical experience with these technologies has increased, these optimal values have been disputed and found to oversimplify what is actually a complex and dynamic relationship between preload, contractility, and afterload.[4,21,24,50] Ongoing work has demonstrated that the patient's RVEF must be taken into consideration when assessing the adequacy of RVEDVI values as a resuscitation endpoint.[4,50]

Cardiac contractility in the critically ill can be described as a series of "ventricular function curves". Each curve has an associated ejection fraction, describing the ventricle's contractility, and an optimal end-diastolic volume, identifying the plateau of the ventricular function curve. Resuscitation to this plateau end-diastolic volume is widely believed to optimize a patient's intravascular volume, cardiac function, and end-organ perfusion (Fig. 5).[47,50]

As demonstrated by Eddy, ventricular function and compliance are constantly changing in the critically ill.[29] As ventricular function changes, the patient "shifts" from one Starling curve to another with identification of a new, optimal plateau end-diastolic volume as a resuscitation endpoint.[4,21,50] Thus, each RVEDVI must be considered in the context of the simultaneous RVEF measurement to determine whether the patient's right ventricular function is increasing, decreasing, or stable. In the presence of unchanging right ventricular contractility and afterload (as evidenced by a stable RVEF), RVEDVI assessment is relatively straightforward, as the target RVEDVI remains unchanged. In the critically ill patient with deteriorating right ventricular contractility or increasing right ventricular afterload, however, RVEDVI assessment becomes more complex.

By the Frank-Starling principle, as RVEF changes as a result of alterations in right ventricular contractility and afterload, the heart shifts to a different ventricular function curve and plateau RVEDVI must change proportionally assuming a constant intravascular volume state (Fig. 5). Thus, whereas an RVEDVI of 100 mL/m^2 is considered normal for a RVEF of 0.40, an RVEDVI of 200 mL/m² would be required for a RVEF of 0.20 assuming intravascular volume has not changed. Thus, since ventricular compliance (and therefore RVEF) is subject to change in the critically ill, there cannot be a single value of RVEDVI that can be considered the goal of resuscitation for all patients. As GEF and GEDVI provide similar information regarding a patient's ventricular contractility and preload, a similar relationship between GEF and GEDVI likely exists, although not yet clinically documented, which must be considered when interpreting GEDVI measurements.

Assuming a normal RVEF, what RVEDVI then is sufficient to optimize a critically ill patient's volume status? Several studies have demonstrated that certain threshold values of RVEDVI correlate with improved patient outcome following surgery or injury. Chang et al prospectively documented a decreased incidence of multiple organ failure and death in patients who received aggressive volume resuscitation with maintenance of an RVEDVI greater than 110 mL/m² (mean RVEF 0.39) compared to those resuscitated to an RVEDVI less than 100 mL/m² (mean RVEF 0.39).[7] Miller et al similarly identified improved visceral perfusion by gastric tonometry and a reduction in organ dysfunction, organ failure, and patient mortality by maintaining an RVEDVI of 120 mL/m² (mean RVEF 0.33) compared to 100 mL/m² (mean RVEF 0.34).[4] Cheatham et al identified significantly higher RVEDVI values in patients who survived IAH and abdominal decompression with a mean RVEDVI of 133 mL/m² for survivors (mean RVEF 0.37) compared to 105 mL/m² for nonsurvivors (mean RVEF 0.36).[5]

Based upon these studies, a reasonable resuscitation protocol is to initially fluid resuscitate patients to a "RVEF-corrected" RVEDVI of 100 mL/m², assuming a normal RVEF of approximately 0.40. In analogy one might postulate to resuscitate patients to a "GEF-corrected" GEDVI of 575 mL/m², assuming normal GEF is approximately 0.30. Patients with lower RVEF or GEF measurements are then resuscitated to proportionally higher RVEDVI or GEDVI values. Figure 6 provides some general guidelines for RVEF and GEF corrected target values for RVEDVI and GEDVI in normal and critically ill conditions. A patient with a RVEF of 0.30, for example, would be considered to have a target RVEDVI of 150 mL/m² and a patient with a RVEF of 0.20, an RVEDVI of 200 mL/m². In analogy, a patient with a GEF of 0.25 might have a target GEDVI value of 650 mL/m² whereas a patient with a GEF of 0.15 might require a GEDVI of 800 mL/m².

If, after achieving such levels of intravascular volume, the patient continues to demonstrate signs of malperfusion, the target RVEDVI or GEDVI should be increased by 20% while simultaneously initiating vasoactive medications as necessary to optimize ventricular contractility and afterload (Table 1). The RVEF or GEF measurement is useful in determining the need for and choice of vasoactive infusion as it defines the current relationship between ventricular contractility and afterload. Patients with a high RVEF or GEF will usually respond to fluid administration alone while those with a low RVEF or GEF will almost invariably benefit from early administration of inotropic support. Restoration of adequate intravascular volume must precede institution of vasoactive medications in order to avoid visceral malperfusion and acidosis.[4]

These target goals should be considered solely to be guidelines for initiating resuscitation rather than definitive endpoints. Each patient should always be resuscitated to the RVEDVI or GEDVI that is identified to restore end-organ function and normalize markers of systemic and regional perfusion adequacy such as urinary output, base deficit, and arterial lactate. It is important to recognize that patients may achieve these endpoints at RVEDVI or GEDVI values below their RVEF- or GEF-corrected target values. Unnecessary over-resuscitation past these values once intravascular volume has been restored has not been demonstrated to benefit the patient and will likely lead to worsening pulmonary function and unnecessary elevations in IAP. In the absence of one of the volumetric monitoring technologies, traditional intracardiac filling pressures such as PAOP and CVP may be used to guide resuscitation with the explicit understanding that transmural estimates of PAOP and CVP must be utilized.

As discussed elsewhere in this textbook, the "critical IAP" that causes end-organ dysfunction varies from patient to patient as a result of both the inciting disease process and preexisting

Table 1. Suggested RVEF- and GEF-corrected target values for RVEDVI and GEDVI in normal and critically ill patients

RVEF	"Normal" RVEDVI (mL/m^2)	"Critically Ill" RVEDVI (mL/m^2)	GEF	"Normal" GEDVI (mL/m^2)	"Critically Ill" GEDVI (mL/m^2)
.20	200	240	.10	875	975
.25	175	210	.15	800	900
.30	150	180	.20	725	825
.35	125	150	.25	650	750
.40	100	120	.30	575	675
.45	75	90	.35	500	600
.50	50	60	.40	450	550

Target RVEDVI and GEDVI values must be considered in light of the patient's current cardiac contractility.

comorbidities. As a result, a single threshold value of IAP cannot be globally applied to the decision making of all critically ill patients. Calculation of the patient's "abdominal perfusion pressure" (APP), defined as mean arterial pressure (MAP) minus IAP, assesses not only the severity of IAP present, but also the adequacy of the patient's systemic perfusion. APP has been demonstrated to be superior to both IAP and global resuscitation endpoints such as arterial pH, base deficit, and arterial lactate in its ability to predict patient outcome and may represent another useful parameter for guiding the resuscitation and management of the patient with IAH or ACS.[63,64] Any resuscitative strategy should incorporate maintenance of an APP > 50 mm Hg in the patient with elevated IAP.

Conclusions

IAH and ACS have been increasingly recognized as causes of significant morbidity and mortality in the critically ill. Cardiovascular dysfunction, as a result of elevations in intra-thoracic pressure and IAP, plays a major role in the organ dysfunction and failure that characterizes IAH/ACS. Aggressive hemodynamic monitoring and optimization of both systemic and regional perfusion is essential to improving patient outcome. Traditional measures of intravascular volume such as PAOP and CVP are commonly erroneous in patients with IAH and reliance on such measurements may lead to inappropriate therapeutic interventions. Volumetric estimates of preload status such as RVEDVI, GEDVI, or ITBVI are especially useful in such patients with changing ventricular compliance and elevated intra-thoracic pressure and IAP. The clinician must be aware of the interactions between intra-thoracic pressure, IAP, PEEP, and intracardiac filling pressures in order to correctly resuscitate these patients.

Commentary

Michael Sugrue

Cheatham and Malbrain have provided an insight into the challenges of the physiological approach to the interaction between cardiovascular haemodynamics and elevated intra-abdominal pressure. The chapter outlines the challenges with haemodynamic monitoring in particular the importance of intra-abdominal hypertension in its role in altering current haemodynamic parameters used for monitoring. Of particular importance and emphasised throughout the chapter is the need for dynamic continuous evaluation of the patient's physiology. Continuous cardiac output measurements are important and should be coupled with continuous abdominal pressure measurements giving rise to a more physiological profusion evaluation, not just of the abdominal cavity itself but also of the global patient. The future lies in further research in these two important topics to improve the outcome for what is a highly morbid problem in the abdominal compartment syndrome.

References

1. Shoemaker WC, Appel PL, Kran HB et al. Prospective trial of supranormal values of survivors as therapeutic goals in high-risk surgical patients. Chest 1998; 94:1176-1186.
2. Boyd O, Grounds RM, Bennett ED. A randomized clinical trial of the effect of deliberate perioperative increase in oxygen delivery on mortality in high-risk surgical patients. JAMA 1993; 270:2699-2707.
3. Wilson J, Woods I, Fawcett J et al. Reducing the risk of major elective surgery: Randomised controlled trial of preoperative optimisation of oxygen delivery. BMJ 318:1099-1103.
4. Miller PR, Meredith JW, Chang MC. Randomized, prospective comparison of increased preload versus inotropes in the resuscitation of trauma patients: Effects on cardiopulmonary function and visceral perfusion. J Trauma 1998; 44:107-113.
5. Cheatham ML, Safcsak K, Block EFJ et al. Preload assessment in patients with an open abdomen. J Trauma 1999; 46:16-22.
6. Rivers E, Nguyen B, Havstad S et al. Early goal-directed therapy in the treatment of severe sepsis and septic shock. NEJM 2001; 345:1368-1377.
7. Chang MC, Meredith JW. Cardiac preload, splanchnic perfusion, and their relationship during resuscitation in trauma patients. J Trauma 1997; 42:577-584.
8. Coombs HC. The mechanism of the regulation of intra-abdominal pressure. Am J Physiol 1922; 61:159-170.
9. Caldwell CV, Ricotta JJ. Changes in visceral blood flow with elevated intraabdominal pressure. J Surg Res 1987; 43:14-20.
10. Kashtan J, Green JF, Parson EQ et al. Hemodynamic effects of increased abdominal pressure. J Surg Res 1981; 30:249-255.
11. Richardson JD, Trinkle JK. Hemodynamic and respiratory alterations with increased intra-abdominal pressure. J Surg Res 1976; 20:401.
12. Lee SL, Anderson JT, Kraut EJ et al. A simplified approach to the diagnosis of elevated intra-abdominal pressure. J Trauma 2002; 52:1169-1172.
13. Malbrain ML. Different techniques to measure intra-abdominal pressure (IAP): Time for a critical reappraisal. Intensive Care Med 2004; 303:357-371.
14. Schein M, Wittmann DH, Aprahamian CC et al. The abdominal compartment syndrome: The physiological and clinical consequences of elevated intra-abdominal pressure. J Am Coll Surg 1995; 180:745-753.
15. Luca A, Cirera I, García-Pagán JC et al. Hemodynamic effects of acute changes in intra-abdominal pressure in patients with cirrhosis. Gastroenterology 1993; 104:222-227.
16. Diamant M, Benumof JL, Saidman LJ. Hemodynamics of increased intra-abdominal pressure. Anesthesiology 1978; 48:23-27.
17. Bloomfield GL, Blocher CR, Fakhry IF et al. Elevated intra-abdominal pressure increased plasma renin activity and aldosterone levels. J Trauma 1997; 42:997-1005.
18. Bloomfield GL, Dalton JM, Sugerman HJ et al. Treatment of increasing intracranial pressure secondary to the acute abdominal compartment syndrome in a patient with combined abdominal and head trauma. J Trauma 1995; 39:1168-1170.
19. Diebel LN, Wilson RF, Dulchavsky SA et al. Effect of increased intra-abdominal pressure on hepatic arterial, portal venous, and hepatic microcirculatory blood flow. J Trauma 1992; 33:279-282.
20. Malbrain MLNG. Intra-abdominal pressure in the Intensive Care Unit: Clinical tool or toy? In: Vincent JL, ed. Yearbook of Intensive Care and Emergency Medicine. Berlin: Springer-Verlag, 2001:547-585.
21. Cheatham ML. Intra-abdominal hypertension and abdominal compartment syndrome. New Horizons 1999; 7:96-115.
22. Malbrain MLNG. Abdominal pressure in the critically ill. Curr Opinion Crit Care 2000; 6:17-29.
23. Chang MC, Miller PR, D'Agostino R et al. Effects of abdominal decompression on cardiopulmonary function and visceral perfusion in patients with intra-abdominal hypertension. J Trauma 1998; 44:440-445.
24. Cheatham ML, Nelson LD, Chang MC et al. Right ventricular end-diastolic volume index as a predictor of preload status in patients on positive end-expiratory pressure. Crit Care Med 1998; 28:1801-1806.
25. Barnes GE, Laine GA, Giam JY. Cardiovascular responses to elevation of intra-abdominal pressure. Am J Physiol 1985; 248:R208.

26. Goodale RL, Beebe DS, McNevin MP et al. Hemodynamic, respiratory, and metabolic effects of laparoscopic cholecystectomy. Am J Surg 1993; 166:533-537.
27. Baxter JN, O'Dwyer PJ. Pathophysiology of laparoscopy. Br J Surg 1995; 82:1-2.
28. MacDonnell SPJ, Lalude OA, Davidson AC. The abdominal compartment syndrome: The physiological and clinical consequences of elevated intra-abdominal pressure (letter). J Am Coll Surg 1996; 183:419-420.
29. Eddy AC, Rice CL, Anardi DM. Right ventricular dysfunction in multiple trauma victims. Am J Surg 1988; 155:712-715.
30. Cullen DJ, Coyle JP, Teplick R et al. Cardiovascular, pulmonary, and renal effects of massively increased intra-abdominal pressure in critically ill patients. Crit Care Med 1989; 17:118-121.
31. Smith PK, Tyson GS, Hammon JW et al. Cardiovascular effects of ventilation with positive end-expiratory airway pressure. Ann Surg 1982; 195: 121-130.
32. Ridings PC, Bloomfield GL, Blocher CR et al. Cardiopulmonary effects of raised intra-abdominal pressure before and after intravascular volume expansion. J Trauma 1995; 39:1071-1075.
33. Connors AF, Speroff T, Dawson NV et al. The effectiveness of right heart catheterization in the initial care of critically ill patients. JAMA 1996; 276:889-897.
34. Trottier SJ, Taylor RW. Physicians' attitudes toward and knowledge of the pulmonary artery catheter: Society of Critical Care Medicine membership survey. New Horiz 1997; 5:201-206.
35. Calvin JE, Driedger AA, Sibbald WJ. Does the pulmonary capillary wedge pressure predict left ventricular preload in critically ill patients? Crit Care Med 1981; 9:437-443.
36. Safcsak K, Fusco MA, Miles WS et al. Does transmural pulmonary artery occlusion pressure (PAOP) via esophageal balloon improve prediction of ventricular preload in patients receiving positive end-expiratory pressure (PEEP)? Crit Care Med 1995; 23(suppl):A244.
37. Diebel LN, Wilson RF, Tagett MG et al. End-diastolic volume: A better indicator of preload in the critically ill. Arch Surg 1992; 127:817-822.
38. Michard F, Teboul JL. Predicting fluid responsiveness in ICU patients. A critical analysis of the evidence. Chest 2002; 121:2000-2008.
39. Malbrain MLNG, Nieuwendijk R, Verbrugghe W et al. Effect of intra-abdominal pressure on pleural and filling pressures. Intensive Care Med 2003; 29:S73.
40. West JB, Dollery CT, Naimark A. Distribution of blood flow in isolated lung: Rrelation to vascular and alveolar pressures. J Appl Physiol 1964; 19:713-724.
41. Iberti TJ, Fischer EP, Leibowitz AB et al. A multicenter study of physicians' knowledge of the pulmonary artery catheter. Pulmonary Artery Catheter Study Group. JAMA 1990; 264:2928-2932.
42. Trottier SJ, Taylor RW. Physicians' attitudes toward and knowledge of the pulmonary artery catheter: Society of Critical Care Medicine membership survey. New Horiz 1997; 5:201-206.
43. Teboul JL, Pinsky MR, Mercat A et al. Estimating cardiac filling pressures in mechanically ventilated patients with hyperinflation. Crit Care Med 2000; 28:3631-3636.
44. Safcsak K, Fusco MA, Miles WS et al. Does transmural pulmonary artery occlusion pressure (PAOP) via esophageal balloon improve prediction of ventricular preload in patients receiving positive end-expiratory pressure (PEEP)? Crit Care Med 1995; 23:A244.
45. Malbrain MLNG, Nieuwendijk R, Verbrugghe W et al. Effect of intra-abdominal pressure on pleural and filling pressures. Intensive Care Med 2003; 29:S73.
46. Diebel LN, Myers T, Dulchavsky S. Effects of increasing airway pressure and PEEP on the assessment of cardiac preload. J Trauma 1997; 42:585-591.
47. Durham R, Neunaber K, Vogler G et al. Right ventricular end-diastolic volume as a measure of preload. J Trauma 1995; 39:218-224.
48. Chang MC, Black CS, Meredith JW. Volumetric assessment of preload in trauma patients: Addressing the problem of mathematical coupling. Shock 1996; 6:326-329.
49. Chaney JC, Derdak S. Minimally invasive hemodynamic monitoring for the intensivist: Current and emerging technology. Crit Care Med 2002; 30:2338-2345.
50. Cheatham ML. Right ventricular end-diastolic volume measurements in the resuscitation of trauma victims. Int J Inten Care 2000; 7:165-176.
51. McNamee JE, Abel FL. Mathematical coupling and Starling's law of the heart. Shock 1996; 6:330.
52. Nelson LD, Safcsak K, Cheatham ML et al. Mathematical coupling does not explain relationship between right ventricular end-diastolic volume (RVEDV) and thermodilution cardiac output (TDCO). Crit Care Med 1999; 27:A151.
53. Haller M, Zollner C, Briegel J et al. Evaluation of a new continuous thermodilution cardiac output monitor in critically ill patients: A prospective criterion standard study. Crit Care Med 1995; 23:860-866.
54. Berton C, Cholley B. Equipment review: New techniques for cardiac output measurement - oesophageal Doppler, Fick principle using carbon dioxide, and pulse contour analysis. Crit Care (London) 2002; 6:216-221.

55. Gödje O, Höke K, Goetz AE et al. Reliability of a new algorithm for continuous cardiac output determination by pulse-contour analysis during hemodynamic instability. Crit Care Med 2002; 30:52-58.
56. Berkenstadt H, Margalit N, Hadani M et al. Stroke volume variation as a predictor of fluid responsiveness in patients undergoing brain surgery. Anesth Analg 2001; 92:984-989.
57. Reuter DA, Kirchner A, Felbinger TW et al. Usefulness of left ventricular stroke volume variation to assess fluid responsiveness in patients with reduced cardiac function. Crit Care Med 2003; 31:1399-1404.
58. Wiesenack C, Prasser C, Rödig G et al. Stroke volume variation as an indicator of fluid responsiveness using pulse contour analysis in mechanically ventilated patients. Anesth Analg 2003; 96:1254-1257.
59. Malbrain MLNG, van Mieghem N, Verbrugghe W et al. PiCCO derived parameters versus "filling pressures" in intra-abdominal hypertension. Intensive Care Med 2003; 29:S130.
60. Hering R, Rudolph J, Spiegel Tv et al. Cardiac filling pressures are inadequate for estimating circulatory volume in states of elevated intraabdominal pressure. Intensive Care Med 1998; 24:S409.
61. Reuter DA, Felbinger TW, Moerstedt K et al. Intrathoracic blood volume index measured by thermodilution for preload monitoring after cardiac surgery. J Cardiothorac Vasc Anesth 2002; 16:191-195.
62. Brienza N, Dambrosio M, Cinnella G et al. Effects of PEEP on intrathoracic and extrathoracic blood volumes evaluated by the COLD system in patients with acute respiratory failure. Minerva Anestesiol 1996; 62:235-242.
63. Cheatham ML, White MW, Sagraves SG et al. Abdominal perfusion pressure: A superior parameter in the assessment of intra-abdominal hypertension. J Trauma 2000; 49:621-627.
64. Malbrain MLNG. Abdominal perfusion pressure as prognostic marker in intra-abdominal hypertension. In: Vincent JL, ed. Yearbook of Intensive Care and Emergency Medicine. New York: Springer, Berlin Heidelberg, 2002:792-814.

CHAPTER 7

Intra-Abdominal Hypertension and the Respiratory System

Ingrid R. A. M. Mertens zur Borg,* Serge J. C. Verbrugge and Claudia I. Olvera

Abstract

Significant increases in intra-abdominal pressure (IAP) are seen in a wide variety of conditions commonly encountered in the intensive care unit.[1-3] Abdominal compartment syndrome (ACS) describes the combination of increased intra-abdominal pressure and organ dysfunction.[4-6] The incidence of intra-abdominal hypertension (IAH) and ACS varies with the case mix studied and the cut-off pressure used to define IAH.[7-9] The World Congress on Abdominal Compartment Syndrome (WCACS) 2004, defined ACS as IAP > 20 mm Hg, associated with single or multiple organ system failure, which was not previously present. Some authors report pulmonary dysfunction as the earliest manifestation of ACS.[2] This chapter will discuss the pulmonary derangements in ACS and describe evidence-based treatment principles for mechanical ventilation and pulmonary treatment in ACS.

Introduction

The majority of studies suggest that the highest incidence of IAH is observed in patients who have undergone emergency laparotomy for abdominal trauma, and that massive fluid resuscitation is a major contributory factor,[1-15] because massive volume infusion produces abdominal distension, lung compression and chest wall stiffening.[15,16] The combination of traumatic lung injury, fluid resuscitation, and supine positioning results in alveolar damage, alveolar collapse (de-recruitment), worsening gas exchange, and declining compliance. ACS has a high mortality and, eventually most deaths result from sepsis and multi-organ failure. The adverse effects of IAH on lung function were first described in the late 1800s.[10] Emerson in 1911 hypothesized a reciprocal relationship between IAP and intra-thoracic pressure, with a steady decline in inspired air with respiratory failure and death occurring with IAP above 27-46 cm H_2O in anesthetized cats and guinea pigs.[11] Baggot, an anaesthetist from Dublin,[12] in 1951 described the clinical effects of abdominal wound closure under tension after a dehiscence or 'abdominal blow-out'. He noted a high mortality with these operations and, referring to earlier investigations, he concluded that death was a result of respiratory dysfunction.

*Corresponding Author: Ingrid R. A. M. Mertens zur Borg—Department of Anesthesiology (Room HN 1279), Erasmus Medical Center, Dr. Molewaterplein 40, 3015 GD Rotterdam, The Netherlands. Email: i.mertenszurborg@erasmusmc.nl

Abdominal Compartment Syndrome, edited by Rao R. Ivatury, Michael L. Cheatham, Manu L. N. G. Malbrain and Michael Sugrue. ©2006 Landes Bioscience.

Pulmonary Dysfunction with Increased Abdominal Pressure

Pulmonary Dysfunction Due to Pneumoperitoneum in Laparoscopic Surgery

Some evidence for the causes of the pathophysiological pulmonary derangements in ACS comes from studies of induced pneumoperitoneum during laparoscopy. The IAP in a normal individual ranges from slightly sub-atmospheric to approximately 6.5 mm Hg, and varies with the respiratory cycle.[17] Patients whose lungs are being mechanically ventilated show a slight increase in IAP due to transmission of pleural pressures across the diaphragm.[18] Pneumoperitoneum (PP) for laparoscopic operations decreases functional residual capacity (FRC) and thoracopulmonary compliance may be reduced by 30 to 50%,[19-23] or even more in obese patients.[24,25] Lung and chest wall mechanical impedances increase with increasing IAP, but these are completely reversible.[26] Nevertheless, diaphragmatic function remains significantly impaired after laparoscopy.[27-29] Recent studies demonstrate that even at the relatively low IAP of 10-15 mm Hg significant alterations in organ function can be seen.[15,30-32]

Moreover, if a patient is immobilized in the supine position for whatever reason, dependent atelectasis is seen after several hours. Sedation and analgesia further exacerbate atelectasis because of cephalad movement of the diaphragm into the thoracic cavity and compression of dorsal/dependent lung regions. General anaesthesia exacerbates this problem, and the addition of neuromuscular blockade worsens it further.[33]

Pulmonary Dysfunction in ACS

Severely elevated IAP, as in ACS, can cause substantial alterations in respiratory system mechanics. Increased IAP displaces the diaphragm into the thoracic cavity, compressing basilar lung segments. Chest radiography may show elevated hemi diaphragms with loss of lung volume.[34] Physiologically, this is manifested as a decrease in FRC, an increase in alveolar dead space (V_{DA}) and ventilation perfusion (V/Q) mismatch.[35] The increase in IAP is translated across the diaphragm, causing a smaller, but proportionate rise in intra-thoracic pressure.[36] As a result, respiratory system and chest wall compliance, and total lung capacity (TLC) and residual volume are reduced.[1,15,19,21,23,35,37] Recent studies show that the decrease in respiratory system compliance is mainly due to a decreased chest wall compliance, the lungs mostly being unaffected. The resultant increase in ventilation-perfusion mismatch and pulmonary dead space leads to hypoxia, hypercapnia and the need for mechanical ventilation.

Because of the changes in compliance of the lungs and chest wall, high peak airway and plateau pressures are needed during mechanical ventilation. The ensuing increase in intra-thoracic pressure and hypoxic pulmonary vasoconstriction due to lung compression can lead to pulmonary hypertension.[38-40] Patients with ACS often encounter a combined ventilatory and hypoxic respiratory failure, which manifests with a low arterial oxygen pressure (PaO_2), an elevated arterial carbon dioxide pressure ($PaCO_2$), as well as an increased difference in partial alveolar and arterial oxygen pressure ($P(A-a)O_2$).[41] Extremely high driving pressures may be required to maintain minimally sufficient tidal volumes, often with loss of delivered tidal volume by distension of ventilatory tubing.

Decreased oxygen delivery, the product of blood oxygen content and the cardiac output, will result in tissue hypoxia. The physiologic responses to tissue hypoxia are, primarily, to increase oxygen delivery by increasing cardiac output and, secondarily, when breathing spontaneously, to increase ventilation. Oxygen therapy will increase alveolar (and thus blood oxygen content) when a low PaO_2 is present and thus will result in increased oxygen delivery. When hypoxia results from ventilatory demands that exceed the ability of the cardiac output to increase oxygen delivery, mechanical ventilation may not only improve $PaCO_2$ but also improve PaO_2 by decreasing oxygen consumption. We will discuss the principles of mechanical ventilation later in this chapter.

Physiological Derangements Due to Pulmonary Dysfunction in ACS

Firstly, as discussed, the pulmonary effects of ACS include mechanical problems by secondary compressive atelectasis and deteriorating lung dynamic compliance[42] with the need to ventilate the patient with high peak inspiratory pressures.[43] The deterioration in lung mechanics may be attributable to both decreased compliance of the lung and to decreased compliance of the thoracic cage.[44] Abdominal decompression by means of laparotomy[8,38,45] or continuous negative abdominal pressure[43,46] may alleviate such mechanical problems. Worsening hypercapnia and respiratory system compliance have been identified as critical indicators of pulmonary failure that warrant emergent abdominal decompression in the setting of IAH.[7] Decompression of the abdominal cavity results in nearly immediate reversal of respiratory failure.[1] Patients identified as being at risk of developing IAH and ACS, should undergo close monitoring. A low threshold for reexploration and decompression of the abdomen if ACS is suspected should be employed. However, the timing, indications and threshold for surgical decompression are controversial, with very few large trials available to give firm guidance.[6,18,47,48] Worsening hypercapnia, deteriorating respiratory system compliance and excessively increased airway pressures often warrant surgical decompression. These disturbances, although severe, may comprise the least complex part of the pulmonary problem in IAH and ACS.

Secondly, as discussed, tissue hypoxia may result due to decreased oxygen delivery. This may contribute to a more complex pulmonary problem in ACS. The resulting intestinal/hepatic ischemia with abdominal wall ischemia/necrosis may provoke the systemic release of pro-inflammatory cytokines which may result in lung neutrophil accumulation and intra-pulmonary oxygen free radical production as shown in animal models.[49] Moreover, IAP elevation to 25 mm Hg for 60 minutes in rats decreased mucosal blood flow which results in bacterial translocation from the gut into the systemic circulation via the mesenteric lymph nodes.[50] Both such mechanisms may contribute to a situation of a systemic inflammatory syndrome which may evolve to multiple organ failure (MOF) of which acute lung injury (ALI) / acute respiratory distress syndrome (ARDS) is a part.[4,51]

Both experimental and human research in the past decade in the field of ARDS/ALI and MOF suggests that if systemic inflammation occurs, the lung becomes an important, causative part of the inflammation-induced systemic disease state that can evolve to MOF. The lung could than act as a propagator of the MOF syndrome[52] rather than being just a simple end-effect organ in ACS/IAH.[53]

Physiological Derangements Caused by Mechanical Ventilation

Animal Models of Ventilation-Induced Lung Injury

As discussed above, compressive atelectasis and deteriorating dynamic respiratory system compliance[42] necessitate to ventilate the patient with ACS with high peak inspiratory pressures. In animal models of ventilation-induced lung injury, especially modes of mechanical ventilation which combine high peak inspiratory pressures with low PEEP settings can induce lung injury which is indistinguishable from the lung injury seen in models of ARDS/ALI.[54,55] The early stages of ventilator-associated lung injury develop at commonly used airway pressures (transalveolar pressure >35 cm H_2O) in animals with normal lungs. The threshold for lung injury may occur at lower pressures in injured lungs. The pathophysiological mechanisms for such ventilation-induced lung injury include endothelial and epithelial breaks due to peak inspiratory overstretching with the loss of barrier function of the alveolar-capillary membrane resulting in the infiltration of protein-rich edema. This may lead to a dose-dependent inactivation of the surfactant system. Furthermore, mechanical ventilation, by itself, may primarily disturb the surfactant system.[54] Surfactant inactivation will further predispose the lungs to alveolar collapse, which promotes the infiltration of protein-rich edema. Ventilation-induced lung injury is, however, not only due to peak inspiratory overstretching but even more so due to "shear forces" which develop at the border zone of open and closed (collapsed) alveolar units.

These "shear forces" tending to disrupt the lung tissue may be as high as 120 cm H_2O at transpulmonary pressures of only 30 cm H_2O.[56] In this way a vicious cycle develops with more lung tissue disrupture, more protein infiltration, and more surfactant changes leading to more alveolar de-recruitment and the need for even higher ventilatory pressures to provide some ventilation. These ventilator-induced changes do not remain confined to the lung but may also have "systemic consequences".

Ventilation-Induced Mediator Translocation

Recent studies suggest that the mode of mechanical ventilation influences the degree of tumour necrosis factor (TNF) alpha translocation from the bloodstream into the lungs as measured by broncho-alveolar lavage in rats challenged intra-abdominally with lipopolysaccharide.[57] Modes of mechanical ventilation that combine high peak inspiratory pressures with low PEEP pressures, especially, resulted in TNF alpha translocation. The application of PEEP significantly reduced TNF alpha translocation. When rats were challenged with LPS into the lung, the same results were obtained for translocation from the lung into the bloodstream. Such findings may be particularly relevant in IAH and ACS as more optimal forms of mechanical ventilation could reduce the extent to which the lung takes part in the systemic inflammatory disease state induced by IAH/ACS.

The release of inflammatory mediators and the production of cytokines by the lung as a result of mechanical ventilation has now been shown in both isolated perfused lungs and in vivo experiments.[58-64] In humans, Ranieri et al showed that mechanical ventilation designed to minimise ventilator-induced lung injury using high PEEP levels and low end-inspiratory stretch could markedly attenuate the cytokine response in ARDS patients compared to conventional ventilation strategy.[65] In contrast to ARDS, inflammatory mediator release as a result of different forms of mechanical ventilation could not be shown in patients with normal pulmonary function.[66] In some studies the lung appears to need a 'first hit'. This first hit might well be provided by the systemic attack in IAH/ACS.

Ventilation-Induced Bacterial Translocation

Based on the observation that mechanically ventilated critically-ill patients often develop pneumonia[67] and septicaemia, the question may be raised whether damaging mechanical ventilation can promote bacteremia and/or sepsis. Also, it is conceivable that mechanical ventilation in the patient with bacteremia/sepsis as a result of IAH/ACS may result in loss of barrier function of the alveolo-capillary barrier with a resultant pneumonia. It has been established that preserving end-expiratory lung volume with PEEP has a beneficial effect on the course of infection in terms of reducing bacterial counts recovered from the lung tissue after prolonged mechanical ventilation of lungs inoculated with bacteria.[68] Moreover, avoiding high peak transpulmonary pressures and preserving end-expiratory lung volume with PEEP has been shown to reduce translocation of *Pseudomonas aeruginosa*,[68] *Escherichia coli*[69] and *Klebsiella pneumoniae*[70] from the lung into the bloodstream. The same principle applies to endotoxin derived from the lung, which may translocate as a result of detrimental forms of mechanical ventilation.[71]

These data suggest that ventilation-induced changes in the barrier function of the lung epithelium and/or endothelium may, to a certain extent, contribute to the development of bacteremia and endotoxemia as it is seen in MOF. The influence of mechanical ventilation in inducing bacterial translocation from the circulation in the direction of the lung has not been shown yet.

Principles of Mechanical Ventilation in ARDS/ALI in ACS

The treatment for ARDS or ALI is primarily supportive with mechanical ventilation, which allows time for treatment of the underlying cause of lung injury and for natural healing. Until recently, most studies of ARDS or ALI reported a mortality rate of 40 to 60%, with death attributed to sepsis or MOF rather than primary respiratory causes.[72,73]

The increase in intra-abdominal pressure in IAH/ACS will inevitably lead to decreased respiratory system-thoracic compliance, decreased FRC, atelectasis and enlargement of the functional right-to-left shunt, hypoxemia with anaerobic metabolism, and metabolic acidosis and pulmonary edema. The application of high inspiratory pressures and volumes with over distension of open alveoli for a long time is associated with an increased risk for barotrauma.[52,74-77] On the other hand, low levels of PEEP may contribute to ventilation-induced lung injury by allowing alveoli to collapse and reopen during each respiratory cycle.[74,75,77-79]

Both experimental and clinical data have demonstrated that ventilation settings that prevent lung injury in both healthy and diseased lungs should prevent alveolar overdistension and recruit all alveoli and prevent their collapse at end-expiration.[75] The lung should be opened and the lung should be kept open with the least possible pressure swings to ensure the required gas exchange. Hemodynamic side effects are thus minimized.[56,80] The open lung is characterized by an optimal gas exchange.[56] The intrapulmonary shunt is ideally less than 10%, which corresponds to a PaO_2 of more than 450 mm Hg on pure oxygen.[81,82]

A rational treatment concept is the following:[56,80]

1. One must overcome a critical opening pressure during inspiration
2. This opening pressure must be maintained for a sufficiently long period of time
3. During expiration, no critical time that would allow collapse of lung units should pass.

The goal of the initial increase in inspiratory pressure is to recruit collapsed alveoli and to determine the critical lung opening pressure. Then, the minimum pressures that prevent the lung from collapse are determined. Finally, after an active reopening man<euvre sufficient pressure is implemented to keep the lung open (Fig. 1).

A clinical study by Amato et al showed that a ventilation strategy aimed at opening atelectatic lungs and keeping them open at all times in combination with a treatment strategy of permissive hypercapnia and a restriction on the size of tidal volume and limited peak inspiratory pressures, resulted in a higher rate of weaning from mechanical ventilation, lower rate of barotrauma, and improved 28 day survival in ARDS patients compared to conventional ventilation.[83] The authors stratified the patients according to PEEP levels and concluded that PEEP levels higher than 12 cm H_2O and especially higher than 16 cm H_2O significantly improved survival of these ARDS patients.[84]

Alveolar recruitment should almost always be possible during the first 48 hours on mechanical ventilation (which may be more difficult if the disease exists for a longer period of time[85]). Even if not all of the lung tissue may be fully recruited for gas exchange, as in consolidating pneumonia, this ventilatory strategy may prevent further damage to the reasserted part of the lung.

Modes of Positive-Pressure Ventilation

There are two basic types of goals for the modes of ventilation: ventilation limited by a pressure target and ventilation limited to the delivery of a specified volume. Formerly, mechanical ventilators could control only one of these parameters during a breath, and the controlled variables were pressure, volume, and (in the case of high-frequency ventilation) time. Newer ventilators are capable of switching between volume and pressure targets, classifying them as "dual-control" modes.

Pressure-Targeted versus Volume-Targeted Modes of Ventilation

There are distinct differences, advantages, and disadvantages in pressure and volume-targeted strategies. Pressuretargeted modes of ventilation allow the clinician to control the peak inspiratory pressure (PIP) and the inspiratory time, or I:E ratio. Flow is delivered in a decelerating fashion and varies from breath to breath. The initial peak flow is rapidly reached at the beginning of the breath and then decreases throughout inspiration, maintaining the peak pressure until a preset inspiratory time is met. Pressuretargeted ventilation allows a more even distribution of ventilation in the lung while using the variable (decelerating) flow profile.[86] It has also

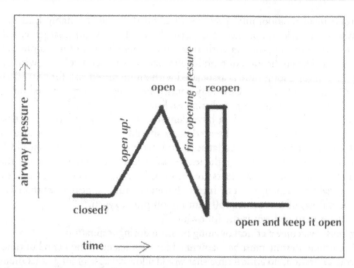

Figure 1. Schematic representation of the opening procedure for collapsed lungs. Note: The imperatives (!) mark the treatment goal of each specific intervention. The bold words mark the achieved state of the lung. At the beginning the precise amount of collapsed lung tissue is not known.

been suggested that patients are more comfortable breathing spontaneously while on pressuretargeted ventilation. This may be due in part to the constant adaptation of peak flows and to the rate of deceleration that occurs between breaths. Because tidal volume is a dependent variable, inconsistent alveolar minute ventilation can occur. When using pressuretargeted ventilation, the clinician must be aware that the tidal volume delivered depends on changes in lung and chest wall compliance and airway resistance.

In volume-targeted modes of ventilation, the controlled variables are tidal volume, which is a function of inspiratory flow (not flow rate), and time. A goal with this strategy is to guarantee a preset minimum minute ventilation, which is usually a function of set tidal volume and set respiratory rate. Along with delivering a preset tidal volume, with certain ventilators the clinician can select the inspiratory flow profile, which dictates whether that flow will be delivered throughout the inspiratory cycle in a constant or decelerating flow pattern. Pressure is the dependent variable in the modes of ventilation in which volume is the target. Because pressures will vary in volume-targeted modes of ventilation, careful monitoring and assessment of respiratory system compliance and resistance is necessary.

Dual-Control Modes

Modes that combine the positive attributes of volume- and pressuretargeted strategies are designed for use in patients with disease processes in which respiratory system mechanics vary and/or in which ventilator dysynchrony occurs. These devices, popularly referred to as dual-control modes, do not control both parameters (pressure and volume) simultaneously; rather, the modes switch from one to the other, based on a measured input variable. The device operates as a timed-cycle, pressurelimited or volume-limited ventilator using a clinician-selected tidal volume- or pressure limit as an input variable to automatically adjust the pressure or volume, based on changes in respiratory system compliance and/or resistance.

Pressure-regulated volume control (PRVC) and Volume-assured pressure-support (VAPS) ventilation are examples of dual-control modes that switches the control parameter "breath to breath" or even "within the breath".

Newer modalities, such as airway pressure release ventilation (APRV), negate the need for paralysis or deep sedation by allowing spontaneous breathing throughout the respiratory cycle.

Spontaneous breathing with supporting ventilation significantly improves ventilation-perfusion matching, cardiac output, CO_2 clearance, and renal blood flow.[87]

As previously discussed, PEEP should be applied to sustain recruitment of as many alveoli as possible in order to maximize gas exchange and improve the distribution of ventilation. Levels of PEEP required to maintain end-expiratory lung volume and limit shear forces may be substantial (>20 cm H_2O).

Despite the existence of different modes of mechanical ventilation, with each mode used, the same principle should be applied: open up the lung and keep the lung open and ensure adequate carbon dioxide exchange with the least possible pressure amplitude. Such goals are commonly achieved by applying modes of mechanical ventilation that use pressure targeted strategies and not by modes using volume-targeted strategies, as these may predispose to regional over inflation of compliant lung areas with under ventilation of noncompliant regions.

Lung Function Monitoring

Lung function measurements should provide basic physiological information on (1) gas transport from the air via the lung into the blood, and (2) should (depending on the level of care) provide techniques which may differentiate between causes of disturbances in gas exchange. Mechanical ventilation should overcome or prevent hypoxia, which is the most important and life-threatening parameter during mechanical ventilation.

The oxygenation index (PaO_2/FiO_2) measured under standard ventilator settings, can be used to define the state of impairment of the lung, although a lower than optimal oxygenation index does not differentiate between:

1. ventilation
2. perfusion
3. diffusion, or
4. ventilation/perfusion (V/Q) problems.

It is, however, the most reliable and routinely available tool to define the state of openness of the lung under standard ventilation conditions.[88]

Peak inspiratory pressure at flow-constant ventilation is a poor parameter to measure alveolar over-stretching as it is influenced by a number of factors independent of alveolar pressure and does not allow to define the state of over-inflation and/or openness of different lung areas.[79]

Total lung volume and functional residual capacity should always be taken into account when interpreting compliance measurements; lung compliance measurements, which are not normalised for lung volume, have only limited information. Thus if FRC measurements are not available, one should at least normalise lung compliance values for lean body weight.

End tidal carbon dioxide ($PetCO_2$) is most commonly used as a noninvasive substitute for $PaCO_2$ in evaluating the adequacy of ventilation, but in patients with cardiopulmonary problems significant increase in $PaCO_2$ are not reflected by comparable increases in $PetCO_2$. $PaCO2$ may be underestimated by $PetCO_2$ if there is a reduction in CO or an increase in V/Q mismatch.[89] Therefore it cannot be recommended to be used in patients with IAH.

Oxygen delivery (DO_2), O_2 consumption (VO_2), and O_2 extraction ratio are the three monitored O_2 transport variables: $DO_2 = CO \times CaO_2$; $VO_2 = CO \times (CaO_2\text{-}CVO_2)$ and the O_2 extraction ratio = VO_2/DO_2. With a closed ventilator system like the Physioflex (Dräger) and the Zeus (Dräger), VO_2 is measured. Together with an arterial and mixed venous blood sample the O_2 extraction ratio is then easily obtained.

Oxygen debt is defined as the amount of O_2 that cells are deprived of as a result of imbalance between DO_2 and VO_2. Correction of this debt is one of the end- points of resuscitation.

Achieving supernormal values of cardiac index (4.5 L/min/m^2), DO_2 index (600 mL/min/m^2), and VO_2 index (170 mL/min/m^2) by fluid infusion and inotropes is a possible way to repay this debt and salvage tissues.[90]

Organ Perfusion and Oxygen Utilisation

Because in patients with ACS especially the abdominal organs are deprived of blood flow, it is important to monitor organ function. Oxygen transport variables, blood lactate level, base deficit, and gastric intramucosal pH (pHi) are considered acceptable markers of organ perfusion and O_2 utilisation.[90] Currently, only the blood lactate level and base deficit and, in some instances, O_2 transport variables are used during surgery. However, with improved technology, alimentary tract pHi, PCO_2 or indocyanine green clearance may also be used in this setting.

Cardiovascular Effects of Mechanical Ventilation in ACS

Positive pressure ventilation (PPV) causes a decrease in cardiac output that can be attributed to at least three mechanisms: (1) decreased venous return (2) right ventricular dysfunction, and (3) alteration of left ventricular distensibility. Decreased venous return is generally the most significant factor causing decreased cardiac output with PPV. Increased intrathoracic pressure results in decreased end-diastolic volume and stroke volume of both ventricles.[91,92]

Augmenting preload with additional intravascular fluid will minimize this effect. In the second mechanism, PPV increases pulmonary vascular resistance and thereby increases right ventricular afterload.[93,94] This effect is most pronounced in patients with preexisting right ventricular dysfunction.[95] The third mechanism by which PPV can cause reduction in cardiac output is alteration of left ventricular distensibility. Elevated pulmonary pressures can cause an elevation in right ventricular end-diastolic volume, resulting in a leftward shift of the intraventricular septum.[96] This shift limits left ventricular distensibility and causes a decrease in cardiac output.[96,97]

Weaning and Extubation

Weaning and extubation can be considered if the cause for the elevated IAP is solved and IAP is normalized. Fluid and electrolyte balance should be optimal. In most of the cases the patients have been ventilated for several days and first a period of weaning is warranted. Again, during the period of weaning it is important that atelectasis is avoided by use of adequate PEEP levels.

Ventilator weaning and extubation are important to decrease the risks associated with mechanical ventilation such as ventilator-associated pneumonia (VAP), airway trauma, and increased costs.[98] These risks must be balanced against the risk of premature extubation, which may lead to difficulty reestablishing endotracheal intubation, hemodynamic instability, and increased mortality.[99] It is estimated that as much as 42% of the time a patient is mechanically ventilated is spent on ventilator weaning.[100] No single approach to ventilator weaning has been established as superior, resulting in the use of different techniques in various institutions.[101] Weaning techniques include intermittent trials of spontaneous breathing or gradually decreasing levels of intermittent mandatory ventilation and pressure support ventilation (see below). These techniques allow the clinician to assess the patient's ability to take on an increasing proportion of the work of breathing.[102,103]

Weaning Modes

Continuous mandatory ventilation (CMV) is synonymous with assist/control ventilation (ACV), in which patients are allowed to "trigger" the ventilator to receive an assisted breath from the device. Intermittent mandatory ventilation (IMV), also referred to as "intermittent demand ventilation," is a partial-support mode. The patient receives mandatory (machine) breaths at a set frequency and volume or a set pressure and inspiratory time. Between mandatory breaths, the patient can breathe spontaneously from either a demand flow or a continuous flow system. The original version of the IMV mode is now considered obsolete. Most modern ventilators operate in an synchronized intermittent mandatory ventilation (SIMV) mode. ACV is the most commonly used mode of mechanical ventilation in the world[104] although many institutions prefer SIMV. In this updated version of IMV, the machine creates timing windows around the scheduled mandatory breaths in order to synchronize each machine's breath with the patient's inspiratory effort, which might vary the machine cycle times slightly. If no inspiratory effort is detected within the time window, the machine delivers a mandatory breath. Although SIMV improves patient/ventilator interaction at low (rate) levels, patients can expend an unanticipated amount of energy, which may contribute to failure in weaning.[105]

Pressure support ventilation (PSV) is a "spontaneous" mode of breathing in which the patient's inspiratory effort is assisted with a set level of inspiratory pressure (pressure support). Basically, PSV is a pressuretargeted mode and is very similar to pressureassist control (pressure control). In a pressurecontrol mode, the variable is time; in a pressuresupport mode, the variable is flow. PSV allows the patient to control his or her respiratory rate, inspiratory time, and inspiratory flow rate. The tidal volume achieved is a function of the respiratory system compliance and resistance. Several factors may influence the effects of pressure support ventilation, including the level of inspiratory pressure support and the pressure rise time. PSV has become a widely used mode during the weaning of patients who require prolonged durations of mechanical ventilation.

Bi-level positive airway pressure (BiPAP) is nearly identical to PSV, but with two sets of pressure level. CPAP is the mode of conventional ventilation that offers the least amount of support. Like PSV, it is classified as a spontaneous breathing mode. CPAP has a set level of pressure that is maintained throughout the respiratory cycle during spontaneous breathing. It has been used synonymously with PEEP, expiratory positive airway pressure (EPAP), and continuous positive-pressure breathing (CPPB). This mode is typically used to assess extubation readiness in an intubated patient.

Clinical trials to determine the best mode of ventilator weaning have been inconclusive. Esteban and colleagues[106] compared techniques of weaning patients from mechanical ventilation. They found that daily spontaneous breathing trials (SBTs) led to extubation three times more quickly than weaning with intermittent mandatory ventilation and about twice as quickly as weaning with pressure support ventilation. In contrast, Brochard and coworkers.[107] Reported that weaning was significantly faster with pressure support ventilation than with intermittent mandatory ventilation or SBT. In summary, weaning protocols should be used daily to identify patients who are ready for extubation, and unless contraindications exist, patients who pass the protocol should proceed to extubation. Patients who fail these protocols should be treated for any reversible causes and reassessed the next day for another weaning attempt.

Conclusions

Manifestations of ACS include pulmonary impairment: an increase in intra-abdominal pressure causes a graded disturbance in pulmonary physiology ranging from mechanical and cardiovascular to systemic derangements. The changes in such patients warrant good observation and the monitoring in patients with ACS has been discussed.

Mechanical problems include decreased FRC, alveolar dead space ventilation, ventilation/perfusion mismatch and increased intrathoracic pressures with a resulting decreased oxygen delivery to the tissues. These changes necessitate mechanical ventilation. Systemic derangements in ACS include a systemic inflammatory response with a release of systemic inflammatory mediators and bacterial translocation into the bloodstream. Mediators and bacteria may translocate across the alveolo-capillary membrane. In this way, the lung may than become a propagator of the systemic inflammatory response and mechanical ventilation has been shown to be an important contributory factor to do so. Protective lung strategies may prevent the transfer of inflammatory mediators, and the transfer of bacteria and bacterial endotoxins from the bloodstream into the lung and vice versa. There are a vast number of ventilatory strategies available and available ventilation techniques have been discussed. All these techniques should comply with one rational concept, which prevents further damage due to artificial ventilation itself. It should produce minimal pressure swings during the ventilation cycle and keep the lung open during the whole ventilatory cycle. Open up the whole lung and keep it open with the least possible influence on the cardio-circulatory system. Such a concept of mechanical ventilation is most readily achieved using pressurecycled modes of mechanical ventilation. Finally, when the signs of ACS resolve, patients should be weaned from mechanical ventilation and the modes of weaning have been discussed here.

Commentary

Manu L. N. G. Malbrain

The interactions between the abdominal and the thoracic compartment pose a specific challenge to the physicians working in the OR and the ICU. Both compartments are linked via the diaphragm and on average a 50% transmission of IAP to the intrathoracic pressure has been noted in previous animal and human studies. Patients with primary ACS will often develop a secondary ARDS and will require a different ventilatory strategy and treatment than patient with primary ARDS. The major problem lays in the diminished total respiratory system compliance resulting in a low FRC and low lung volumes mimicking a form of restrictive lung disease. Together with the alterations caused by secondary ARDS this will lead to the so-called "baby-lungs". Some key-issues to remember are:

1. The presence of IAH decreases total respiratory system compliance by a decrease in chest wall compliance, lung compliance being unchanged (except in case of concomitant primary ARDS)
2. Best PEEP can be set to counteract IAP whilst in the same time avoiding over-inflation of already well-aerated lung regions (consider applying weights on the chest at PEEP-levels above 20 cm H2O)
3. The ARDS consensus definitions should take into account PEEP and IAP values
4. During lung protective ventilation, the plateau pressures should be limited to transmural plateau pressures (ΔPplat: plateau pressure minus IAP/2) below 35cm H2O instead of the classical alveolar plateau pressures measured by the ventilator
5. The wedge criterion in ARDS consensus definitions is futile in case of IAH and should be adapted since most patients with IAH and secondary ARDS will have filling pressures above the 18 mm Hg definition cut-off
6. The presence of IAH dramatically increases lung edema especially in case of direct lung injury or capillary leak, within this concept monitoring of EVLWI seems warranted
7. The combination of capillary leak, positive fluid balance and raised IAP poses the patient at an exponential danger for lung edema
8. Body position affects IAP, putting an obese patient in the upright position can cause ACS, for instance in case of preexisting aerophagia related to noninvasive ventilation. Conversely the abdomen should hang freely during prone positioning, while the anti-Trendelenburg position may improve respiratory mechanics, however it can decrease splanchnic perfusion.
9. The use of curarisation should be balanced against the beneficial effect on abdominal muscle tone resulting in decrease in IAP and improvement of APP, however the more cranial position of the diaphragm during curarisation (especially in conditions of IAH or ACS) may worsen lung mechanics resulting in atelectasis and infection.
10. The presence of IAH will lead to pulmonary hypertension via increased intrathoracic pressures with direct compression on lung vessels and via the diminished left and right ventricular compliance. In this case the administration of inhaled NO or ilomedine (prostacyclin) may be justified.

References

1. Kron IL, Harman PK, Nolan SP. The measurement of intra-abdominal pressure as a criterion for abdominal reexploration. Ann Surg 1984; 199(1):28-30.
2. Eddy V, Nunn C, Morris Jr JA. Abdominal compartment syndrome. The Nashville experience. Surg Clin North Am 1997; 77(4):801-812.
3. Ivatury RR, Porter JM, Simon RJ et al. Intra-abdominal hypertension after life-threatening penetrating abdominal trauma: Prophylaxis, incidence, and clinical relevance to gastric mucosal pH and abdominal compartment syndrome. J Trauma 1998; 44(6):1016-1021, (discussion 1021-1013).
4. Sugerman HJ, Bloomfield GL, Saggi BW. Multisystem organ failure secondary to increased intraabdominal pressure. Infection 1999; 27(1):61-66.
5. Schein M, Ivatury R. Intra-abdominal hypertension and the abdominal compartment syndrome. Br J Surg 1998; 85(8):1027-1028.
6. Burch JM, Moore EE, Moore FA et al. The abdominal compartment syndrome. Surg Clin North Am 1996; 76(4):833-842.
7. Ivatury RR, Diebel L, Porter JM et al. Intra-abdominal hypertension and the abdominal compartment syndrome. Surg Clin North Am 1997; 77(4):783-800.
8. Meldrum DR, Moore FA, Moore EE et al. Prospective characterization and selective management of the abdominal compartment syndrome. Am J Surg 1997; 174(6):667-672, (discussion 672-663).
9. Ertel W, Oberholzer A, Platz A et al. Incidence and clinical pattern of the abdominal compartment syndrome after "damage-control" laparotomy in 311 patients with severe abdominal and/or pelvic trauma. Crit Care Med 2000; 28(6):1747-1753.
10. Wendt E. Ueber den einfluss des intraabdominalen druckes auf die absondererungsgeschwindigkeit des harness. Arch Physiologische Heilkunde 1876; 57:525-527.
11. Emerson H. Intra-abdominal pressures. Arch Intern Med 1911; 7:754-784.
12. Baggot MG. Abdominal blow-out: A concept. Current Research in Anesthesia and Analgesia 1951; 30:295-298.
13. Malbrain MLNG. Intra-abdominal pressure in the intensive care unit. In: Vincent JL, ed. Yearbook of Intensive Care and Emergency Medicine. Berlin: Springer-Verlag, 2001:547-585.
14. Morris Jr JA, Eddy VA, Blinman TA et al. The staged celiotomy for trauma. Issues in unpacking and reconstruction. Ann Surg 1993; 217(5):576-584, (discussion 584-576).
15. Ridings PC, Bloomfield GL, Blocher CR et al. Cardiopulmonary effects of raised intra-abdominal pressure before and after intravascular volume expansion. J Trauma 1995; 39(6):1071-1075.
16. Mutoh T, Lamm WJ, Embree LJ et al. Volume infusion produces abdominal distension, lung compression, and chest wall stiffening in pigs. J Appl Physiol 1992; 72(2):575-582.
17. Sanchez NC, Tenofsky PL, Dort JM et al. What is normal intra-abdominal pressure? Am Surg 2001; 67(3):243-248.
18. Schein M, Wittmann DH, Aprahamian CC et al. The abdominal compartment syndrome: The physiological and clinical consequences of elevated intra-abdominal pressure. Journal of the American College of Surgeons 1995; 180(6):745-753.
19. Bardoczky GI, Engelman E, Levarlet M et al. Ventilatory effects of pneumoperitoneum monitored with continuous spirometry. Anaesthesia 1993; 48(4):309-311.
20. Fahy BG, Barnas GM, Flowers JL et al. The effects of increased abdominal pressure on lung and chest wall mechanics during laparoscopic surgery. Anesth Analg 1995; 81(4):744-750.
21. Hirvonen EA, Nuutinen LS, Kauko M. Ventilatory effects, blood gas changes, and oxygen consumption during laparoscopic hysterectomy. Anesthesia and Analgesia 1995; 80(5):961-966.
22. Obeid F, Saba A, Fath J et al. Increases in intra-abdominal pressure affect pulmonary compliance. Arch Surg 1995; 130(5):544-547.
23. Oikkonen M, Tallgren M. Changes in respiratory compliance at laparoscopy: Measurements using side stream spirometry. Can J Anaesth 1995; 42(6):495-497.
24. Dumont L, Mattys M, Mardirosoff C et al. Changes in pulmonary mechanics during laparoscopic gastroplasty in morbidly obese patients. Acta Anaesthesiol Scand 1997; 41(3):408-413.
25. Sprung J, Whalley DG, Falcone T et al. The impact of morbid obesity, pneumoperitoneum, and posture on respiratory system mechanics and oxygenation during laparoscopy. Anesth Analg 2002; 94(5):1345-1350.
26. Fahy BG, Barnas GM, Nagle SE et al. Changes in lung and chest wall properties with abdominal insufflation of carbon dioxide are immediately reversible. Anesth Analg 1996; 82(3):501-505.
27. Shulman SM, Chuter T, Weissman C. Dynamic respiratory patterns after laparoscopic cholecystectomy. Chest 1993; 103(4):1173-1177.
28. Erice F, Fox GS, Salib YM et al. Diaphragmatic function before and after laparoscopic cholecystectomy. Anesthesiology 1993; 79(5):966-975, (discussion 927A-928A).

29. Sharma RR, Axelsson H, Oberg A et al. Diaphragmatic activity after laparoscopic cholecystectomy. Anesthesiology 1999; 91(2):406-413.
30. Gargiulo NJ IIIrd SR, Leon W, Machiedo GW et al. Hemorrhage exacerbates bacterial translocation at low levels of intra-abdominal pressure. Arch Surg 1998; 133(12):1351-1355.
31. Malbrain MLNG. Effects of abdominal compression and decompression on cardiorespiratory function and regional perfusion. Intensive Care Med 2000; 26:S264.
32. Bongard F, Pianim N, Dubecz S et al. Adverse consequences of increased intra-abdominal pressure on bowel tissue oxygen. J Trauma 1995; 39(3):519-524, (discussion 524-515).
33. Brismar B, Hedenstierna G, Lundquist H et al. Pulmonary densities during anesthesia with muscular relaxation—a proposal of atelectasis. Anesthesiology 1985; 62(4):422-428.
34. Williams M, Simms H. Abdominal compartment syndrome: Case reports and implications for management in critically ill patients. Am Surg 1997; 63(6):555-558.
35. Puri GD, Singh H. Ventilatory effects of laparoscopy under general anaesthesia. Br J Anaesth 1992; 68(2):211-213.
36. Kelman GR, Swapp GH, Smith I et al. Caridac output and arterial blood-gas tension during laparoscopy. Br J Anaesth 1972; 44(11):1155-1162.
37. Johannsen G, Andersen M, Juhl B. The effect of general anaesthesia on the haemodynamic events during laparoscopy with CO2-insufflation. Acta Anaesthesiologica Scandinavica 1989; 33(2):132-136.
38. Cullen DJ, Coyle JP, Teplick R et al. Cardiovascular, pulmonary, and renal effects of massively increased intra-abdominal pressure in critically ill patients. Critical Care Medicine 1989; 17(2):118-121.
39. Richardson J, Trinkle J. Hemodynamic and respiratory alterations with increased intra-abdominal pressure. J Surg Res 1976; 20(5):401-404.
40. Saggi B, Sugerman H, Ivatury R et al. Abdominal compartment syndrome. J Trauma 1998; 45(3):597-609.
41. Ayres SM GA, Holbrook PR, Shoemaker WC. Textbook of Critical Care Medicine. 3rd ed. Philadelphia: WB Saunders, 1995.
42. Sugrue M, Jones F, Janjua KJ et al. Temporary abdominal closure: A prospective evaluation of its effects on renal and respiratory physiology. J Trauma 1998; 45(5):914-921.
43. Bloomfield G, Saggi B, Blocher C et al. Physiologic effects of externally applied continuous negative abdominal pressure for intra-abdominal hypertension. J Trauma 1999; 46(6):1009-1014, (discussion 1014-1006).
44. Quintel M, Pelosi P, Caironi P et al. An increase of abdominal pressure increases pulmonary edema in oleic acid-induced lung injury. Am J Respir Crit Care Med 2004; 169(4):534-541.
45. Biffl WL, Moore EE, Burch JM et al. Secondary abdominal compartment syndrome is a highly lethal event. Am J Surg 2001; 182(6):645-648.
46. Valenza F, Bottino N, Canavesi K et al. Intra-abdominal pressure may be decreased noninvasively by continuous negative extra-abdominal pressure (NEXAP). Intensive Care Med 2003; 29(11):2063-2067.
47. Sugrue M. Intra-abdominal pressure: Time for clinical practice guidelines? Intensive Care Med 2002; 28(4):509-514.
48. Mayberry J. Prevention of the abdominal compartment syndrome. Lancet 1999; 354(9192):1749-1750.
49. Rezende-Neto JB, Moore EE, Melo de Andrade MV et al. Systemic inflammatory response secondary to abdominal compartment syndrome: Stage for multiple organ failure. J Trauma 2002; 53(6):1121-1128.
50. Diebel LN, Dulchavsky SA, Brown WJ. Splanchnic ischemia and bacterial translocation in the abdominal compartment syndrome. The Journal of Trauma 1997; 43(5):852-855.
51. Balogh Z, McKinley BA, Holcomb JB et al. Both primary and secondary abdominal compartment syndrome can be predicted early and are harbingers of multiple organ failure. J Trauma 2003; 54(5):848-859, (discussion 859-861).
52. Verbrugge SJ, Lachmann B. Mechanisms of ventilation-induced lung injury: Physiological rationale to prevent it. Monaldi Arch Chest Dis 1999; 54(1):22-37.
53. Demling RH. Adult respiratory distress syndrome: Current concepts. New Horiz 1993; 1(3):388-401.
54. Verbrugge SJ, Sorm V, Lachmann B. Mechanisms of acute respiratory distress syndrome: Role of surfactant changes and mechanical ventilation. J Physiol Pharmacol 1997; 48(4):537-557.
55. Verbrugge SJ, Gommers D, Lachmann B. Conventional ventilation modes with small pressure amplitudes and high positive end-expiratory pressure levels optimize surfactant therapy. Crit Care Med 1999; 27(12):2724-2728.
56. Lachmann B. Open up the lung and keep the lung open. Intensive Care Med 1992; 18(6):319-321.

57. Haitsma JJ, Uhlig S, Goggel R et al. Ventilator-induced lung injury leads to loss of alveolar and systemic compartmentalization of tumor necrosis factor-alpha. Intensive Care Med 2000; 26(10):1515-1522.
58. Cheng KC, Zhang H, Lin CY et al. Ventilation with negative airway pressure induces a cytokine response in isolated mouse lung. Anesth Analg 2002; 94(6):1577-1582, (table of contents).
59. Chiumello D, Pristine G, Slutsky AS. Mechanical ventilation affects local and systemic cytokines in an animal model of acute respiratory distress syndrome. Am J Respir Crit Care Med 1999; 160(1):109-116.
60. Ricard JD, Dreyfuss D, Saumon G. Production of inflammatory cytokines in ventilator-induced lung injury: A reappraisal. Am J Respir Crit Care Med 2001; 163(5):1176-1180.
61. Tremblay L, Valenza F, Ribeiro SP et al. Injurious ventilatory strategies increase cytokines and c-fos m-RNA expression in an isolated rat lung model. J Clin Invest 1997; 99(5):944-952.
62. Verbrugge SJ, Uhlig S, Neggers SJ et al. Different ventilation strategies affect lung function but do not increase tumor necrosis factor-alpha and prostacyclin production in lavaged rat lungs in vivo. Anesthesiology 1999; 91(6):1834-1843.
63. von_Bethmann AN, Brasch F, Nusing R et al. Hyperventilation induces release of cytokines from perfused mouse lung. American Journal of Respiratory and Critical Care Medicine: An Official Journal of the American Thoracic Society, Medical Section of the American Lung Association 1998; 157(1):263-272.
64. Haitsma JJ, Uhlig S, Verbrugge SJ et al. Injurious ventilation strategies cause systemic release of IL-6 and MIP-2 in rats in vivo. Clin Physiol Funct Imaging 2003; 23(6):349-353.
65. Ranieri VM, Suter PM, Tortorella C et al. Effect of mechanical ventilation on inflammatory mediators in patients with acute respiratory distress syndrome: A randomized controlled trial. Jama 1999; 282(1):54-61.
66. Wrigge H, Zinserling J, Stuber F et al. Effects of mechanical ventilation on release of cytokines into systemic circulation in patients with normal pulmonary function. Anesthesiology 2000; 93(6):1413-1417.
67. Fabregas N, Torres A, El-Ebiary M et al. Histopathologic and microbiologic aspects of ventilator-associated pneumonia. Anesthesiology 1996; 84(4):760-771.
68. Tilson MD, Bunke MC, Smith GJ et al. Quantitative bacteriology and pathology of the lung in experimental Pseudomonas pneumonia treated with positive end-expiratory pressure (PEEP). Surgery 1977; 82(1):133-140.
69. Nahum A, Hoyt J, Schmitz L et al. Effect of mechanical ventilation strategy on dissemination of intratracheally instilled Escherichia coli in dogs. Crit Care Med 1997; 25(10):1733-1743.
70. Verbrugge SJ, Sorm V, van 't Veen A et al. Lung overinflation without positive end-expiratory pressure promotes bacteremia after experimental Klebsiella pneumoniae inoculation. Intensive Care Med 1998; 24(2):172-177.
71. Murphy DB, Cregg N, Tremblay L et al. Adverse ventilatory strategy causes pulmonary-to-systemic translocation of endotoxin. Am J Respir Crit Care Med 2000; 162(1):27-33.
72. Suchyta MR, Clemmer TP, Elliott CG et al. The adult respiratory distress syndrome. A report of survival and modifying factors. Chest 1992; 101(4):1074-1079.
73. Milberg JA, Davis DR, Steinberg KP et al. Improved survival of patients with acute respiratory distress syndrome (ARDS): 1983-1993. Jama 1995; 273(4):306-309.
74. Webb HH, Tierney DF. Experimental pulmonary edema due to intermittent positive pressure ventilation with high inflation pressures. Protection by positive end-expiratory pressure. Am Rev Respir Dis 1974; 110(5):556-565.
75. Dreyfuss D, Saumon G. Ventilator-induced lung injury: Lessons from experimental studies. Am J Respir Crit Care Med 1998; 157(1):294-323.
76. Dreyfuss D, Basset G, Soler P et al. Intermittent positive-pressure hyperventilation with high inflation pressures produces pulmonary microvascular injury in rats. Am Rev Respir Dis 1985; 132(4):880-884.
77. Dreyfuss D, Soler P, Basset G et al. High inflation pressure pulmonary edema. Respective effects of high airway pressure, high tidal volume, and positive end-expiratory pressure. Am Rev Respir Dis 1988; 137(5):1159-1164.
78. Ashbaugh DG, Petty TL, Bigelow DB et al. Continuous positive-pressure breathing (CPPB) in adult respiratory distress syndrome. J Thorac Cardiovasc Surg 1969; 57(1):31-41.
79. Tremblay LN, Slutsky AS. Role of pressure and volume in ventilator-induced lung injury. Appl Cardiopulm Pathophysiology 1997; 6:179-190.
80. Lachmann B, Jonson B, Lindroth M et al. Modes of artificial ventilation in severe respiratory distress syndrome. Lung function and morphology in rabbits after wash-out of alveolar surfactant. Crit Care Med 1982; 10(11):724-732.

81. Sjostrand UH, Lichtwarck-Aschoff M, Nielsen JB et al. Different ventilatory approaches to keep the lung open. Intensive Care Med 1995; 21(4):310-318.
82. Kesecioglu J. Mechanical ventilation in ARDS. Adv Exp Med Biol 1996; 388:533-538.
83. Amato MB, Barbas CS, Medeiros DM et al. Effect of a protective-ventilation strategy on mortality in the acute respiratory distress syndrome. N Engl J Med 1998; 338(6):347-354.
84. Barbas CSV, Medeiros D, Magaldi RB. High PEEP levels improved survival in ARDS patients. AM J Resp Crit Care Med 2002; 165:A 218.
85. Grasso S, Mascia L, Del Turco M et al. Effects of recruiting maneuvers in patients with acute respiratory distress syndrome ventilated with protective ventilatory strategy. Anesthesiology 2002; 96(4):795-802.
86. Alvarez A, Subirana M, Benito S. Decelerating flow ventilation effects in acute respiratory failure. J Crit Care 1998; 13(1):21-25.
87. Putensen C, Rasanen J, Lopez FA. Ventilation-perfusion distributions during mechanical ventilation with superimposed spontaneous breathing in canine lung injury. Am J Respir Crit Care Med 1994; 150(1):101-108.
88. Bohm S, Lachmann B. Pressure-control ventilation. Putting a mode into perspective. Int J Intensive Care 1996; 3:12-27.
89. Bhavani-Shankar K, Moseley H, Kumar AY et al. Capnometry and anaesthesia. Can J Anaesth 1992; 39(6):617-632.
90. Porter JM, Ivatury RR. In search of the optimal end points of resuscitation in trauma patients: A review. J Trauma 1998; 44(5):908-914.
91. Leithner C, Podolsky A, Globits S et al. Magnetic resonance imaging of the heart during positive end-expiratory pressure ventilation in normal subjects. Crit Care Med 1994; 22(3):426-432.
92. Huemer G, Kolev N, Kurz A et al. Influence of positive end-expiratory pressure on right and left ventricular performance assessed by Doppler two-dimensional echocardiography. Chest 1994; 106(1):67-73.
93. Robotham JL, Lixfeld W, Holland L et al. The effects of positive end-expiratory pressure on right and left ventricular performance. Am Rev Respir Dis 1980; 121(4):677-683.
94. Pinsky MR, Desmet JM, Vincent JL. Effect of positive end-expiratory pressure on right ventricular function in humans. Am Rev Respir Dis 1992; 146(3):681-687.
95. Brooks H, Kirk ES, Vokonas PS et al. Performance of the right ventricle under stress: Relation to right coronary flow. J Clin Invest 1971; 50(10):2176-2183.
96. Jardin F, Farcot JC, Boisante L et al. Influence of positive end-expiratory pressure on left ventricular performance. N Engl J Med 1981; 304(7):387-392.
97. Terai C, Uenishi M, Sugimoto H et al. Transesophageal echocardiographic dimensional analysis of four cardiac chambers during positive end-expiratory pressure. Anesthesiology 1985; 63(6):640-646.
98. Ely EW, Baker AM, Evans GW et al. The distribution of costs of care in mechanically ventilated patients with chronic obstructive pulmonary disease. Crit Care Med 2000; 28(2):408-413.
99. Epstein SK, Ciubotaru RL, Wong JB. Effect of failed extubation on the outcome of mechanical ventilation. Chest 1997; 112(1):186-192.
100. Esteban A, Alia I, Ibanez J et al. Modes of mechanical ventilation and weaning. A national survey of Spanish hospitals. The Spanish Lung Failure Collaborative Group. Chest 1994; 106(4):1188-1193.
101. Venus B, Smith RA, Mathru M. National survey of methods and criteria used for weaning from mechanical ventilation. Crit Care Med 1987; 15(5):530-533.
102. Luce JM, Pierson DJ, Hudson LD. Intermittent mandatory ventilation. Chest 1981; 79(6):678-685.
103. Nathan SD, Ishaaya AM, Koerner SK et al. Prediction of minimal pressure support during weaning from mechanical ventilation. Chest 1993; 103(4):1215-1219.
104. Esteban A, Anzueto A, Alia I et al. How is mechanical ventilation employed in the intensive care unit? An international utilization review. Am J Respir Crit Care Med 2000; 161(5):1450-1458.
105. Esteban A, Frutos F, Tobin MJ et al. A comparison of four methods of weaning patients from mechanical ventilation. Spanish Lung Failure Collaborative Group. N Engl J Med 1995; 332(6):345-350.
106. Esteban A FF, Tobin MJ et al. A comparison of four methods of weaning from mechanical ventilation. N Engl J Med 1995; 332:345-350.
107. Brochard L, Rauss A, Benito S et al. Comparison of three methods of gradual withdrawal from ventilatory support during weaning from mechanical ventilation. Am J Respir Crit Care Med 1994; 150(4):896-903.

CHAPTER 8

Intra-Abdominal Hypertension and the Kidney

Michael Sugrue, * **Ali Hallal and Scott D'Amours**

Abstract

Intra-abdominal hypertension (IAH) has been associated with renal impairment for over 150 years. It is only recently however, that a clinically recognised relationship has been found. An increasing number of large clinical studies have identified that IAH ($\geq$15 mm Hg) is independently associated with renal impairment and increased mortality. The evidence comes from both animal and human experiments. It is related to a multi-factorial effect, predominately related to renal perfusion, coupled with reduced abdominal perfusion pressure, reduced cardiac output and increased systemic vascular resistance. Further to this there is an alteration in humoral and neurogenic factors aggravating renal function. The risk of renal impairment with IAH is further exacerbated by hypovolaemia and other factors such as sepsis.

In summary, intra-abdominal pressures of $\geq$15 mm Hg exert a clinically significant effect on renal function.

Introduction

Renal impairment and renal failure are one of the commonest causes of surgical admissions to ICU. The treatment of renal impairment and its prevention has evolved in the last two decades with an increasingly physiological approach to patient resuscitation and the greater awareness of the importance of intra-abdominal hypertension (IAH).[1,2] While we have known about the adverse effects of raised intra-abdominal pressure (IAP) on renal function for over 150 years, there are many practicing clinicians who remain unaware or unconvinced of this relationship[3] and we still have patients who have tight abdominal closures (Fig. 1). While it is understandable that the older textbooks in surgery would not be expected to discuss the role of intra-abdominal hypertension in renal impairment, contemporary peer reviewed articles of postoperative renal impairment have neglected intra-abdominal hypertension and the abdominal compartment syndrome (ACS) as a predisposing cause of renal failure.[4,5]

While the literature contains a number of published articles relating to abdominal hypertension only a few prospective studies of large numbers of patients exist.[6-8]

A key question that needs to be answered is whether IAH is in fact an independent cause of renal impairment.

To determine this we need to answer the following questions: Is there supportive evidence from human experiments? Does it make scientific sense? Is there a strong association that is consistent from study to study? Is the temporal relationship right? Is there a dose-response relationship and is it reversible? Is the association independent of other confounding factors?

*Corresponding Author: Michael Sugrue—Department of Trauma, Liverpool Hospital, Liverpool NSW 2170, Australia. Email: michael.sugrue@swsahs.nsw.gov.au

Abdominal Compartment Syndrome, edited by Rao R. Ivatury, Michael L. Cheatham, Manu L. N. G. Malbrain and Michael Sugrue. ©2006 Landes Bioscience.

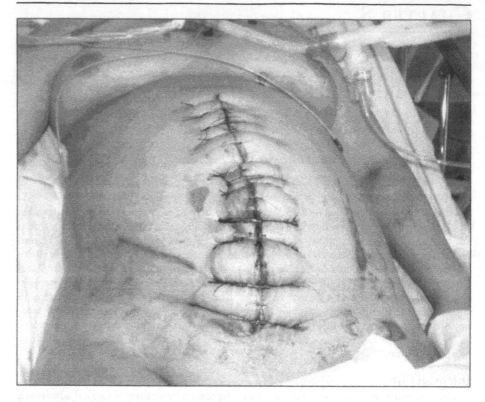

Figure 1. Tight abdominal closure and resultant IAH seen occasionally.

This chapter will attempt to answer these questions while exploring the complex and evolving relationship between IAH and its progression into ACS on renal function.

Is There Supportive Evidence from Human Experiments?

There are a number of interesting human experiments from the last century, particularly Bradley's 1947 landmark paper on the association of renal impairment and increased IAP in human volunteers.[9] In general however, reports have been sporadic, small in numbers and often retrospective. Two large studies[6,7] have enrolled prospectively over 350 general ICU patients and support the association between IAH and renal impairment. In the larger of these prospective studies of 276 patients, (mean APACHE II of 14.6 ± 7.7, range 1-37 admitted to ICU following abdominal surgery (emergency in 174/263), IAH was found be an independent causal factor for renal impairment. The pattern of IAH is shown in Figure 2 and the relationship of IAP and renal function is shown in Figure 3. IAH of 18 mm Hg or greater occurred in 41% of postoperative general ICU patients. Of the 107 patients with intra-abdominal hypertension 35/107 (32.7%) developed renal impairment compared to 22/156 (14.1%) of those with normal IAP.

Biancafiore and colleagues, looking prospectively at the relationship between IAH and renal function in subjects undergoing orthotopic liver transplantation found, that 34/108 patients had IAP of 25 mm Hg or more. Renal impairment was observed in 32% of patient with IAP more than 25 mm Hg.[8] Acute renal failure developed in 17/108 patients (16%), 11 (65%)of whom had IAH (p< 0.01) with a mean of 27.9 ± 9.9 mm Hg versus 18.6 ± 5.2 mm Hg in those without acute renal failure (p< 0.001). Their logistic regression analysis showed that intraoperative transfusions of more than 15 units, respiratory failure and intra-abdominal pressure of >24 mm Hg or more were independent risk factors for renal failure.

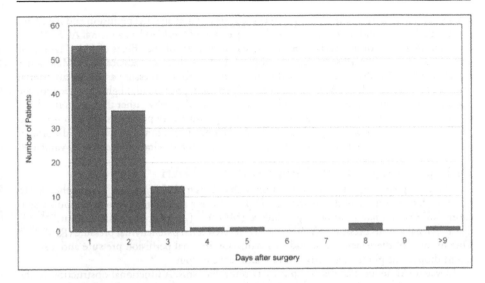

Figure 2. First day at which IAH occurred after abdominal surgery. Figure reprinted with permission from Arch Surg 1999; 134:1082-1085.

Does It Make Physiological Sense?

We have known about compartment syndromes in various body cavities for centuries, of particular importance raised intracranial pressure and limb compartment syndromes. Increasingly however, unusual compartment syndromes are being reported, including ocular compartment syndromes[10] and visceral compartment syndromes including isolated hepatic compartment syndrome following trauma.[11] It makes sound physiological sense that like other

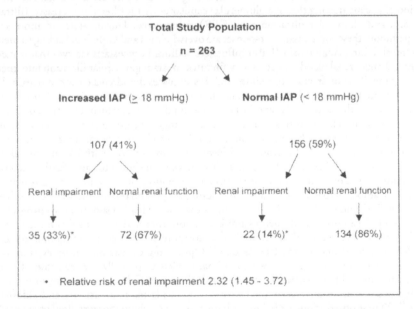

Figure 3. Overall breakdown of relationship of IAP and renal function. Figure reprinted with permission from Arch Surg 1999; 134:1082-1085.

areas in the body a rise in IAP will have adverse effects not just on renal perfusion and function, but on all the intra-abdominal organs, with progression of local IAH to a formal ACS. Newer concepts are being continuously proposed as scientific basis of the effects of IAP. The recent concept of abdominal perfusion pressure proposed by Cheatham and advocated by Malbrain is interesting.[2,12] Cheatham has defined the abdominal perfusion pressure as the mean arterial pressure minus IAP (MAP-IAP). The concept of perfusion pressure highlights the need for the abdomen to be considered part of the whole person physiologically rather than as an insolated cavity. It makes sense that the abdominal perfusion pressure is important, similar to the renal perfusion pressure proposed by Ulyatt.[13,14] Cheatham's suggestion that using an abdominal perfusion pressure of 50 mm Hg or more as an endpoint of resuscitation needs to be validated.

Pathophysiology of Renal Impairment and IAH

The pathophysiology of renal dysfunction with IAH is multifactorial and is hypothesized to be through one or more of the following: (1) part of the systemic inflammatory response syndrome (SIRS) and late multiple organ failure (MOF),[15] (2) reduced cardiac output,[16,17] (3) elevated renal venous pressure (RVP)[18] and (4) elevated renal parenchymal pressure (RPP).[19,20] One or more of these factors can lead to a reduction in renal perfusion pressure and a subsequent drop in the glomerular filtration rate and urine output.

Elevated IAP reduces cardiac output by two mechanisms. A functional obstruction to the vena cava and increased intrathoracic pressure reduce venous return to the heart thereby decreasing preload and cardiac output. The second mechanism involves the direct increase in systemic vascular resistance leading to an increase in afterload and a subsequent fall in cardiac output. A drop in cardiac output will cause a drop in the renal blood flow (RBF),[21] which will cause a drop in the renal perfusion pressure and a corresponding drop in afferent arteriolar pressure. The juxtaglomerular cells perceive this change as a decreased stretch exerted on the afferent arteriolar walls, which results in the release of more renin into the renal circulation. This results in the formation of angiotensin I, which is converted in the kidney and peripherally to angiotensin II by angiotensin converting enzyme. Angiotensin II induces preferential constriction of efferent arterioles. As a result, intraglomerular pressure is maintained and the fraction of plasma flowing through glomerular capillaries that is filtered is increased (filtration fraction), and glomerular filtration rate (GFR) is preserved. During states of more severe hypoperfusion, these compensatory responses are overwhelmed and GFR falls, leading to prerenal acute renal failure. Angiotensin II also influences sodium homeostasis via two major mechanisms: it changes renal blood flow so as to maintain a constant glomerular filtration rate, thereby changing the filtration fraction of sodium and, it stimulates the adrenal cortex to release aldosterone. Elevated plasma levels of aldosterone enhance renal sodium retention and thus result in the expansion of the extracellular and intravascular fluid volume, which, in turn, dampens the stimulus for renin release. In this context, the renin-angiotensin-aldosterone system regulates volume by modifying renal haemodynamics and tubular sodium transport (Fig. 4). Increases in plasma renin activity and aldosterone level have been shown to rise with IAP in animal models[18,22-25] while administration of ACE inhibitors have been shown to cause further deterioration in urine function. Other humoral factors such as endothelin may affect renal function by acting in an autocrine/paracrine fashion constricting the pre and post glomerular vessels thereby decreasing RBF and GFR. This decrease reduces the filtered load of sodium, favouring sodium reabsorption. In addition, endothelin stimulates aldosterone and acts directly on the proximal tubule to increase sodium reabsorption.[26] Catecholamine release occurs in response to increasing IAP in patients undergoing laparoscopy.[27] Optimizing cardiac output by expanding the intravascular volume has been shown in animal models to partially reverse some of the derangement caused by the elevated intra-abdominal pressure. The improvement in urine production in such an instance is coupled with a drop in plasma renin activity and aldosterone levels.[21,28] These observations are in line with the reasoning behind the crystalloid based, preload driven, goal oriented resuscitation for shock recommended as standard of care in North American

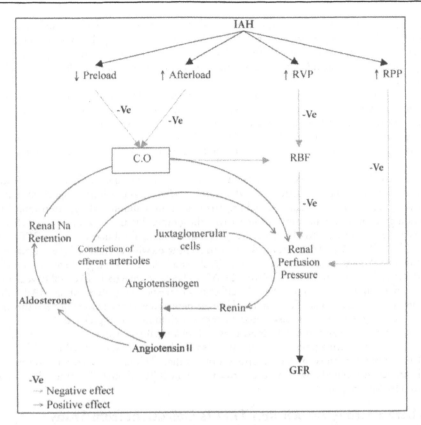

Figure 4. Possible mechanism of renal derangement in patients with IAH.

trauma centers.[29,30] The concept that intravenous fluid resuscitation improves urinary output in the presence of IAH is not new, as Thorington very eloquently demonstrated in animal experiments in 1923, that with rising IAP, increasing degrees of renal failure occurred but this responded in part to an intravenous fluid challenge.[31] It has been shown by other investigators that the haemodynamic changes in animals with IAH are made worse by hypovolaemia.[32] In patients with IAH, despite significantly elevated central venous pressures, continued fluid resuscitation commonly results in improved cardiac output[33] On the other hand some studies would suggest that the above strategy is detrimental in IAH as further expansion of the intravascular volume with crystalloid will lead to more gut edema increasing further the intra-abdominal pressure causing full blown ACS.[34]

Because normalizing cardiac output with volume expansion fails to fully correct the renal derangements observed with IAH, renal vein compression and renal parenchymal compression have been implicated as primary mediators. Bloomfield et al were able to show an increase in the renal venous pressure (RVP) coupled with an increase in IAP induced by ethylene glycol instillation in the peritoneum, associated with a significant drop in the urine output, an increase in plasma renin activity and aldosterone levels. These then normalized after decompression.[22] Doty et al showed that isolated elevation of the RVP by temporary occlusion of the renal vein with a vessel loop in swine leads to the same altered renal function.[18] Whether the increase in renal parenchymal pressure (RPP) plays a major role in the pathophysiology of renal impairment with IAH is controversial. Stone showed that decapsulated ischaemic kidneys in rhesus monkeys had better renal function than capsulated kidneys.[35] Renal decapsulation was widely practiced but fell out of favour because of its morbidity. On the other hand Doty et al showed that high renal parenchymal pressure induced by compressing the kidneys in between two acrylic plates alone did not lead to similar results, and concluded that renal parenchymal compression plays a much less role than renal venous compression as a mediator of renal derangement in ACS.[19]

Little is known about the molecular events that mediate IAH. Barish recently has cast some light on the subject.[36] In a rat model of IAH they demonstrated that there is a dynamic renal gene expression response to early IAH. The molecular changes are observed as early as 30 minutes after induction of IAH. Further characterization of the genes up-and down regulated by IAH may help in the future to develop a better understanding of this particular pathophysiology.

The most likely direct effect of increased IAP is an increase in the renal vascular resistance coupled with a moderate reduction in cardiac output. Ulyatt has previously suggested that the filtration gradient (FG) is a key to renal impairment in intra-abdominal hypertension.[13] The filtration gradient is the mechanical force across the glomerulus and is equal to the difference between the glomerular filtration pressure (GFP) and proximal tubular pressure (PTP), thus FG = GFP - PTP. Where IAP is elevated PTP can be equated with IAP and GFP is estimated by the difference between mean arterial pressure (MAP) and IAP. Filtration gradient can therefore be calculated by the formula FG = MAP - 2 (IAP). Therefore changes in IAP will have a much greater effect on urine formation than the effect of a corresponding alteration in MAP.[13] Pressure on the ureter is not a key factor in renal impairment as renal stents have not been shown to improve urinary output.[31] Shenasky used an external abdominal compression device on dogs to demonstrate decreases in renal blood flow.[37] McDougall has demonstrated a decrease in renal venous flow during pneumoperitoneum also.[38] These decreases in renal blood flow and GFR have been confirmed by laser Doppler flow studies.[39] Chiu et al reported a decrease in renal parenchymal blood flow from the cortex to the medulla although this redistribution was not seen by McDougall.[38]

Is There a Strong Association That Is Consistent from Study to Study?

The first prospective study of 88 patients by Sugrue and colleagues showed a strong association between increased IAP and renal impairment. It was not designed to show a direct causal relationship however[6] and did not account for confounding variables which could have affected renal function. Their second study involving a new cohort of 263 not only confirmed the prevalence of IAH in post laparotomy patients in ICU (40.7%), but also showed an independent causal relationship. Ivatury and colleagues reported the incidence of IAH to be in excess of 50% in patients undergoing fascial closure.[40] In the first multi-center study of IAH reported recently, the prevalence of IAH (defined as a maximal IAP of greater or equal to 12 mm Hg) and ACS was 50.5% and 8.2% respectively.[41] The literature does not contain any reports of series of patients where increased IAP is associated with improved renal function.

Is the Temporal Relationship Right?

Demonstration of a temporal relationship between IAH and renal impairment is a clinical challenge. This is in part due to our current limitation in clinically useful tests to evaluate renal function. A rise in creatinine takes time following renal insult. Sugrue in a review of 263 patients found that in 35 patients who developed renal impairment, the impairment (elevated serum creatinine) occurred at a mean lag period of 2.7 days (range 0-35) after. The effect of IAH on renal function would appear to be gradual rather than immediate.

Table 1. An analysis of renal function for different categories of intra-abdominal pressure

IAP n=263	Renal Impairment n=57	Normal Renal Function n=206
< 18 mm Hg n=156	22 (14%)	134 (86%)*
18-25 mm Hg n=86	22 (26%)	64 (74%)*
> 25 mm Hg n=21	13 (62%)	8 (38%)*

x^2=26.06; df=3; p< 0.001

Is There a Dose Response Relationship and Is It Reversible?

Studies have supported the concept that renal impairment is dose related, with the incidence of renal impairment doubling once IAP goes above 25 mm Hg as shown in Table 1.[7] Other cut-off values for IAH have been used ranging from 12mm to 20 mm Hg.[7,20,41] Many have advocated earlier abdominal decompression to avoid irreversible renal impairment.[42] It has been claimed that abdominal decompression reverses the sequelae of increased IAP.[43] Many series of small numbers of patients have shown that decompression is associated with improved renal and cardiovascular physiology in patients with IAH and ACS. This response however is not universal, Meldrum et al reported a 100% response with decompression.[44] Sugrue and colleagues reported a success of only 20%, although the cohort in their study was not well defined.[45] In addition, the series reported by Sugrue et al, like many others, was not an intention to treat study. Few conclusions, therefore, can be established from the current literature, other than to confirm that timely decompression of the abdomen may help restore renal function in some subgroups (as yet to be defined). It is important to be cautious about an over-enthusiastic approach to abdominal decompression as renal function will not always be improved, although current literature suggest that 85% of patients will be improved.[46]

Is the Association Independent of Other Confounding Factors?

The evidence that IAP is an independent causal factor of renal impairment is supported by the strong clinical association between IAH and renal impairment. Hypotension, sepsis, and age >60 and IAH are well established causes of renal impairment.[7] Hypertension, diabetes and aortic clamping while on bivariate analysis were associated with renal impairment, failed on multivariate analysis to achieve independent significance. The prevalence of risk factors and results of multivariate analysis are shown in Tables 2 and 3 respectively. Intra-abdominal hypertension is currently the fourth most important cause of renal impairment in postoperative ICU patients.

Is There Evidence from Human Experiments?

Initial reports of IAH were often following aortic surgery with postoperative hemorrhage from the graft suture line.[47] These may constitute a different subgroup of patients from those with secondary and tertiary peritonitis where tissue oedema and intra-abdominal sepsis, rather than free intraperitoneal or retroperitoneal blood are the dominant causes of IAH. Even though the side effects of IAP are increasingly reported,[8] the number of large clinical series remains limited.

Table 2. *Comparison of the prevalence of risk factors for renal impairment in patients with normal and impaired renal function*

Risk Factors Assessed	Normal Renal Function (n=206)	Renal Impairment (n=57)	Unadjusted Odds Ratio (95% CI)
Increased IAP	72 (35%)	35 (61%)*	3.0 (1.62-5.42)[#]
Hypovolemia	44 (21%)	20 (35%)	2.0 (1.05-3.77)
Aminoglycosides	77 (37%)	33 (58%)*	2.3 (1.27-4.18)
Radiocontrast	23 (11%)	12 (21%)	2.1 (0.98-4.58)
Sepsis	76 (37%)	37 (65%)*	3.2 (1.71-5.84)[#]
Hypotension	84 (41%)	42 (74%)*	4.1 (2.11-7.80)[#]
CCF	7 (3%)	3 (5%)	1.6 (0.40-6.31)
Hypertension	49 (24%)	24 (42%)*	2.3 (1.26-4.31)
Diabetes	15 (7%)	5 (9%)	1.2 (0.43-3.53)
Age 60+	114 (55%)	47 (82%)*	3.8 (1.82-7.92)[#]
NSAIDS	27 (13%)	13 (23%)	2.0 (0.94-4.10)
ACE Inhibitors	18 (9%)	6 (11%)	1.2 (0.46-3.26)
Diuretics	23 (11%)	13 (23%)	2.4 (1.10-5.00)
Gout	7 (3%)	3 (5%)	1.6 (0.40-6.31)
Aortic Clamping	35 (17%)	19 (33%)*	2.4 (1.26-4.73)
Dehydration	7 (3%)	2 (4%)	1.0 (0.21-5.12)

* $p < 0.01$ (bivariate, chi squared); [#] $p < 0.05$ (unadjusted); Table reprinted with permission from Arch Surg 1999; 134:1082-1085.

Table 3. *Forced entry logistic regression model of clinical factors associated with renal impairment*

Variable	Wald Statistic	Significance	Adjusted Odds Ratio (95% CI)
Sepsis	7.69	0.006*	2.88 (1.37-6.07)
Age 60+	4.68	0.03*	2.70 (1.10-6.62)
Hypotension	4.35	0.04*	2.25 (1.05-4.80)
IAH	4.10	0.004*	2.11 (1.04-4.27)
Hypertension	2.89	0.09	1.95 (0.91-4.18)
Aorta Clamping	2.38	0.12	1.89 (0.84-4.25)
Diuretics	1.78	0.18	1.84 (0.75-4.53)
Aminoglycosides	1.36	0.2	1.53 (0.75-3.14)
Radiocontrast	0.46	0.5	1.44 (0.51-6.07)
Hypovolaemia	0.46	0.8	1.09 (0.50-2.39)

Table reprinted with permission from Arch Surg 1999; 134:1082-1085.

In conclusion, there is irrefutable evidence to support that intra-abdominal hypertension is a direct and independent causal factor leading to renal impairment Renal failure is one of the main expressions of ACS. Given the prevalence of intra-abdominal hypertension of around 40% and ACS of 5-10% in postoperative and trauma patients in ICU it is imperative that further research be undertaken.

Commentary

Rao R. Ivatury

One of the most dramatic sequelae of increased IAP is the effect on renal function and urine output. Sugrue and colleagues review the current evidence that establishes this relationship in this chapter. An increasing number of large clinical studies have identified that IAH (≥ 15 mm Hg) is independently associated with renal impairment and increased mortality. The etiology of these changes are not entirely well established, however it may be multifactorial: reduced renal perfusion, reduced cardiac output and increased systemic vascular resistance and alterations in humoral and neurogenic factors. The risk of renal impairment with IAH is further exacerbated by hypovolaemia and other factors such as sepsis. It has been also demonstrated in the case of cirrhotic patients with ascites that renal function may be improved by paracentesis of the ascitic fluid and reduction in the IAP. The benefits of prompt reduction of IAP are also quite dramatic in patients with primary and secondary IAH after trauma. It therefore behooves us as clinicians to be cognizant of the elevated IAP and its effect on renal function: often the first sign of impending ACS.

References

1. Sugrue M. Intra-abdominal pressure: Time for clinical practice guidelines? Int Care med 2002; 28:389-391.
2. Malbrain ML. Is it wise not to think about intra-abdominal hypertension in the ICU? Curr Opin Crit Care 2004; 10(2):132-45.
3. Fischer M. Raised intra-abdominal pressure, renal failure, and the bumble bee. Int Care Med 1990; 16:285-286.
4. Carmichael p, Carmichael AR. Acute renal failure in the surgical setting. ANZ J Surg 2003; 73:144-153.
5. Sugrue M, Balogh Z, Malbrain M. Intra-abdominal hypertension and renal failure. ANZ J Surg 2004; 74:78.
6. Sugrue M, Buist MD, Hourihan F et al. Prospective study of intra-abdominal hypertension and renal function after laparotomy. Br J Surg 1995; 82:235-238.
7. Sugrue M, Jones F, Deane SA et al. Intra-abdominal hypertension is an independent cause of post-operative renal impairment. Arch Surg 1999; 134:1082-1805.
8. Biancofiore G, Bindi ML, Romanelli AM et al. Postoperative intra-abdominal pressure and renal function after liver transplantation. Arch Surg 2003; 138:703-6.
9. Bradley SE, Bradley GP. The effect of increased intra-abdominal pressure on renal function in man. J Clin Invest 1947; 26:1010-1022.
10. Priluck IA, Blodgett DW. The effect of increased intra-abdominal pressure on the eyes. Nebraska M J 1996; 8-9.
11. Pearl LB, Trunkey DD. Comaprtment sydnrome of the liver. J Trauma 199; 47:796-798.
12. Cheatham ML, White MW, Sagraves SG et al. Abdominal perfusion pressure; A superior parameter in assessment of Intra-abdominal hypertension. J Trauma 2000; 49:621-627.
13. Ulyatt DB. Elevated intra-abdominal pressure Australian Anaes 1992; 108-114.
14. Malbrain ML. Abdominal perfusion pressure as a prognostic marker in intra-abdominal hypertension. In: Vincent JL, ed. Yearbook of Intensive care and Emergency Medicine. Berlin: Springer-Verlag, 2002; 792-814.
15. Rezende-Neto JB, Moore EE, Melo de Andrade MV et al. Systemic inflammatory response secondary to abdominal compartment syndrome: Stage for multiple organ failure. J Trauma 2002; 53:1121-8.
16. Ridings PC, Bloomfield GL, Blocker CR et al. Cardiopulmonary effects of raised intra-abdominal pressure before and after intra-vascular volume expansion. J Trauma 1995; 39:1071-1075.
17. Robotham JL, Wise RA, Bromberger-Barnea B. Effects of changes in abdominal pressure on left ventricular performance and regional blood flow. Crit Care Med 1985; 13:803-809.
18. Doty JM, Saggi BH, Sugerman HJ et al. Effect of increased renal venous pressure on renal function. J Trauma 1999; 47:1000-3.
19. Doty JM, Saggi BH, Blocher CR et al. Effects of increased renal parenchymal pressure on renal function. J Trauma 2000; 48:874-877.
20. Platell CF, Hall J, Clarke G et al. Intra-abdominal pressure and renal function after surgery to the abdominal aorta. Aust NZ J Surg 1990; 60:213-216.

21. Lindstrom P, Wadstrom J, Ollerstam A et al. Effects of increased intra-abdominal pressure and volume expansion on renal function in the rat. Nephrol Dial Transplant 2003; 18:2269-77.
22. Bloomfield GL, Blocher CR, Fakhry IF et al. Elevated intra-abdominal pressure increases plasma renin activity and aldosterone levels. J Trauma 1997; 42:997-1005.
23. Kotzampassi K, Metaxas G, Paramythiotis D et al. The influence of continuous seven-day elevated intra-abdominal pressure in the renal perfusion in cirrhotic rats. J Surg Res 2003; 115:133-8.
24. Vargas JC, Fields D, Razvi I. Direct parenchymal compression to 15 mm Hg produces oliguria. J Urol 1995; 153:514.
25. Caldwell CB, Ricotta JJ. Changes in visceral blood flow with elevated intra-abdominal press. J Surg Res 1987; 43:14-20.
26. Hamilton BD, Chow GK, Inman SR et al. Increased intra-abdominal pressure during pneumoperitoneum stimulates endothelin release in a canine model. J Endourol 1998; 12:193-197.
27. Mikami O, Fujise K, Matsumoto S et al. High intra-abdominal pressure increases plasma catecholamine concentrations during pneumoperitoneum for laparoscopic procedures. Arch Surg 1998; 133:39-43.
28. London ET, Ho HS, Neuhaus AM et al. Effect of intravascular volume expansion on renal function during prolonged CO2 pneumoperitoneum. J Trauma 2000; 231:195-201.
29. Rivers E, Nguyen B, Havstad S et al. Early goal-directed therapy collaborative group. Early goal-directed therapy in the treatment of severe sepsis and septic shock. N Engl J Med 2001; 8(345):1368-77.
30. Miller PR, Meredith JW, Chang MC. Randomized, prospective comparison of increased preload versus inotropes in the resuscitation of trauma patients: Effects on cardiopulmonary function and visceral perfusion. J Trauma 1998; 44:107-13.
31. Thorington JM, Schmidt CF. A study of urinary output and blood pressure changes resulting in experimental ascites. Am J Med Sc 1923; 165:880-889.
32. Toomasian JM, Glavinovich G, Johnson MN. Haemodynamic changes following pneumoperitoneum and graded haemorrhage in the dog. Sur Forum 1978; 29:32-33.
33. Malbrain ML, Cheatham ML. Cardiovascular effects and optimal preload markers in intra-abdominal hypertension. In: Vinent JL, ed. Yearbook of intensive Care and emergency medicine. Berlin: Springer-Verlag, 2004, in press.
34. Balogh Z, McKinley BA, Cocanour CS et al. Patients with impending abdominal compartment syndrome do not respond to early volume loading. Am J Surg 2003; (6):602-8.
35. Stone Hh, Fulenwider JT. Renal decapsulation in the prevention of post ischaemic oliguria. Ann Surg 1977; 186:343-355.
36. Edil BH, Tuggle DW, Puffinbarger NK et al. The impact of intra-abdominal hypertension on gene expression in the kidney. J Trauma 2003; 55:857-9.
37. Shenasky JH, Gillenwater JY. The renal hemodynamic and functional effects of external counterpressure. SGO 1972; 134:253-258.
38. Mc Dougall Em, Monk TG, Wolf JS. The effect of prolonged pneumoperitoneum on renal function in an animal model. J Am Coll Sur 1996; 182:317-328.
39. Chiu AW, Azadzoi KM, Hatzichristou DG. Effects of intra-abdominal pressure on renal perfusion during laparoscopy. J Endourol 1994; 8:99-103.
40. Ivatury RR, Simon RJ, Islam S et al. A prospective randomized study of end points of resuscitation after major trauma: Global oxygen transport indices versus organ-specific gastric mucosal pH. J Am Coll Surg 1996; 183:145-154.
41. Malbrain ML, Chiumello D, Pelosi P et al. Prevalane of intra-abdominal hypertension in critically ill patients. A multicentre epidemiology study. Inten Care Med2004; 30:822-829.
42. Ivatury RR, Porter JM, Simon RJ et al. Intra-abdominal hypertension after life-threatening penetrating abdominal trauma: Prophylaxis, incidence, and clinical relevance to gastric mucosal pH and abdominal compartment syndrome. J Trauma 1998; 44:1016-1021.
43. Kopelman T, Harris C, Miller R et al. Abdominal compartment syndrome in patients with isolated extraperitoneal injuries. J Trauma 2000; 49:744-749.
44. Meldrum DR, Moore FA, Moore EE et al. Prospective characterization and selective management of the abdominal compartment syndrome. Am J Surg 1997; 174:667-672.
45. Sugrue M, Jones F, Janjua KJ et al. Temporary abdominal closure. A prospective evaluation of its effects on renal and respiratory physiology. J Trauma 1998; 45:914-921.
46. Sugrue M, D'Amours S. Abdominal compartment syndrome in patients with isolated extraperitoneal injuries. J Trauma 2001; 51:419.
47. Fietsam R, Villalba M, Glover JL et al. Intra-abdominal compartment syndrome as a complication of ruptured abdominal aortic aneurysm repair. Am Surg 1989; 55:396-402.

CHAPTER 9

Intra-Abdominal Hypertension and the Splanchnic Bed

Rao R. Ivatury* and Lawrence N. Diebel

Abstract

Intra-abdominal hypertension has profound effects on splanchnic organs, causing diminished perfusion, mucosal acidosis and setting the stage for multiple organ failure. If uncorrected, IAH will result in abdominal compartment syndrome and increase morbidity and mortality. The pathologic changes are more pronounced after sequential insults of ischemia-reperfusion and IAH. It appears that IAH and ACS may serve as the second insult in the two-hit phenomenon of the causation of multiple-organ dysfunction syndrome.

Intra-abdominal hypertension (IAH), as elucidated throughout this book, may result in profound physiologic effects that may culminate in organ dysfunction and failure.[1-10] The abdominal compartment syndrome (ACS) is a constellation of these physiologic sequelae of IAH. The effects of IAH on the splanchnic circulation has been known a long time. But only recently, they have been identified as a potential mechanism for the Multiorgan Dysfunction Syndrome (MODS) following IAH and ACS.

IAH and Splanchnic Flow

In animal experiments Diebel and associates[11] showed a decline in the mesenteric and gastro-intesinal mucosal blood flow with an IAP above 20 mm Hg. Intestinal mucosal blood flow diminished to 61% of the baseline at an IAP of 20 mm Hg and 28% of the baseline at an IAP of 40 mm Hg (Fig. 1). Corresponding to these changes, the intestinal mucosa, as studied by tonometer, showed severe acidosis. These changes were disproportionate to the reduction in cardiac output associated with increasing IAP. These investigators also noted that, in anesthetized pigs, an IAP of 10 mm Hg caused a significant decrease in hepatic arterial blood flow (HABF) and hepatic microvascular blood flow (HMVBF). Despite a constant cardiac output and mean arterial pressure, at an IAP of 20 mm Hg the HABF was reduced to 45% of the control, the HMVBF to 71% and the portal venous blood flow to 65% of the control. The decreases were exaggerated with higher levels of IAP.[12] Similar results were seen by Bongard and colleagues[13] as they created IAH by insufflating the peritoneal cavity with helium to an IAP of 15 and 20 mm Hg for 60 minutes. Tissue oxygen partial pressure (TPO$_2$) in the bowel, measured with fluorescence quenching catheters, fell progressively as the IAP was increased while the subcutaneous TPO$_2$ remained unchanged. These changes were independent of changes in cardiac output (Fig. 2). Engum and associates[13] studied puppies with placement of an intra-abdominal inflatable balloon to simulate ACS. Baseline pressures were 2 to 5 cm H$_2$O in the stomach and bladder catheters, 1 to 3 mm Hg in the intra-abdominal catheter, and correlated with a gastric tissue pH level of 7.4. Significantly high correlation coefficients were

*Corresponding Author: Rao R. Ivatury—VCURES, Virginia Commonwealth University, 1521 West Hospital, P.O. Box 980454, Richmond, Virginia, U.S.A. Email: rivatury@hsc.vcu.edu

Abdominal Compartment Syndrome, edited by Rao R. Ivatury, Michael L. Cheatham, Manu L. N. G. Malbrain and Michael Sugrue. ©2006 Landes Bioscience.

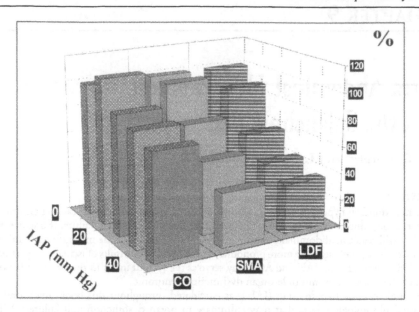

Figure 1. Effect of increasing abdominal pressure on cardiac output (CO), superior mesenteric arterial flow (SMA) and laser doppler mucosal flow in the gut (LDF). Data from Diebel et al, 1992. (Reprinted from: Ivatury RR, Cayten CG: The Textbook of Penetrating Trauma, Williams and Wilkins, Baltimore, 1996.)

observed between these various pressures. Gastric tissue pH level dropped to 7.0 with a BP and GP of 20 cm H_2O and IAP of 10 mm Hg, to 6.8 at 30 cm H_2O and 20 mm Hg, and 6.5 at 40 cm H_2O and 30 mm Hg, respectively. The authors suggested that changes in gastric tissue pH in association with increased IAP may be an early indicator of impending abdominal compartment syndrome. In fact, Pusajo and colleagues[15] and Sugrue et al[16] prospectively evaluated postoperative patients with IAP and gastric mucosal pH (pHi) monitoring. Compared to patients with normal pHi, patients with a pHi < 7.32 were 11 times more likely to have an elevated IAP. Ivatury et al noted that IAH was associated with gut mucosal acidosis[3,4] in their series of patients with catastrophic penetrating abdominal trauma.

Similar data of splanchnic dysfunction were evident form a porcine model of intra-abdominal hypertension. IAP of 30 mm Hg produced changes consistent with medium grade liver necrosis and medium grade mucosal damage in the bowel.[17] A fascinating study[18] described a device, an abdominal cavity chamber, to observe the changes caused by increased intra-abdominal pressure on the microcirculation.of animals. Intra-abdominal pressure was increased by intra-abdominal insufflation of gas. By using a fluorescent marker, the authors quantitatively assessed mucosa perfusion index, functional capillary density, red blood cell velocity, capillary diameters, and flow motion during increased intra-abdominal pressure by intravital video microscopy. When compared with controls, animals subjected to an intra-abdominal pressure of 10 and 15 mm Hg showed a significant stepwise decrease in mucosa perfusion index, functional capillary density, and red blood cell velocity, indicating a progressive impairment of mucosal microcirculation. Capillary diameter and flow motion did not change with respect to intra-abdominal pressure.

An example of the clinical relevance of these physiologic aberrations was provided by a study by Kologlu et al[19] who studied the effect of elevated IAP on healing of colonic anastomoses. Thirty rats, all with right colonic anastomoses, were divided into five groups. Group 1 was the control group, and group 2 had fecal peritonitis. IAP was maintained between 4 to 6 mm Hg in group 3, 8 to 12 mm Hg in group 4, and 14 to 18 mm Hg in group 5 until all rats were sacrificed on day 4. Bursting pressures and tissue hydroxyproline concentrations of

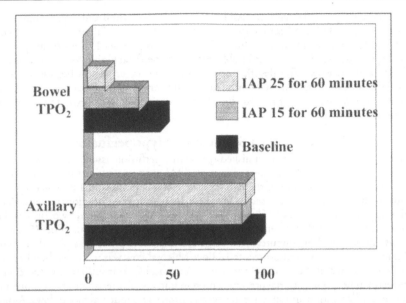

Figure 2. Effects of increased abdominal pressure on bowel tissue oxygenation. Data from Bongard et al, 1995. (Reprinted from: Ivatury RR, Cayten CG: The Textbook of Penetrating Trauma, Williams and Wilkins, Baltimore, 1996.)

anastomoses were then analyzed and compared. The bursting pressure and hydroxyproline concentrations had good correlation ($P<0.001$, $r = 0.76$). 4 to 6 mm Hg IAP delayed healing as much as fecal peritonitis. More elevated IAP delayed healing even more than fecal peritonitis.

Laparoscopy and IAH

The widespread application of laparoscopic surgery also spurred a clinical interest in the effects of increased IAP on cardiopulmonary and visceral function during induced pneumo-peritoneum.[20-26] Eleftheriadis and coauthors[20] studied 16 women undergoing cholecystectomy, eight by an open technique and eight laparoscopically. In all patients hepatic microcirculation was measured by a single-fiber laser-Doppler microde introduced into the hepatic parenchyma. Intestinal pH was measured by a gastric tonometer. Compared to the open cholecystectomy group, laparoscopy patients exhibited a significant decrease in hepatic microcirculatory flow and gastric mucosal pH. Both of these reverted to normal levels after the pneumoperitoneum was deflated. Schwarte and associates[22] similarly showed attenuation of microvascular oxygen saturation in gastric mucosa in 16 patients undergoing elective diagnostic laparoscopy. The increase in IAP from baseline to 8 mm Hg decreased microvascular oxygen saturation in gastric mucosa from 69 +/- 7% (mean +/- SD) to 63 +/- 8% at 8 mm Hg IAP ($P < 0.05$), with a further significant reduction to 54 +/- 13% at 12 mm Hg IAP. The mucosal microvascular oxygen saturation recovered rapidly to baseline level after release of increased IAP. In striking contrast to regional mucosal oxygen saturation, systemic oxygenation did not change with either of the interventions. Knolmayer et al[21] also demonstrated that increasing levels of IAP were correlated with decreased arterial pH, increased mixed venous CO_2, decreased intramucosal pH, and increased arterial CO_2. Gastric pHi differed significantly from baseline at IAP levels of 16 mm Hg and 18 mm Hg, even though no significant effects were observed on cardiac output or arterial lactate.

Windberger et al[23] investigated the IAP range for laparoscopic procedures that elicits only moderate splanchnic and pulmonary hemodynamic and metabolic changes. They increased the IAP to 7 and 14 mm Hg, each for 30 minutes in 10 healthy pigs. Portal and hepatic venous pressure increased in parallel with the IAP but the transmural portal and hepatic venous pressures decreased ($p < 0.01$), indicating decreased venous filling. Portal flow was maintained at 7 mm Hg but decreased at 14 mm Hg from 474 +/- 199 to 395 +/- 175 mL/min ($p < 0.01$), whereas hepatic arterial flow remained stable. Hepatic superficial blood flow decreased during insufflation and increased after desufflation. Intestinal, portal as well as hepatic venous pH decreased significantly at an IAP of 14 mm Hg. The authors concluded that the hemodynamic and metabolic derangement in the splanchnic beds is dependent on the extent of carbon dioxide pneumoperitoneum.

Inotropes and IAH-Induced Splanchnic Hypoperfusion

The exact mechanism of the diminished splanchnic perfusion associated with IAH is not known. but may involve a direct effect of increased IAP on mesenteric arterial resistance, humoral factors or a combination of the two.[1-23] The changes in intestinal mucosal blood flow associated with IAH could be reversed by low dose dobutamine but not dopamine, as demonstrated by Agusti and associates.[27] They studied 25 pigs. IAH was induced by peritoneal insufflation of CO_2 to an IAP of 15 mm Hg. Low dose Dopamine or Dobutamine (5 mcg,mL/min), was administered 60 minutes later. A perivascular flow probe was placed around the superior mesenteric artery to measure arterial flow. Mucosal flow was measured by a laser Doppler probe positioned in the lumen of the ileum. Peritoneal CO_2 insufflation induced significant increases in heart rate, arterial pressure, and systemic vascular resistance with concomitant decreases in cardiac output and superior mesenteric arterial and mucosal blood flows. Although dobutamine infusion reversed the decrease in cardiac output, it failed to restore superior mesenteric artery blood flow. Intestinal mucosal blood flow, however, returned to baseline levels. Dopamine also attenuated the decrease in cardiac output, but had no beneficial effect on splanchnic hemodynamic variables.

The aforementioned studies only investigated the effect of IAH, as a single insult, on splanchnic flow and visceral organ function : a scenario quite different from what is usually observed in the trauma patient, who undergoes sequential insults of initial hypovolemic shock and resuscitation and subsequent IAH. Several authors investigated whether these sequential insults may amplify the ill effects of IAH and also lower the level of IAH critical to detrimental organ function. Simon et al[28] noted that in animals subjected to a 20% hemorrhage followed by resuscitation and then an increase in the IAP to 10 and 20 mm Hg, the PaO_2 / FiO_2 ratios were significantly lower than in a control group of animals without prior shock and resuscitation. A follow-up study suggested a similar synergestic adverse result on superior mesenteric artery flow with ischemia-reperfusion followed by IAH.[29] In a porcine model of hemorrhagic shock and abdominal compartment syndrome (ACS), Varela and associates[30] studied continuous near-infrared spectroscopy (NIRS)-derived gastric tissue oxygen saturation ($GStO_2$) and muscle tissue oxygen saturation ($MstO_2$). A significant decrease in SMA flow, $GStO_2$, SvO_2, and $MStO_2$ was observed after hemorrhage in Group 1 (hemorrhage, no ACS) and in Group 2 (hemorrhage + ACS). $GStO_2$ correlated well with SMA flow and mesenteric oxygen delivery as did $MStO_2$ with SvO_2 and systemic oxygen delivery. The authors concluded that NIRS measurement of $GStO_2$ and $MStO_2$ reflected changes in mesenteric and systemic perfusion respectively during hemorrhage and ACS.

Oxygen free radical production and bacterial translocation associated with increased IAP attracted the attention of other investigators.[31-35] In one study, after the IAP was maintained at 15 mm Hg for 60 minutes, the mean arterial pressure was unchanged; the jejunal mucosal blood flow (measured by laser doppler flowmetry) was significantly decreased; the gut metabolic activity, as indicated by oxygen extraction (measured from portal vein and aortic oxygen content), was significantly increased. Thirty minutes after abdominal deflation free radical production (measured by levels of malondialdehyde) was increased in the intestinal mucosa, liver, spleen and lung. In contrast to controls, the animals with increased IAP showed significant E.coli counts in the mesenteric lymph nodes, liver and spleen three hours after abdominal deflation. The authors argued that these findings were an example of ischemia-reperfusion injury, and that increased IAP caused significant intestinal ischemia, followed by reperfusion injury after abdominal decompression.[31] Diebel and associates[32] also noted bacterial translocation during IAH in a murine experiment. Bacteria translocated primarily to the mesenteric lymph node in the animals with increased IAP, whereas bacterial translocation did not occur in the sham control group (p < 0.05). The most common bacterial species cultured from the rats with increased IAP was *Escherichia coli*. Other organisms recovered from the tissues harvested included Enterobacter, Enterococcus, Pseudomonas, and Staphylococcus. Gargiulo and colleagues[34] used a rodent model of hemorrhage, resuscitation and elevated IAP to 10 mm Hg to study the phenomenon of bacterial translocation. Hemorrhage and resuscitation alone did not increase bacterial translocation to the mesenteric lymph nodes, liver, or spleen. An increase in IAP to 10 mm Hg resulted in a significant level of translocation to the nodes and liver. Hemorrhage and resuscitation did increase the level of translocation to the liver and spleen when IAP was increased to 10 mm Hg. The authors concluded that hemorrhage and resuscitation, in association with an IAP of 10 mm Hg, increased bacterial translocation. Other authors could not demonstrate evidence of translocation in similar clinical and experimental studies.[33]

Against this background of conflicting data, we[35] hypothesized that the failure to demonstrate bacterial translocation in these experimental models may be related to culture techniques and that the demonstration by PCR of bacterial DNA products may be more sensitive. Nineteen swine were divided into two groups. In the experimental group, group 1 (n = 10), animals were hemorrhaged to a mean arterial pressure (MAP) of 25-30 mm Hg for a period of 30 minutes and resuscitated to baseline MAP. Subsequently, intra-abdominal pressure (IAP) was increased to 30 mm Hg above baseline by instilling sterile normal saline into the peritoneal cavity. The IAP was maintained at this level for 60 minutes. Acid/base status, gastric mucosal ph (pHi), superior mesenteric artery (SMA) blood flow, and hemodynamic parameters were measured and recorded. Blood samples were analyzed by polymerase chain reaction (PCR) for the presence of bacteria. Spleen, lymph node, and portal venous blood cultures were obtained at 24 hours. The second group was the control. These animals did not have the hemorrhage, resuscitation, or intra-abdominal hypertension (IAH) but were otherwise similar to the experimental group in terms of laparotomy and measured parameters. SMA blood flow in group 1 (baseline of 0.87 +/- 0.10 L/min) decreased in response to hemorrhage (0.53 +/- 0.10 L/min, p = 0.0001) and remained decreased with IAH (0.63 L/min +/- 0.10, p = 0.0006) as compared to control and returned towards baseline (1.01 +/- 0.5 L/min) on relief of IAH. pHi (baseline of 7.21 +/- 0.03) was significantly decreased with hemorrhage (7.04 +/- 0.03, p = 0.0003) and decreased further after IAH (6.99 +/- 0.03, p = 0.0001) in group 1 compared to control, but returned toward baseline at 24 hours (7.28 +/- 0.04). The mean arterial pH decreased significantly from 7.43 +/- 0.01 at baseline to 7.27 +/- 0.01 at its nadir within group 1 (p = 0.0001) as well as when compared to control (p = 0.0001). Base excess was also significantly decreased between groups 1 and 2 during hemorrhage (3.30 +/- 0.71 vs. 0.06 +/- 0.60, p = 0.001) and IAH (3.08 +/- 0.71 vs. -1.17 +/- 0.60, p = 0.0001). In group 1, 8 of the 10 animals had positive lymph node cultures, 2 of the 10 had positive spleen cultures, and 2 of the 10 had positive portal venous blood cultures for gram-negative enteric bacteria. Only 2 of the 10 animals had a positive PCR. In group 2, five of the nine animals had positive lymph node cultures, zero of the nine had positive spleen cultures, and one of the nine had positive portal venous blood cultures. Two of the nine animals had positive PCRs. There was no significant difference in cultures or PCR results between the two groups (Fisher's exact test, p = 0.3). These data showed that in this clinically relevant model, hemorrhage-reperfusion and IAH caused significant GI mucosal acidosis and ischemia as well as systemic acidosis. However, we could not show any evidence for increased bacterial translocation based on PCR or tissue or blood cultures in the experimental as compared with the control group.

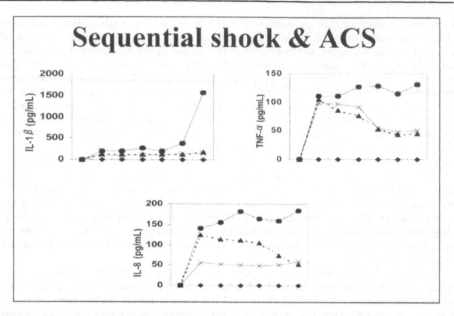

Figure 3. Effects of sequential hemorrhage-reperfusion (circles) and IAH on cytokines. These were all higher (p < 0.01) as compared with the either of the insults alone. Data from Oda et al.[41]

IAH and Multi-Organ Dysfunction Syndrome (MODS)

Recent clinical and laboratory studies suggest that splanchnic ischemia may have a pivotal role in the causation of MODS.[36-40] Uncontrolled inflammatory response manifested by the release of cytokines has been implicated as an important pathogenic component. The syndrome of multiple organ dysfunction is also seen in the clinical scenario of massive trauma leading to hemorrhagic shock and subsequent resuscitation, either as a single-hit or as a second-hit phenomenon. In fact, one of these second hits may be IAH and the ensuing ACS.[3,36-40] In a recent study, we hypothesized that sequential hemorrhagic shock (HS) and ACS would result in greater cytokine activation and polymorphonuclear neutrophil (PMN)-mediated lung injury than with either insult alone.[41] Twenty Yorkshire swine (20-30 kg) were studied. Group 1 (n = 5) was hemorrhaged to a mean arterial pressure of 25 to 30 mm Hg for 60 minutes and resuscitated to baseline mean arterial pressure. Intra-abdominal pressure was then increased to 30 mm Hg above baseline and maintained for 60 minutes. Group 2 (n = 5) was subjected to HS alone and Group 3 (n = 5) to ACS alone. Group 4 (n = 5) had sham experiment without HS or ACS. Central and portal venous interleukin-1beta, interleukin-8, and tumor necrosis factor-alpha levels were serially measured. Bronchoalveolar lavage (BAL) for protein and PMNs was performed at baseline and 24 hours after resuscitation. Lung myeloperoxidase was evaluated at 24 hours after resuscitation. Portal and central vein cytokine levels were equivalent but were significantly higher in Group 1 than in other groups (Fig. 3). BAL PMNs, lung myeloperoxidase activity and BAL protein were all higher (p < 0.01) in Group 1 compared with the other groups, suggesting that, in this clinically relevant model, sequential insults of ischemia-reperfusion (HS and resuscitation) and ACS were associated with significantly increased portal and central venous cytokine levels and more severe lung injury than HS or ACS alone, as a possible explanation for the causation of multi-organ failure.

Rezende-Nato and associates[42] induced IAH in Sprague-Dawley rats which were divided into 5 groups, 10 animals each. Intra-abdominal pressure (IAP) was increased to 20 mm Hg for 60 and 90 minutes in two different groups. In a third group the abdomen was decompressed for 30 minutes following IAP of 20 mm Hg before samples were collected. The other animals were used as controls. Hemodynamic response was monitored throughout the procedure. Cytokine levels were assessed in the plasma. Remote organ injury was assessed by histopathology and myeloperoxidase activity. IAH caused a significant increase in the levels of TNF-alpha and IL-6, 30 minutes after abdominal decompression. Plasma concentration of IL-1beta was elevated after 60 minutes of IAH. Abdominal decompression, however, did not cause a significant increase in the levels of this cytokine. Lung neutrophil accumulation was significantly elevated only after abdominal decompression. Histopathological findings showed intense pulmonary inflammatory infiltration including atelectasis and alveolar edema. The authors concluded that IAH provoked the release of pro-inflammatory cytokines which may serve as a second insult for the induction of MOF. In a subsequent study,[43] the same group of investigators attempted to develop a clinically relevant two-event animal model of postinjury MOF using the ACS as a second insult. Male adult rats underwent hemorrhagic shock (30 mm Hg x 45 min) and were resuscitated with crystalloids and shed blood. The timing of postshock systemic neutrophil (PMN) priming was determined by the surface expression of CD11b via flow cytometry. Finding maximal PMN priming at 8 h, but no priming at 2 h (early) and 18 h (late), ACS (25 mm Hg x 60 min) was introduced at these time points. At 24 h postshock, lung injury was assessed by lung elastase concentration and Evans blue dye extravasation in bronchoalveolar lavage. Liver and renal injuries were determined by serum alanine aminotransferase, serum creatinine, and blood urea nitrogen. The ACS during the time of maximal systemic PMN priming (8 h) provoked lung and liver injury, but did not if introduced at 2 or 18 h postshock when there was no evidence of systemic PMN priming. The 24-h mortality of this two-event model was 33%. These findings corroborate the potential for the ACS to promote multiple organ injury when occurring at the time of systemic PMN priming. This tme period is consistent with the clinical course of a severly injured patient developing MODS with ACS as the second precipitating event.

In summary, these clinical and experimental observations illustrate the profound implications of IAH on the splanchnic bed, even at very low levels of IAP. Constant monitoring of IAP and aggressive intervention to prevent IAH are crucial to avoid the development of ACS, MODS and mortality.

Commentary

Michael Sugrue

Ivatury and Diebel have given us an insight into the science behind the sequelae of Intra-Abdominal Hypertension (IAH) and the Abdominal Compartment Syndrome (ACS) on visceral perfusion. It is clear that this is physiologically and clinically relevant. Intra-abdominal pressures beyond 10 mm Hg will exert adverse effect at a cellular, organ and compartment level. Not only are the effects of raised intra-abdominal pressure due to increased arterial resistance but also increased venous pressures in part related to increased central venous pressure. Apart from the phenomena of increased mesenteric vascular resistance other factors that count include tissue oedema, sepsis, alteration in the inflammatory cascades further aggravating a critical balance resulting in hypoxia.

The phenomenon of ACS may be local exerting an effect on individual organs such as the kidney and liver or may have a broader effect on the entire abdominal cavity. The concept of perfusion is crucial. Abdominal perfusion in the early phase of IAH before it advances to ACS may fall before there is evidence of a drop in systemic perfusion. Hence the importance of abdominal perfusion pressure measurement. What remains a significant challenge is prevention and treatment of IAH and ACS. A current dilemma is how much fluids to give- enough to maintain perfusion yet not too much too cause ACS.

In addition, we need to have a greater understanding of the potential role of decompression, negative pressure and finally identify patients who are at higher risk of developing adverse effects.

References

1. Ivatury RR, Diebel L, Porter JM et al. Intra-abdominal hypertension and the abdominal compartment syndrome. Surg Clin North Am 1997; 77:783-800.
2. Ivatury RR, Simon RJ. Intraabdominal hypertension: The abdominal compartment syndrome. In: Ivatury RR, Cayten CG, eds. The Textbook of penetrating trauma. Williams & Wilkins, 1996:939-951.
3. Ivatury RR, Porter JM, Simon RJ et al. Intra-abdominal hypertension after life threatening abdominal trauma: Incidence, prophylaxis and clinical relevance to gastric mucosal pH and abdominal compartment syndrome. J Trauma 1998; 44:1016-1021.
4. Ivatury RR, Simon RJ, Islam S et al. A prospective randomized study of end points of resuscitation after major trauma. Global oxygen transport indices versus organ specific gastric mucosal pH. J Amer Coll Surg 1996; 183:145-154.
5. Saggi BH, Sugerman HJ, Ivatury RR et al. Abdominal compartment syndrome. J Trauma 1998; 45:597-609.
6. Schein M, Rucinski JR, Wise L. The abdominal compartment syndrome in the critically ill patient. Current opinion in critical care 1996; 2:287-294.
7. Schein M, Wittmann DH, Aprahamian CC et al. The abdominal compartment syndrome: The physiological and clinical consequences of elevated intra-abdominal pressure. J Amer Coll Surg 1995; 180:747-753.
8. Burch JM, Moore EE, Moore FA et al. The abdominal compartment syndrome. Surg Clin North Amer 1996; 76:833-842.
9. Widergren JT, Battisella FD. The open abdomen treatment for intraabdominal compartment syndrome. J Trauma 1994; 37:158.
10. Meldrum DR, Moore FA, Moore EE et al. Prospective characterization and selective management of the abdominal compartment syndrome. Am J Surg 1997; 174:667-672.
11. Diebel LN, Dulchavsky SA, Wilson RF. Effect of increased intra-abdominal pressure on mesenteric arterial and intestinal mucosal blood flow. J Trauma 1992; 33:45-49.
12. Diebel LN, Wilson RF, Dulchavsky SA. Effect of increased intr-abdominal pressure on hepatic arterial, portal venous, and hepatic microcirculatory blood flow. J Trauma 1992; 33:279-283.
13. Bongard FB, Ryan M, Dubecz. Adverse consequences of increased intraabdominal pressure on bowel tissue oxygen. J -Trauma 1995; 39:519-525.
14. Engum SA, Kogon B, Jensen E et al. Gastric tonometry and direct intraabdominal pressure monitoring in abdominal compartment syndrome. Pediatr Surg 2002; 37:214-8.
15. Pusajo JF, Bumaschny E, Agurrola A et al. Post-operative intra-abdominal pressure. Its relation to splanchnic perfusion, sepsis, multiple organ failure and surgical reintervention. Intensive & Crit Care Digest 1994; 13:2-4.
16. Sugrue M, Jones F, Lee A et al. Intraabdominal pressure and gastric mucosal pH: Is there an association? World J Surg 1996; 20:988-991.
17. Toens C, Schachtrupp A, Hoer J et al. A porcine model of the abdominal compartment syndrome shock. 2002; 18:316-21.
18. Stephan ST, Neufang T, Mueller A et al. A new abdominal cavity chamber to study the impact of increased intra-abdominal pressure on microcirculation of gut mucosa by using video microscopy in rats. Critical Care Medicine 2002; 30:1854-1858.
19. Kologlu M, Sayek I, Kologlu LB et al. Effect of persistently elevated intraabdominal pressure on healing of colonic anastomoses. Am J Surg 1999; 178:293-7.
20. Eleftheriadis E, Kotzampassi K, Botsios D et al. Splanchnic ischemia during laparoscopic cholecystectomy. Surg Endosc 1996; 10:324-326.
21. Knolmayer TJ, Bowyer MW, Egan JC et al. The effects of pneumoperitoneum on gastric blood flow and traditional hemodynamic measurements. Surg Endosc 1998; 12:115-8.
22. Schwarte LA, Scheeren TWL, Lorenz CC et al. Moderate increase in intraabdominal pressure attenuates gastric mucosal oxygen saturation in patients undergoing laparoscopy. Anesthesiology 2004; 100:1081-1087.
23. Windberger UB, Auer R, Keplinger F et al. The role of intra-abdominal pressure on splanchnic and pulmonary hemodynamic and metabolic changes during carbon dioxide pneumoperitoneum. Gastrointest Endosc 1999; 49:84-91.
24. Dorsay DA, Greene FL, Baysinger CL. Hemodynamic changes during laparoscopic cholecystectomy monitored with transesophageal echocardiography. Surg Endosc 1995; 9:128-134.
25. Shuto K, Kitano S, Yoshida T et al. Hemodynamic and arterial blood gas changes during carbon dioxide and helium pneumoperitoneum in pigs. Surg Endosc 1995; 9:1173-1178.

26. Junghans T, Bohm B, Grundel K et al. Does pneumoperitoneum with different gases, body positions, and intraperitoneal pressures influence renal and hepatic blood flow? Surgery 1997; 121:206-211.
27. Agusti M, Elizalde JI, Adalia R et al. Dobutamine restores intestinal mucosal blood flow in a porcine model of intra-abdominal hyperpressure. Critical Care Medicine 2000; 28:467-472.
28. Simon RJ, Friedlander MH, Ivatury RR et al. Hemorrhage lowers the threshold for intra-abdominal hypertension (IAH) induced pulomnary dysfunction. J Trauma 1997; 42:398-405.
29. Friedlander M, Simon RJ, Ivatury RR et al. The effect of hemorrhage on SMA flow during increased intraabdominal pressure. J Trauma 1998; 45:433-439.
30. Varela JE, Cohn SM, Giannotti GD et al. Near-infrared spectroscopy reflects changes in mesenteric and systemic perfusion during abdominal compartment syndrome. Surgery 2001; 129:363-70.
31. Eleftheriadis E, Kotzampassi K, Papanotas K et al. Gut ischemia, oxidative stress and bacterial translocation in elevated abdominal pressure in rats. World J Surg 1996; 20:11-16.
32. Diebel LN, Dulchavsky SA, Brown HJ. Splanchnic ischemia and bacterial translocation in abdominal compartment syndrome. J Trauma 1997; 43:852-855.
33. Jacobi CA, Ordemann J, Bohm B et al. Does laparoscopy increase bacteremia and endotoxemia in a peritonitis model? Surg Endosc 1997; 11:235-238.
34. Gargiulo IIIrd NJ, Simon RJ, Leon W et al. Hemorrhage exacerbates bacterial translocation at low levels of intra-abdominal pressure. Arch Surg 1998; 133:1351-5.
35. Doty J, Oda J, Ivatury RR et al. The effect of hemorrhage followed by intraabdominal hypertension on bacterial translocation. J Trauma 2002; 52:13-7.
36. Biffl WL, Moore EE, Burch JM et al. Secondary abdominal compartment syndrome is a highly lethal event. Am J Surg 2001; 182:542-6.
37. Raeburn CD, Moore EE, Biffl WL et al. The abdominal compartment syndrome is a morbid complication of postinjury damage control surgery. Am J Surg 2002; 182:542-546.
38. Offner PJ, de Souza AL, Moore EE et al. Avoidance of abdominal compartment syndrome in damage-control laparotomy after trauma. Arch Surg 2001; 136:676-81.
39. Balogh Z, McKinley BA, Cox Jr CS et al. Abdominal compartment syndrome: The cause or effect of post injury multiple organ failure. Shock 2003; 20:483-492.
40. Balogh Z, McKinley BA, Holcomb JB et al. Both primary and secondary abdominalcompartment syndrome can be predicted early and are harbingers of multiple organ failure. Journal of Trauma-Injury Infection & Critical Care 2003; 54:848-861.
41. Oda J, Ivatury RR, Blocher CR et al. Amplified cytokine response and lung injury by sequential hemorrhagic shock and abdominal compartment syndrome in a laboratory model of ischemia-reperfusion. J Trauma 2002; 52:625-31.
42. Rezende-Neto JB, Moore EE, Melo de Andrade MV et al. Systemic inflammatory response secondary to abdominal compartment syndrome: Stage for multiple organ failure. J Trauma 2002; 53:1121-8.
43. Rezende-Neto JB, Moore EE, Masuno T et al. The abdominal compartment syndrome as a second insult during systemic neutrophil priming provokes multiple organ injury. Shock 2003; 20:303-8.

CHAPTER 10

Intra-Abdominal Hypertension and the Liver

Julia Wendon,* Gianni Biancofiore and Georg Auzinger

There is increasing awareness of the deleterious consequences of intra-abdominal hypertension (IAH) and the abdominal compartment syndrome (ACS) on end organ function in critically ill patients. The pathophysiological implications of IAH and ACS in regard to functional impairment of the renal, cardiovascular and respiratory system, as well as the brain and intestine, are well reported, however our knowledge regarding hepatic complications is limited.

This is surprising given that the incidence of significant elevations in intra-abdominal pressure (IAP) in patients with end stage liver disease,[1] acute liver failure, hepatic trauma[2] and following liver transplantation[3] is high.

Patients with end stage cirrhosis due to chronic liver disease frequently present with large volume tense ascites. Despite a significantly raised IAP these patients might not be symptomatic due to adaptive processes of the abdominal wall occurring over a period of weeks or months. However any additional insult such as variceal hemorrhage or septic shock (i.e., spontaneous bacterial peritonitis) requiring aggressive fluid resuscitation is likely to increase IAP acutely by virtue of capillary leak and consequent tissue edema. This can quickly lead to the development of ACS in an already stressed noncompliant system. At the same time an increase in IAP by 10 mm Hg was found to augment variceal pressure, radius, wall tension and variceal volume, with a consequently high risk of variceal rupture and hemorrhage,[4] therefore potentially perpetuating a vicious cycle. Relieving raised IAP in turn via total paracentesis can reverse these negative effects and will lead to a significant drop in portal pressure, hepatic venous pressure gradient and azygos blood flow while increasing cardiac output.[5]

Hyperacute liver failure is more likely to present with the acute complication of IAH and ACS due to frequently encountered high cardiac output, low peripheral resistance status and massive capillary leak, requiring large fluid volume administration, similar to patients with septic shock or severe burns. A major complicating factor is severe coagulopathy, which might preclude or at least complicate therapeutic measures to reduce IAH such as paracentesis or surgical decompression. Similarly grade IV or V liver trauma, regardless of its management, operative or non operative, will often be complicated by IAH and ACS due to the combination of intra-abdominal haemorrhage, tight packing of the liver and massive volume resuscitation leading to intestinal and abdominal wall edema.

More recently Biancofiore et al[3] reported on a high incidence of IAH in patients following liver transplantation (see below).

Effects of an increase in IAP on liver blood flow and liver cell function will be discussed below, however most of the results are based on animal trials or studies on "healthy" patients undergoing laparoscopic surgery and therefore difficult to apply to the critically ill patient.

*Corresponding Author: Julia Wendon—Institute of Liver Studies, Kings College Hospital, London, U.K. Email: Julia.wendon@kcl.ac.uk

Abdominal Compartment Syndrome, edited by Rao R. Ivatury, Michael L. Cheatham, Manu L. N. G. Malbrain and Michael Sugrue. ©2006 Landes Bioscience.

Intra-Abdominal Pressure (IAP) and Portosystemic Visceral Dysfunction

Blood flow abnormalities resulting from IAH may severely affect kidney, gut and liver function. For example IAP levels of 15-20 mm Hg can cause oliguria, which is likely to progress to anuria if the pressure reaches 30 mm Hg; an IAP of 20 mm Hg has been shown to reduce glomerular filtration rate (GFR) by up to 25%.[6] From an intestinal point of view, it has been demonstrated that abdominal pressure values of 20-40 mm Hg reduce mesenteric blood flow by 40-70%, and that IAP levels of less than 20 mm Hg can reduce it by 30-40%.[7] Impaired bowel perfusion has been linked to abnormalities in the physiological mucosal barrier function of the gut, resulting in an increased risk of bacterial translocation and consequently contributing to septic complications and multi organ dysfunction or failure. In another study the adverse effects of IAH on hepatic blood flow have been pointed out; the investigators used ultrasonic flow probes to continuously monitor hepatic arterial and portal venous blood flow, the latter being more susceptible to increases in IAP: in this pig model changes in hepatic artery flow were seen at 10 mm Hg IAP whereas portal venous flow was decreased at 20 mm Hg IAP despite "normal" cardiac output and systemic blood pressure.[8] In any case, at 20 mm Hg IAP, the hepatic artery and portal venous blood flow were decreased to 45% and 65% respectively compared to their control values. As with the gut, a reduced blood flow negatively influences metabolic liver function. In a rabbit model, using indocyanine green clearance and the arterial ketone body ratio as indicators of liver function, an IAP of 20 mm Hg caused a slight decrease in sinusoidal blood flow but did not affect hepatic energy status. An IAP of 30 mm Hg however reduced the hepatic mitochondrial redox status and a decreased energy level became evident despite sufficient oxygenation of arterial blood.[9] Such derangements are likely to be accentuated if patients are hypovolemic. Therefore the combination of hypotension or a low effective intravascular blood volume and IAH can be extremely detrimental to hepatic perfusion and metabolism. Finally, elevated IAP has been questioned as the possible cause of "unexplained" postoperative liver function test disturbances after laparoscopic cholecystectomy. The charts of patients undergoing open or laparoscopic cholecystectomy were reviewed and compared retrospectively, and altered alanine transaminase levels were found to be present in 34% of the patients undergoing the procedure via the laparoscopic approach; a finding not observed in the "open" group. The biochemical derangements did not translate into any clinically significant complication. The reason for this phenomenon was attributed to pneumoperitoneum-induced IAH, as it was the only variable not present in the open approach group.[10]

Although measurment and monitoring of IAP is gaining increasing interest in clinical practice and IAH is recognised as a serious postoperative risk factor compromising patient homeostasis; no consensus has yet been reached concerning the IAP level that should be considered critical in absolute terms or in terms of organ dysfunction. We know from animal studies that abdominal pressure values of 20-40 mm Hg reduce mesenteric blood flow by 40-70%, and even IAP levels of less than 20 mm Hg can reduce it by 30-40%.[8] It has also been shown that relatively low IAP levels (10 mm Hg) can lead to a significant reduction in hepatic blood flow to as low as 39% of normal.[8] Regarding renal function, IAP levels of 15-20 mm Hg can cause oliguria, which can progress to anuria when pressure reaches 30 mm Hg,[11] and an IAP of 20 mm Hg can reduce GFR by up to 25%.[8] From a clinical point of view a ROC curve analysis showed that the critical IAP values (i.e., those with the best sensitivity/specificity) in OLT recipients were 23 mm Hg for respiratory failure (p< 0.05), 24 mm Hg for renal failure (p< 0.05), and 25 mm Hg for death (p< 0.01).[3]

Pathophysiology of IAH in Acute Liver Failure

The pathophysiological consequences of IAH on various organ systems make it a challenging problem in patients with acute liver failure.

The incidence of raised intracranial pressure is high in this patient population[12] and significant increase in IAP can further raise ICP to critical levels. High IAP augments intrathoracic pressure, which in turn can impede venous outflow from the brain. In addition, cerebral perfusion pressure is often low due to reduced preload, inadequate stroke volume and peripheral vasodilatation all of which contribute to arterial hypotension.

Increase of IAP above 10 mm Hg reduces cardiac output (CO) due to an increase in afterload and reduction in preload. This might not become apparent in acute liver failure patients due to the frequently present significant peripheral vasodilation. Venous pooling within the splanchnic bed might cause a degree of congestion within the liver. A direct compressive effect on the abdominal segment of the inferior vena cava has also been reported which might result in a Budd Chiari like syndrome.[13] Any drop in CO and hence forward flow can render the already damaged liver acutely ischaemic. Similar complications can be observed in severe hepatic trauma where expanding hematomata or tight surgical packing can lead to acute organ ischaemia especially in the presence of inadequate CO.

IAP above 15-20 mm Hg can significantly reduce thoraco abdominal compliance by virtue of a cephalad movement of the diaphragm. This can be accentuated by massive fluid resuscitation and development of abdominal and chest wall edema. The consequent reduction in FRC and rise in pulmonary vascular resistance can lead to significant ventilation perfusion mismatch. Ventilation with low tidal volumes and limitation of airway pressures according to the results of the ARDS network trial, in the absence of transpulmonary pressure gradient measurements, is potentially harmful in patients with acute liver failure. Permissive hypercapnia is not practical in patients with raised ICP and high FiO_2 requirements are a commonly accepted contraindication for liver transplantation. We tend to routinely measure extravascular lung water index (EVLWI) in this patient group, which might help distinguish between severe lung injury and restricted thoracoabdominal compliance as the main cause for critical hypoxia.[14] The latter tends to improve rapidly after relief of intra-abdominal pressure in the operating theatre or the ICU.

Many patients with acute liver failure present with acute renal failure (ARF) before significant IAH develops, due to mainly prerenal causes. However, especially in the group of patients undergoing liver transplantation, perioperative rise in IAP is often a contributing factor to ARF.

Intra-Abdominal Pressure in Liver Transplant Recipients: Incidence and Clinical Relevance

Orthotopic liver transplantation (OLT) is the treatment of choice for end-stage liver disease. Patients undergoing OLT are at risk of increased intra-abdominal pressure both preoperatively and after the transplant procedure.

In the preOLT period many patients with hepatic cirrhosis develop tense ascites leading to IAH and its accompanying cardiovascular and renal complications. In 1988 Savino et al could clearly show that a decrease in IAP following paracentesis, can increase cardiac index, stroke index and left ventricular stroke work with positive therapeutic effects on renal function.[11] Thus, in critically ill cirrhotic patients, the measurement of IAP is not only a useful tool to monitor the manipulation of ascitic fluid pressure quantitatively in order to optimise hemodynamic and renal function, but it may help clinicians in their therapeutic efforts and provide a quantitative correlate for the clinical concept of tense ascites.

After OLT, intra-abdominal pressure is frequently elevated as a consequence of certain specific complications related to the transplant procedure. These include the frequent occurrence of intraperitoneal hemorrhage that may be of surgical nature or due to coagulopathy, use of perihepatic or retroperitoneal packs to control bleeding, bowel congestion due to portal hypertension and/or the requirement for massive fluid resuscitation and use of a pneumatic anti-shock garment. In a series of 104 OLT recipients who spent at least 4 days in the Intensive Care Unit, IAP was measured every 6 hours for at least 72 hours following surgery using the urinary bladder technique. With the primary aim to emphasize the possible deleterious effects of elevated intra-abdominal pressure, IAH was defined as an abdominal pressure ≥ 25 mm Hg on at least two consecutive measurements. The mean IAP of the study population after OLT was 23, 21 and 20 mm Hg (range 4-65 mm Hg) respectively in the first 24 hours after the procedure and on the following 2 days. Thirty-four patients (31.5% of the entire population) were found to have elevated IAP as per study criteria. Their mean IAP values were 35, 26 and 24 mm Hg during the 3 days of observation as opposed to 18, 18 and 17 mm Hg in the rest of the study patients with normal or only sporadically elevated abdominal pressure (p< 0.05 and < 0.001). Among the studied individuals, those who developed renal failure had a mean IAP of 27.9 ± 9.9 mm Hg whereas it measured 18.6 ± 5.2 mm Hg in subjects without renal impairment. IAH was associated with a relative risk for ARF of 9.8, which was higher than the risk of renal failure due to sepsis, respiratory failure, congestive cardiac failure, relaparotomy, abdominal packing and significant intraoperative blood transfusion requirements (p < 0.0001).[3] In this setting the calculation of the Filtration Gradient, as proposed by Sugrue et al,[15] was a useful discriminator for renal dysfunction as its values were significantly higher in those with normal IAP (p< 0.001 on the first two days and p< 0.01 on the third). Although the incidence of primary graft dysfunction did not differ in patients with elevated and normal IAP, the latter group showed better hepatic function, as assessed by the aPTT ratio. Finally, IAH was associated with a lower PaO_2/FiO_2 ratio, measured before patients were weaned from mechanical ventilation, less frequent early extubation following the transplant procedure and a higher mortality. Although the incidence of IAH appears to be similar in OLT patients and other critically ill patient groups, the intra-abdominal pressure values recorded in transplant patients were significantly higher when compared both to surgical and medical patients, where IAP was frequently found to be less then 20 mm Hg:[16,17] the accumulation of blood and clots in the abdominal cavity, intestinal edema or congestion due to portal hypertension and the frequent requirement for massive infusion of fluids and blood components is the most likely reason for this finding.

Haemodynamic Monitoring in Patients with Liver Disease and IAH

The use of filling pressures, such as central venous pressure (CVP) or pulmonary artery occlusion pressure (Paop) as markers of preload can be misleading in patients on positive pressure ventilation with high positive end expiratory pressure (PEEP), in the presence of impaired or rapidly changing myocardial compliance or in the context of elevated IAP. All of these conditions are commonly present in the patient with liver failure or following hepatic trauma.

Volumetric indicators of preload such as right ventricular end diastolic volume (RVEDV), measured with a modified pulmonary artery catheter equipped with a fast response thermistor,[18] or intrathoracic blood volume index (ITBVI) measurements using the transpulmonary thermodilution technique appear to be more accurate then pressure derived preload markers under above circumstances.

We could recently show in a group of 17 patients, with mixed aetiology (acute and chronic liver failure, liver trauma and following liver transplantation), that ITBVI correlated better with stroke volume – and cardiac index (SVI, CI) than CVP. Nine patients had significantly raised IAP (21 ± 4 mm Hg) and in this subgroup only ITBVI correlated with indices of blood flow.[19] Following a fluid bolus there was also a significant correlation between percentage change in ITBVI and SVI (unpublished data).

In a different group of 12 patients with elevated IAP, median 18.5 mm Hg (16-34), 62 simultaneous recordings of CVP, ITBVI, CI, SVI and stroke volume variation (SVV – a dynamic marker of preload) were performed. SVV correlated best with CI and SVI, followed by ITBVI, no correlation for CVP and CI or SVI could be found.[20] However there are limitations in the practical usefulness of this new technology. SVV measurements are only reliable in the absence of cardiac arrhythmias, patients need to be heavily sedated and/or paralized (i.e., without any spontaneous respiratory efforts) and measurements are influenced by the size of the tidal volume administered.[21] The accuracy of SVV as a functional marker of preload has thus far not been validated in patients with multiple organ failure or shock.

Conclusion

The liver appears to be particularly susceptible to injury in the presence of elevated IAP. Animal and human studies have shown impairment of hepatic cell function and liver perfusion even with only moderately elevated intra-abdominal pressure. Furthermore, acute liver failure, decompensated chronic liver disease and liver transplantation are frequently complicated by IAH and the ACS. Significant IAH correlates with extra-hepatic organ dysfunction and mortality in patients undergoing liver transplantation. Thus close monitoring and early recognition of IAH, followed by aggressive treatment of IAH and ACS may confer an outcome benefit in patients with liver disease.

Commentary

Michael Sugrue

Perfusion is key to the liver's physiological performance. Liver dysfunction seen in abdominal compartment syndrome, may be related to alterations in global systemic perfusion, abdominal perfusion and end organ perfusion itself. Obviously cardiac output, preload and afterload are crucial as systemic regulators of liver perfusion. Local intra-abdominal hypertension (IAH) progressing to the abdominal compartment syndrome may result in tissue hypoxia. Increasingly we are recognising that within the capsule of the liver itself, local haematoma formation may have an adverse affect on tissue perfusion. With increasing IAP there is decreased hepatic arterial flow, decreased venous portal flow and increase in the portacollateral circulation. All exert physiological effects with decreased lactate clearance, glucose metabolism and mitochondrial function. This is evidenced by cytochrome p450 abnormalities and alteration in gene expression. The exciting recent work from Biancofiore and colleagues from Pisa in Italy has identified how crucial IAH is in patients undergoing liver transplantation.

References

1. Malbrain MLNG, Wyffels E, Wilmer AP et al. Effects of raised intra-abdominal pressure (IAP) and subsequent abdominal decompression on cardiovascular and renal function in medical ICU patients. Canada, Ottawa: Abstract book of the 7th World Congress of Intensive Care Medicine, 1997:75.
2. Chen RJ, Fang JF, Chen MF. Intra-abdominal pressure monitoring as a guideline in the nonoperative management of blunt hepatic trauma. J Trauma 2001; 51:44-50.
3. Biancofiore G, Bindi ML, Romanelli AM et al. Intra-abdominal pressure monitoring in liver transplant recipients: A prospective study. Intensive Care Med 2003; 1:30-36.
4. Escorsell A, Gines A, Llach J et al. Increasing intra-abdominal pressure increases pressure, volume and wall tension in esophageal varices. Hepatology 2002; 36:936-940.
5. Luca A, Feu F, Garcia-Pagan JC et al. Favorable effect of total paracentesis on splanchnic hemodynamics in cirrhotic patients with tense ascites. Hepatology 1994; 20:30-33.
6. Saggi BH, Sugerman HJ, Ivatury RR et al. Abdominal compartment syndrome. J Trauma 1998; 45:597-609.
7. Cheatham ML, White MW, Sagraves SG et al. Abdominal perfusion pressure: A superior parameter in the assessment of intra-abdominal hypertension. J Trauma 2000; 49:621-627.
8. Diebel LN, Wilson RF, Dulchavsky SA et al. Effect of increased intra-abdominal pressure on hepatic arterial, portal venous and hepatic microcirculatory blood flow. J Trauma 1992; 33:279-283.
9. Nakatami T, Sakamoto Y, Kaneko I et al. Efects of intra-abdominal hypertension on hepatic energy metabolism in a rabbit model. J Trauma 1998; 44:446-453.
10. Ve A, Schein M, Margolis M et al. Liver enzymes are commonly elevated following laparoscopic cholecystectomy: Is elevated intra-abdominal pressure the cause? Dig Surg 1998; 15:256-259.
11. Savino JA, Cerabona T, Agarwal N et al. Manipulation of ascitic fluid pressure in cirrhotics to optimize hemodynamic and renal function. Ann Surg 1988; 208:504-511.

12. Jalan R. Intracranial hypertension in acute liver failure: Pathophysiological basis of rational management. Semin Liver Dis 2003; 23:271-282.
13. Wachsberg RH. Narrowing of the upper abdominal inferior vena cava in patients with elevated intraabdominal pressure: Sonographic observations. J Ultrasound Med 2000; 19(3):217-22.
14. Auzinger G, Sizer L, Bernal W et al. Incidence of lung injury in acute liver failure - diagnostic role of extravascular lung water index. Critical Care 2004; 8(Suppl 1):P40.
15. Sugrue M, Jones F, Deane SA et al. Intra-abdominal hypertension in an indipendent cause of postoperative renal impairment. Arch Surg 1999; 134:1082-1085.
16. Malbrain MLNG. Abdominal pressure in the critically ill: Measurement and clinical relevance. Intensive Care Med 1999; 25:1453-1458.
17. Pelosi P, Malacrida R, Oggioni M et al. Intra abdominal pressure in the criticaly ill patients: A prospective, observational, multicentre study. Procedings 9th ESA Annual Meeting.
18. Cheatham ML, Safcsak K, Block EF et al. Preload assessment in patients with an open abdomen. J Trauma 1999; 46:16-22.
19. Sutcliffe R, Meares H, Auzinger G et al. Preload assessment in severe liver disease associated with intraabdominal hypertension. Intensive Care Medicine 2002; 28(Suppl 1):S177A688.
20. Auzinger GM, Tilley R, Sizer L et al. Markers of preload in patients with severe liver disease and intraabdominal hypertension. Intensive Care Medicine 2002; 28(Suppl 1):S7A12.
21. Reuter DA, Bayerlein J, Goepfert MS et al. Influence of tidal volume on left ventricular stroke volume variation measured by pulse contour analysis in mechanically ventilated patients. Intensive Care Med 2003; 29(3):476-80.

CHAPTER 11

Intra-Abdominal Hypertension and the Central Nervous System

Giuseppe Citerio* and Lorenzo Berra

Abstract

In animal studies, increases in intra-abdominal pressure (IAP) raise central venous pressure (CVP) and pleural pressure (PP) and, eventually, result in elevation of intracranial pressure (ICP) and decrease of cerebral perfusion pressure (CPP). Clinical studies documented similar correlations. Particularly, in patients with an intracranial hypertension (HICP), in which the compensatory capacities of accepting intracranial volumes are exhausted, the effect of high IAP may induce a further harmful increase in ICP.

In head trauma victims with associated intra-abdominal lesions accurate monitoring of IAP is recommended, particularly if HICP is recorded. The cornerstone for treating intra-abdominal hypertension (IAH) and abdominal compartment syndrome (ACS) is the identification of patients at risk and the early recognition and treatment of its harmful effects. Thus, decompressive laparotomy can be a useful adjunct in the treatment of refractory HICP, after the exclusion of other removable causes, while the use of laparoscopy should be considered an absolute contraindication in HICP patients and should be avoided in patients with recent head injury.

Further laboratory and clinical investigation and a strict monitoring of the IAP in HICP patients will allow us a better understanding and treatment of this pathology to reduce the burden of IAP on CNS.

Introduction

The effects of elevated intra-abdominal pressure (IAP) have been investigated by several historical studies. In the 19th century, Marey and Burt described the respiratory effects of intra abdominal hypertension (IAH). In 1911, Emerson highlighted the cardiovascular derangements in various animal models of intra-abdominal hypertension and, in 1913, Wendt described the association of IAH and renal dysfunction.

Nevertheless, only in 1994, Josephs et al[1] at the Boston University School of Medicine evaluated, in a porcine model, the effect of raising the IAP with a pneumoperitoneum on intracranial pressure (ICP) and cerebral perfusion pressure (CPP). They demonstrated that IAH causes an ICP increase; specifically, pneumoperitoneum during laparoscopy raises ICP (Fig. 1). The authors concluded that laparoscopy should not be used in patients with severe head injuries.

A year later, in 1995, Irgau et al[2] and Bloomfield et al[3] published two interesting case reports. Irgau reported a case of a patient with an intracranial mass lesion in which ICP aug-

*Corresponding Author: Giuseppe Citerio—UO Neuroanestesia e Neurorianimazione, Dipartimento di Medicina Perioperatoria e Terapie Intensive, H San Gerardo, Via Donizetti, 106, 20052 Monza (Mi), Italy. Email: g.citerio@hsgerardo.org

Abdominal Compartment Syndrome, edited by Rao R. Ivatury, Michael L. Cheatham, Manu L. N. G. Malbrain and Michael Sugrue. ©2006 Landes Bioscience.

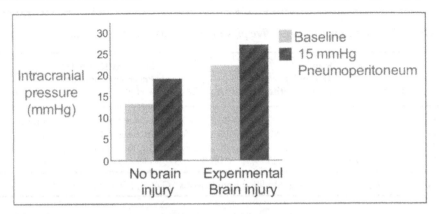

Figure 1. Effect of 15 mm Hg pneumoperitoneum in an animal model with and without intracranial hypertension. Modified from Josephs LG et al, J Trauma 1994; 36(6):815-819.[1]

mented abruptly when the peritoneal cavity was insufflated during laparoscopic cholecystectomy. Bloomfield et al showed the successful management of a patient with severe multisystem injury in whom abdominal decompression, indicated only by the clinical conditions, dramatically reduced high ICP, previously unresponsive to medical measures.

Since then, many laboratory and clinical investigators are tentatively trying to address the role of the IAH and abdominal compartment syndrome (ACS) on the central nervous system (CNS), and to answer relevant questions, as:

1. What is the physiological response of the CNS to variation of IAP?
2. How is the IAP variation transmitted to the ICP and CPP?
3. What is the role of the IAP, IAH, and ACS on CNS in presence or absence of head injury?
4. What are the maneuvers that may cause a secondary damage to the CNS in presence of ACS?
5. How can we prevent an increase in ICP and a decrease in CPP in patients with ACS?
6. And, what should we know about the CNS approaching a patient with ACS with or without head injury?

Regardless of improvements in the understanding and treatment of ACS, many questions remain still unanswered, especially on the effects of IAH and ACS on the CNS. At present, this area of research is one of the most challenging and fascinating topics for many laboratory and clinical investigators.

In the first part of the chapter we present the findings of laboratory studies and clinical investigations reported in the literature; thereafter, we will conclude with some practical indications for a better management of this unusual relationship between two distant organ systems.

Animal Studies

Animal studies have clearly shown the effects of IAH on the ICP and CPP and illustrated the pathophysiological pathways of these derangements.

Evidence that IAP Increments Increase the ICP

Josephs et al[1] were the first group of investigators to evaluate the effect of increased IAP on intracranial pressure dynamics, using a porcine model. Five pigs were enrolled in this animal protocol at the Boston University Medical Center. Animals were anesthetized, tracheostomized, ventilated and an ICP pressure transducer was inserted over the left parietal cortex. A balloon catheter was positioned in the epidural space for reproducing high ICP. Heart rate, MAP, ICP and arterial blood gas (ABG) measurements were recorded every 5 minutes for 30 minutes

Table 1. Average values from the five pigs for each experimental period

	Basal	Standard 15 mm Hg Pneumoperitoneum	High ICP	Standard 15 mm Hg Pneumoperitoneum + High ICP	p Value
ICP	13.46 ± 0.68	18.72 ± 1.5	22.6 ± 1.75	27.4 ± 0.93	≤ 0.0001
MAP	81.92 ± 9.81	86.78 ± 7.75	85 ± 8.26	81.64 ± 7.87	ns
PaCO$_2$	37.82 ± 2.23	40.52 ± 0.95	41.20 ± 1.43	39.00 ± 1.10	ns
PaO$_2$	99.2 ± 13.22	105.35 ± 8.85	108.8 ± 9.5	115 ± 9.89	≤ 0.02

All values are expressed in mm Hg. Modified from Josephs LG et al; J Trauma 1994; 36(6):815-8.

before, during, and after establishment of 15 mm Hg CO_2 pneumoperitoneum. The ICP was then raised by inflating the epidural balloon to an ICP between 20-25 mm Hg and measurements were repeated before, during, and after pneumoperitoneum. Josephs showed that a standard pneumoperitoneum increases ICP; in particular in the uninjured model mean ICP increased from a baseline of 13.46 ± 1.01 mm Hg to 18.72 ± 1.50 mm Hg during pneumoperitoneum (p = 0.0001). In the head injury model (epidural balloon inflated) ICP moved from 22 ± 1.75 mm Hg to 27.40 ± 0.93 mm Hg (p = 0.0001). CPP decreased not significantly from 62.46 to 55.02 mm Hg (Table 1).

The authors hypothesized, according to the modified Monro-Kellie doctrine, that the mechanism through which pneumoperitoneum increases ICP is simply mechanical. The Monro-Kellie doctrine recognizes three main contents in the cranial space: vascular, cerebrospinal fluid (CSF) and parenchyma. The doctrine states that, in adults, changes in one or more of these contents result in reciprocal changes in the remaining compartment. The imbalance of the contents produces, when buffers mechanisms are exhausted, an ICP increase. In other words, ICP reflects the relationship between the volume of intracranial contents (vascular, CSF, and parenchyma) and the volume of the cranial vault. In their brain injury model, Josephs et al[1] increased directly the intracranial volume by inflating the epidural balloon meanwhile, inducing the pneumoperitoneum, they indirectly increased the vascular contents by reducing the cerebral blood outflow, due to the decreased thoracoabdominal compliance, thus producing an ICP rise.

Evidence that IAP Increments Increase Central Venous Pressure, Pleural Pressure, ICP and Decrease Cardiac Index and CPP

The confirmation of these aforementioned hypothesis came a few years later. Using a porcine model of acutely elevated IAP, Bloomfield et al[4]

1. Clarified the mechanisms by which IAH increases ICP and decreases CPP,
2. Evaluated the effect of intravascular volume expansion upon ICP and CPP,
3. Studied the relationship between IAP, pleural pressure (PP), central venous pressure (CVP) and ICP.

They measured the effects of elevated IAP upon ICP and CPP before and after intravascular volume resuscitation. IAP was increased in 5 anesthetized swine by inflating an intraperitoneal balloon to 25 mm Hg above baseline. Intravascular volume was then expanded and, finally, abdominal decompression was performed (Fig. 2).

Changes in ICP and systemic and pulmonary hemodynamic parameters, secondary to increasing IAP, were measured. PaO$_2$ and PaCO$_2$ were maintained relatively constant.

Bloomfield et al[4] showed that elevated IAP significantly increased ICP (7.6 ± 1.2 vs. 21.4 ± 1.0 mm Hg), PP and CVP whereas cardiac index (CI) and CPP (82.2 ± 6.3 vs. 62.0 ± 10.0 mm Hg) decreased significantly. Intravascular volume expansion further significantly increased ICP (27.8 ± 1.0 mm Hg), and increased both mean arterial pressure (MAP, 83.4 ± 14.0 versus

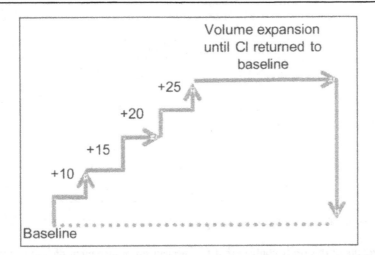

Figure 2. Study design for elucidating effects of elevated IAP upon ICP and CPP before and after intravascular volume resuscitation. Details described in the text. Modified from Bloomfield GL et al, J Trauma 1996; 40: 936-941.[5]

103.4 ± 8.9 mm Hg) and CPP (75.6 ± 9.0 mm Hg). Abdominal decompression returned ICP toward baseline (11.2 ± 1.8 mm Hg) and further increased CPP (79.8 ± 9.7 mm Hg) (Fig. 3).

The authors concluded that elevated IAP increases CVP, PP, ICP and decreases CI, MAP and CPP. Moreover, volume expansion, in the presence of elevated IAP, further raised the ICP and, because of a larger increase in MAP, CPP. An interesting finding was that an IAP greater than or equal to 25 mm Hg produces a statistically significant decrease in CPP, even in animals without head injuries.

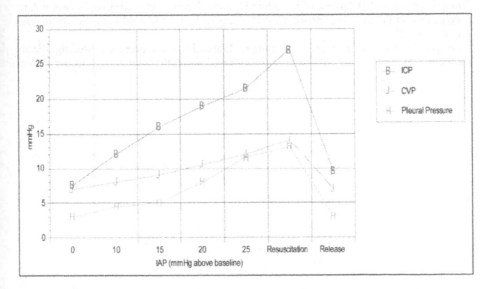

Figure 3. Effect of increasing intra-abdominal pressure (IAP) upon intracranial pressure (ICP), central venous pressure (CVP) and pleural pressure before and after intravascular volume expansion (Resuscitation) and after abdominal decompression (Release). Modified from Bloomfield GL et al. J Trauma 1996; 40(6):936-943.[5]

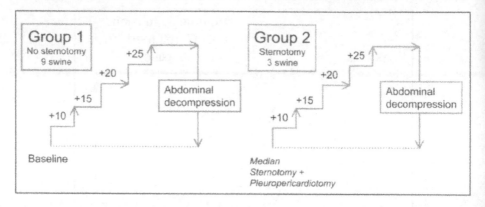

Figure 4. Scheme of the Bloomfield's study design.

The results of this study confirm that the mechanism of increment of ICP is purely mechanical (at $PaCO_2 \leq 45$ mm Hg, and arterial pH ≥ 7.35). According to the Monro-Kellie doctrine, the authors demonstrated that cerebral venous outflow via the jugular venous system is impeded by a significant rise in the CVP. This phenomenon was amplified by volume expansion, due to the additional rise in CVP. They suggested that the functional obstruction of the jugular venous system and the ensuing obligatory increase in the volume of the cerebral vascular space, are the mechanisms responsible for the increased ICP caused by elevated IAP.

Evidence that Increased IAP Produces a Raise in CVP, PP, ICP and Decreases CPP

Furthermore in 1997, Bloomfield et al[5] in order to better clarify the relationship between the PP, ICP and CPP in presence of IAH, repeated the animal-study in two groups of pigs. In Group 1 animals had IAP increased to 25 mm Hg above baseline, then released. In Group 2, to prevent a rise in PP, animals underwent a sternotomy and pleuropericardotomy before increasing IAP (Fig. 4).

As predicted, in the first group IAP rising to 25 mm Hg above baseline caused significant increases in ICP, PP, PAOP, CVP and decreases in CI and CPP. Interestingly, in the second

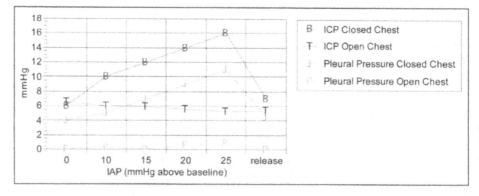

Figure 5. Effect of increasing intra-abdominal pressure (IAP) in closed and open-chest animals on ICP (ICP= filled boxes closed chest, open boxes animals with thoracotomy) and pleural pressure (PP= closed circles animals without sternotomy, open triangles animals with sternotomy). Modified from Bloomfield: Crit Care Med 1997; 25(3):496-503.[5]

group, sternotomy and pleuropericardotomy abolished all the effects of increased IAP, except the decreased CI (Fig. 5).

In conclusion, these laboratory studies clearly showed that an IAP rise decreases the thoracic-abdominal compliance and increases the mean intrathoracic pressures. These in turn reduce cerebral venous outflow, causing an ICP elevation.

Clinical Studies

Despite those laboratory investigations providing evidence about the relationship between the IAH, ICP and CPP, only recently some investigations focused on the clinical impact of such findings.

As previously mentioned, in 1995, two case reports simultaneously confirmed the laboratory results. Yet, no clinical trial was performed in the clinical ward to asses the effect of an increased IAP on CNS.

In 2001, the first clinical study evaluating IAP and ICP was carried out by our group[6] at San Gerardo Hospital, Monza, Italy. We designed a prospective, sequential, nonrandomized study to systematically measure the effects of artificially increased IAP in 15 head trauma patients and to clarify the pressure transmission modalities between different body compartments (abdomen, chest and head). IAP was increased by positioning a soft, 15-L water bag on the patient's abdomen. Strict inclusion criteria were implemented: intubated and mechanically ventilated head injury adult patients were considered eligible for the study at the end of the acute phase, after the evacuation of surgical masses and when no intracranial hypertension was recorded (ICP < 20 mm Hg and CPP > 70 mm Hg) throughout the 24 hours preceding the enrollment. Many parameters were monitored: IAP,[7] MAP, CVP, ICP, CPP, jugular bulb pressure (IJP), jugular bulb oxygen saturation (SjO$_2$), cerebral oxygen extraction (CEO$_2$), intracranial compliance measured as pressurevolume index (PVI[8]), compliance of the respiratory system divided into its pulmonary and chest wall components and gas exchange. Measurements were carried out before and 20 minutes after the IAP rise. MAP, ICP, IAP, CVP and IJP were recorded continuously, as shown in Figure 6.

We found that placing weights upon the abdomen generated a significant increase in IAP, which rose from 4.7 ± 2.9 to 15.5 ± 4.1 mm Hg (p <.001) (Fig. 7). The rise in IAP caused concomitant and rapid increases in CVP from 6.2 ± 2.4 to 10.4 ± 2.9 mm Hg (p <.001), IJP from 11.9 ± 3.2 to 14.3 ± 2.4 mm Hg (p <.001), and ICP from 12.0 ± 4.2 to 15.5 ± 4.4 mm Hg (p <.001).

All these changes required only seconds to reach a plateau and remained increased till the IAP returned to baseline after the weight removal.

A curious result from this study was the MAP increase from 94 ± 11 to 100 ± 13 mm Hg (p <.01), which allowed the maintenance of a stable CPP (82.4 ± 10.3 vs. 84.7 ± 11.5 mm Hg; p = NS), despite the ICP increase. As noticed, an increase in MAP and a stable CPP do not agree with the previous animal findings, and they speculated that the difference lies upon:

1. A different level of IAP. IAP was raised to ≥ 25 mm Hg in animals vs. 15 mm Hg in humans;
2. Animals were heavily sedated with high dose of pentobarbital, with possible vasoplegic effects. In the clinical setting propofol (3-6 mg/K/h) and fentanyl (1.5 γ/K/h) were used;
3. As demonstrated by physiologic studies, the rise in the intrathoracic pressure may facilitate the systolic ejection, although decreasing the venous return.

Interestingly, we found that respiratory system compliance decreased in all patients (from 58.9 ± 9.8 to 44.9 ± 9.4 mL/cm H$_2$O; p <.001) (Fig. 8). However, thoracic transmural pressure (TTP = CVP - esophageal pressure) remained constant during the study time, while chest wall compliance decreased significantly (from 204.7 ± 37.1 to 123.6 ± 38.0 mL/cm H$_2$O; p <.001); lung compliance did not change.

These findings let us confirm the hypothesis that HIAP displaces the diaphragm upward, reducing the chest wall compliance, hence respiratory system compliance. The pressure in the abdominal compartment (IAP) is directly transmitted to the thoracic compartment, raising

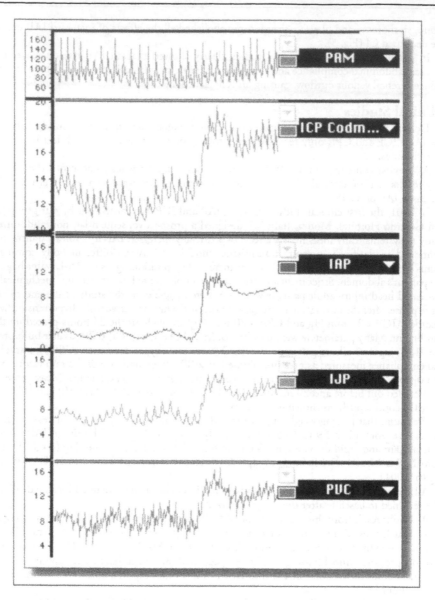

Figure 6. Effect of IAP rise, obtained with a weight application (dotted line). Computerized tracing are from top to bottom: PAM= arterial pressure; ICP= intracranial pressure; IAP= intra-abdominal pressure; IJP= jugular pressure; PVC= central venous pressure.

intrathoracic pressures (CVP, esophageal pressure) and, thereafter, to the cerebral compartment (IJP, ICP). In other words, the ICP rise appears to be the result of an obstruction to the cerebral venous drainage, causing elevation of pressure in the intracranial compartment.

Utilizing a simplified model of cerebral circulation, as presented by Huseby,[9] the rise in jugular vein pressure produces the transmission of the pressure to superior sagittal sinus. This pressure increase is transferred to the Starling's resistor at the cortical bridging vein level. The increase in the outflow pressure in the Starling resistor requires, for maintaining a constancy of

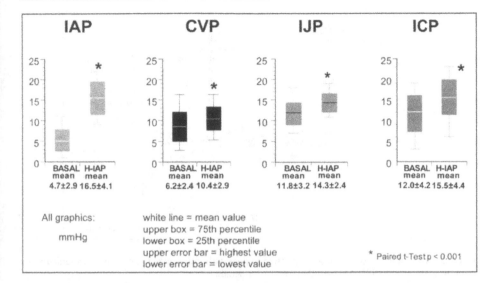

Figure 7. Effect of weight application. Basal and high IAP values. All graphics white line = mean value, upper box = 75th percentile, lower box = 25th percentile, upper error bar = highest value, lower error bar = lowest value. * p< 0.001.

the cerebral blood downflow, a parallel rise in the cerebral venous pressure, thus a rise in ICP (Fig. 9).

This explains also the continued effect on ICP till the increased IAP is released, thus removing the effect on the Starling's resistor. Furthermore, all patients enrolled in the study were in a stable condition, with a starting ICP < 20 mm Hg, and a normal intracranial compliance. This suggests that the starting ICP of all patients was on the flat portion of the Starling pressure/volume curve (point a in Fig. 10) and this may explain the significant but not clinically relevant ICP increase.

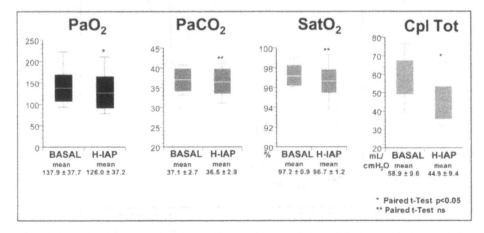

Figure 8. Effect of raised IAP on PaO_2, PaO_2, oxygen saturation, total respiratory compliance. All graphics white line = mean value, upper box = 75th percentile, lower box = 25th percentile, upper error bar = highest value, lower error bar = lowest value. * p< 0.05, ** ns.

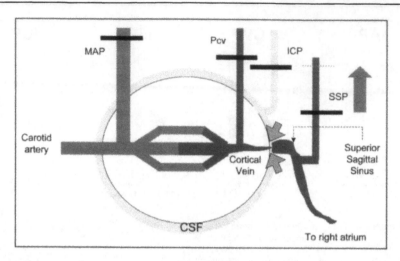

Figure 9. Effect of PEEP on intracranial pressure in dogs with intracranial hypertension. Modified from Huseby, J Neurosurg 1981; 55:704-707. See text for details.

We speculated that probably in the presence of an intracranial hypertension the effect of high IAP may induce a more profound and harmful increase in ICP due to a different starting ICP position on the Starling curve owing to the reduction/absence of the compensatory capacity.

Based on those results, our recommendations were:

1. Routine assessment of IAP could help clinicians to identify remediable causes of increased ICP,
2. Laparoscopic techniques causing IAP rise should be used with caution in patients with concomitant head and abdominal injury.[11]

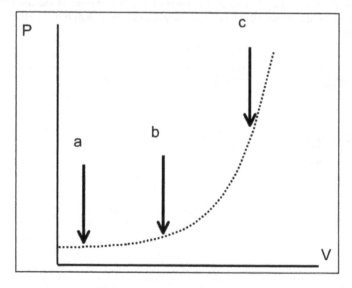

Figure 10. Pressure-volume curve of the craniospinal compartment. It illustrates the principle that in the physiologic range, i.e., near the origin of the x-axis on the graph (point a), intracranial pressure remains normal in spite of small additions of volume until a point of decompensation (point b), after which each subsequent increment in total volume results in an ever larger increment in intracranial pressure (point c).

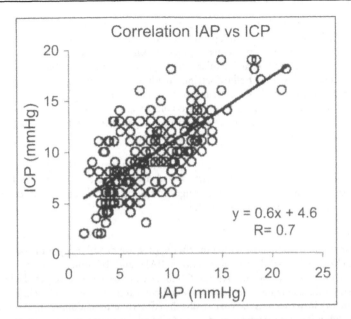

Figure 11. Scattergram of IAP versus ICP (positive correlation). From: Deeren DH, Dits H, Malbrain ML. Correlation between intra-abdominal and intracranial pressure in nontraumatic brain injury. Intensive Care Med 2005; 31(11):1577-81.

These suggestions are of a great importance for patient care, because abdominal trauma is commonly associated with head trauma. The Major Trauma Outcome Study[11] documented that up to 40% of patients with major abdominal trauma had an associated head injury. The Boston University Medical Center Trauma Registry described approximately 30% of the patients who are victims of blunt trauma have an intracranial injury. An Italian data collection on head trauma victims[12] showed the presence of associated severe abdominal complications in 8% of the ICU admitted patients (90 patients out of 1086).

Recently, an interesting paper has been published by the Adams Cowley Shock Trauma Center of University of Maryland on decompressive laparotomy to treat intractable intracranial hypertension after traumatic brain injury.[13] Seventeen head injured patients underwent decompressive laparotomy for intractable intracranial hypertension unresponsive to maximal therapy. Before decompression, mean ICP and mean IAP were respectively 30 ± 8.1 mm Hg and 27.5 ± 5.2 mm Hg. After abdominal decompression ICP dropped by at least 10 mm Hg to a mean of 17.5 ± 3.2 mm Hg. In 6 patients that died this decrease was transient. The 11 surviving had persistent decreases in ICP.

Malbrain et al[14] evaluated the effects of IAP on ICP and CPP in patients with nontraumatic brain injury. The prevalence of intra-abdominal hypertension is high in critically ill patients. An epidemiological survey studied patients in medical and surgical ICUs and it was found that 54.4% of medical and 65% of surgical ICU patients have an IAP of 12 mm Hg or more.[16] The aims of this study were:

    a. to confirm with a large number of measurements the positive correlation between IAP and ICP in patients with nontraumatic brain injury, and

    b. to establish the changes in ICP and CPP that accompany changes in IAP. Eleven patients were enrolled in the study with ICP-monitoring because of ischemic (4), hemorrhagic (5) and metabolic (2) encephalopathy. Mean ICP was 9.8 ± 3.3 mm Hg, IAP 8.1 ± 3.7 mm Hg, CPP 82.5 ± 16.6 mm Hg. They found a tight association between increases in IAP and increases in ICP even at only slightly elevated IAP levels, as shown in Figure 11

Specifically, Deeren et al[15] found significant correlations between IAP and ICP (r = 0.70; p < 0.01), and ΔIAP and ΔICP (r = 0.69; p < 0.01), IAP and CPP (r = -0.31; p < 0.01), and ΔIAP and ΔCPP (difference between two consecutive CPP measurements) (r = -0.16; p = 0.03).

Overall, the clinical reports and clinical investigations confirmed the laboratory studies, validating the modified Monro-Kellie doctrine. In adults, intracranial pressure reflects the relationship between intracranial contents: osseous, vascular, cerebrospinal fluid and parenchymal.[16,17] Importantly, the intracranial pressure-volume curve is not linear. In the physiologic range, small volume increases do not cause substantial pressure increases until a point of decompensation, after which each small increase in volume results in a large increase in intracranial pressure.

If traumatic or nontraumatic brain injury causes an increase of one of the intracranial compartments or if it adds an extra volume, intracranial compliance can be reduced. In that situation, it seems plausible that even minor congestion with a small increase in cerebral blood volume may lead to a marked increase in ICP.

Summary

Summarizing the evidence presented in this chapter, we can identify some points:
- The effects of intra-abdominal hypertension (IAH) and abdominal compartment syndrome (ACS) on the central nervous system (CNS) have not been extensively studied to date, and remain a challenging area for laboratory and clinical investigators. Only in the last decade laboratory investigations tried to describe the effects of intra-abdominal hypertension (IAH) and abdominal compartment syndrome (ACS) on the central nervous system (CNS). This data, utilizing mainly a pneumoperitoneum model, demonstrated a mechanical effect of high IAP on ICP, generating an increase in ICP. This effect requires for its realization the pressure transmission through an intact rib cage. The increase in intrathoracic pressure is transmitted to CVP and, therefore, to the cerebral venous outflow, obstructing it, generating, at the end, a rise in ICP.
- Seminal clinical study demonstrated the same phenomena in head injured patients and in non traumatic brain damaged population. Due to the Monro-Kellie doctrine, the ICP rise depends in its expression intensity upon the position of the patient on the PV curve of the CNS. If the compliance of the system is reduced, small changes in volume produce a higher impact on ICP. The opposite in a normal, compliant, ICP system.

For these reasons:
- According to this preliminary data, accurate monitoring of IAP in head trauma victims with associated intra-abdominal lesions is worthwhile. Appears, at least daily, wise to consider in each individual patient with brain damage, mainly if high intracranial pressure is recorded, the presence of high IAP as additional extracranial cause of HICP. This "secondary insult" could be easily identified and corrected thus removing its effect on intracranial dynamics. The key to managing IAH and ACS is, in fact, the identification of at-risk patients and early recognition of the harmful effects. It is better to prevent ACS than to allow it to occur and manage the sequelae. The literature supports frequent determinations of IAP via bladder measurements in patients at increased risk for ACS, beginning on arrival to the ICU and continuing until the risk of IAH is eliminated. An IAP near 15-20 cm H_2O is gaining acceptance as a pressure at which those caring for the critically ill should take note. Decompression must be strongly considered if the IAP continues to rise or if clinical deterioration occurs. Prior to decompression, aggressive attempts should be made to correct coagulation deficits, hypothermia, acidosis and hypovolemia.
- For these considerations, the use of laparoscopy in the acute post-traumatic phase appears to be more foe than friend. Laparoscopic techniques cause IAP pressure levels comparable to the experimental high IAP, thus producing an impact on ICP. Recent head injury should be considered an absolute contraindication for laparoscopic procedures.

We will welcome further experimental and clinical studies that will address the impact of increased IAP on cerebral dynamics. At this point we know that IAP must be taken in account as a possible extracranial cause of intracranial hypertension in critically ill patients; the pathophysiologic mechanisms have been already established but need some clarifications. Decompressive laparotomy has been recently pointed out as a measure to reduce the burden of high IAP on CNS but further work is needed to identify its correct timing in the care of our everyday patient.

Commentary

Michael L. Cheatham

While the effects of elevated intra-abdominal pressure (IAP) on cardiac, pulmonary, and renal function are well-recognized, the significant implications of intra-abdominal hypertension (IAH) on intracranial pressure (ICP), cerebral blood flow, and cerebral perfusion pressure (CPP) are frequently not. As stated by Drs. Citerio and Berra, the abdomen and brain are commonly considered to be two distant and unrelated organ systems. The evidence presented in this chapter, however, demonstrates that this is not the case. IAH, through cephalad elevation of the diaphragm and increased intrathoracic pressure (ITP), directly limits cerebral venous outflow, decreasing CPP and raising ICP. These potentially dangerous changes, especially in the presence of brain injury and decreased cerebral compliance, can have significant detrimental effects on patient morbidity and mortality.

Due to the potential for IAH-induced secondary injury, clinicians should have a low threshold for initiating serial IAP measurements in the head injured patient. Such measurements are clearly indicated in the patient with concomitant abdominal injuries and should be strongly considered in the isolated head injury patient with refractory intracranial hypertension. Patients with evidence of IAH and inadequate CPP despite appropriate measures to raise mean arterial pressure should undergo either immediate decompressive laparotomy or percutaneous abdominal fluid drainage depending upon the etiology of the patient's IAH. Therapies that raise ITP such as positive end-expiratory pressure (PEEP) and modes of mechanical ventilation that result in elevated inspiratory pressures should be judiciously applied. Closure of an open abdomen in the head injured patient must be cautiously considered and performed only when the patient is clinically ready to tolerate any resultant increase in IAP and therefore ICP. These measures will serve to minimize IAP, improve cerebral venous outflow, and optimize CPP thereby improving patient outcome.

References
1. Josephs LG, Este-McDonald JR, Birkett DH et al. Diagnostic laparoscopy increases intracranial pressure. J Trauma 1994; 36(6):815-8.
2. Irgau I, Koyfman Y, Tikellis JI. Elective intraoperative intracranial pressure monitoring during laparoscopic cholecystectomy. Arch Surg 1995; 130(9):1011-3.
3. Bloomfield GL, Dalton JM, Sugerman HJ et al. Treatment of increasing intracranial pressure secondary to the acute abdominal compartment syndrome in a patient with combined abdominal and head trauma. J Trauma 1995; 39(6):1168-70.
4. Bloomfield GL, Ridings PC, Blocher CR et al. Effects of increased intra-abdominal pressure upon intracranial and cerebral perfusion pressure before and after volume expansion. J Trauma 1996; 40(6):936-41, (discussion 941-3).
5. Bloomfield GL, Ridings PC, Blocher CR et al. A proposed relationship between increased intra-abdominal, intrathoracic, and intracranial pressure. Crit Care Med 1997; 25(3):496-503.
6. Citerio G, Vascotto E, Villa F et al. Induced abdominal compartment syndrome increases intracranial pressure in neurotrauma patients: A prospective study. Crit Care Med 2001; 29(7):1466-71.
7. Iberti TJ, Kelly KM, Gentili DR et al. A simple technique to accurately determine intra-abdominal pressure. Crit Care Med 1987; 15:1140-1145.
8. Marmarou A, Maset AL, Wood JD et al. Contribution of CSF and vascular factors to elevation of ICP in severely head-injured patients. J Neurosurg 1987; 66:883-890.

9. Huseby JS, Luce JM, Cary JM et al. Effects of positive end-expiratory pressure on intracranial pressure in dogs with intracranial hypertension. J Neurosurg 1981; 55(5):704-5.
10. Athanassiou L, Citerio G, Pesenti A. Laparoscopy is contraindicated in neurotrauma patients? There is certainly a doubt! Crit Care Med 2002; 30(10):2402-3, (author reply 2403).
11. Gennarelli TA, Champion HR, Copes WS et al. Comparison of mortality, morbidity, and severity of 59,713 head injured patients with 114,447 patients with extracranial injuries. J Trauma 1994; 37(6):962-8.
12. Citerio G, Stocchetti N, Cormio M et al. Neuro-Link, a computer-assisted database for head injury in intensive care. Acta Neurochir (Wien) 2000; 142(7):769-76.
13. Joseph DK, Dutton RP, Aarabi B et al. Decompressive laparotomy to treat intractable intracranial hypertension after traumatic brain injury. J Trauma 2004; 57(4):687-95.
14. Malbrain ML, Chiumello D, Pelosi P et al. Prevalence of intra-abdominal hypertension in critically ill patients: A multicentre epidemiological study. Intensive Care Med 2004; 30(5):822-9.
15. Deeren DH, Dits H, Malbrain ML. Correlation between intrabdominal and intracranial pressure in nontraumatic brain injury. Intensive Care Med 2005; 31(11):1577-81.
16. Andrews PJ, Citerio G. Intracranial pressure. Part one: Historical overview and basic concepts. Intensive Care Med 2004; 30(9):1730-3.
17. Citerio G, Andrews PJ. Intracranial pressure Part two: Clinical applications and technology.Intensive. Care Med 2004; 30(10):1882-5.

CHAPTER 12

Abdominal Compartment Syndrome Provokes Multiple Organ Failure:
Animal and Human Supporting Evidence

Christopher D. Raeburn* and Ernest E. Moore

Abstract

Damage control surgery has undoubtedly increased the survival of severely injured patients; however, a subset of these salvaged patients go on to develop the devastating complication of the abdominal compartment syndrome (ACS). Clinical studies have demonstrated a clear association of the ACS with multiple organ failure (MOF). In animals, the ACS increases systemic levels of pro-inflammatory cytokines, primes neutrophils for cytotoxicity and results in remote organ injury. Both clinical and human studies confirm that the ACS causes a disproportionate decrease in mesenteric perfusion, which can occur even in the absence of hypotension and decreased cardiac output. We hypothesize that this ischemia/reperfusion injury to the gut serves as a second-insult in a two-event model of MOF and further propose that mesenteric lymph is the conduit by which gut-derived proinflammatory agents induce remote organ injury.

Background

The deleterious effects of increased intra-abdominal pressure (IAP) have been known for over a century; however, it was not until the advent of damage control surgery (DCS) that abdominal compartment syndrome (ACS) was recognized as a significant clinical entity.[1] In the 1980s damage control surgery was instituted as a life saving strategy in severely injured patients who developed the constellation of hypothermia, acidosis and coagulopathy, "the bloody vicious cycle" (Fig. 1). Patients with severe liver injury and historically fatal coagulopathy were being salvaged by placing packs around the injured liver and abbreviating the laparotomy to allow for correction of the factors promoting the coagulopathy.[2] Despite the undisputed benefit of postinjury DCS, it became apparent that some of these patients developed a new set of morbid complications, namely the abdominal compartment syndrome and multiple organ failure (MOF). The observed association of damage control surgery, ACS and MOF has initiated debate as to whether ACS is a cause of MOF or merely an end result.[3]

Despite the large volume of literature published on the abdominal compartment syndrome, no uniform definition for ACS exists. In short, ACS can be defined as an increase in intra-abdominal pressure to a level sufficient to produce an adverse physiologic response.[4] Intra-abdominal pressure is most commonly determined by measurement of urinary bladder pressure[5,6] and in hospitalized patients is normally about 6.5 mm Hg.[7] Both animal and

*Corresponding Author: Christopher D. Raeburn—Department of Surgery, University of Colorado Health Sciences Center, 4200 E. 9th Avenue, Denver, Colorado, 80262 U.S.A. Email: christopher.raeburn@uchsc.edu

Abdominal Compartment Syndrome, edited by Rao R. Ivatury, Michael L. Cheatham, Manu L. N. G. Malbrain and Michael Sugrue. ©2006 Landes Bioscience.

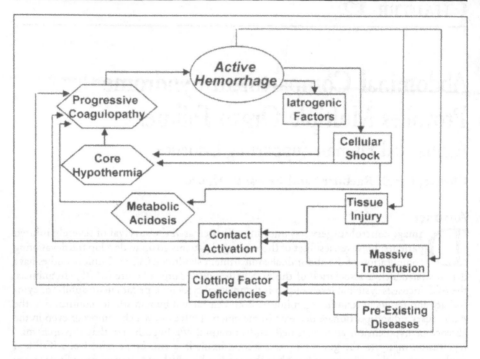

Figure 1. The bloody vicious cycle. Hemorrhagic shock induces many factors which lead to metabolic acidosis, core hypothermia and progressive coagulopathy which combine to provoke further hemorrhage.

human studies demonstrate that an increase in IAP to 15 mm Hg is sufficient to produce remote organ dysfunction.[8-10] Increased IAP is graded as follows: I = 10-15 mm Hg, II = 16-25 mm Hg, III = 26-35 mm Hg and IV = >35 mm Hg.[10] As the grade increases so does the number of organs adversely effected. Animal studies have demonstrated that antecedent hemorrhage lowers the threshold at which increased IAP induces remote organ dysfunction.[11-13] Overt impairment of renal, pulmonary and cardiac function in the setting of increased IAP is virtually diagnostic of ACS; however, impairment of mesenteric blood flow may be an earlier indication of underlying ACS.[9,14]

The etiology of ACS is primarily related to the accumulation of intra-abdominal blood, ascites and visceral edema to the degree that abdominal wall compliance is overcome. Primary ACS is most commonly a result of abdominal trauma but can also occur following pancreatitis, ruptured abdominal aneurysms, retroperitoneal hemorrhage, pneumoperitoneum, neoplasm and ascites.[15-17] Secondary ACS results from protracted circulatory shock due to an extra-abdominal inciting event such as thoracic or extremity trauma, burn injury or sepsis.[18-22] Regardless of the etiology of ACS, the pattern of organ dysfunction, incidence of MOF and mortality are similar.[19]

The incidence of ACS depends largely on the population examined. Approximately 1% of all patients admitted to a trauma ICU develop ACS.[19,23] However, up to 36% of patients requiring damage control surgery will develop ACS.[9,10,24,25] Patients requiring DCS are at high risk for ACS because they commonly require massive crystalloid resuscitation due to protracted hemorrhagic shock, "the salt water vicious cycle,"[3] (Fig. 2) and require intra-abdominal packing, which is an independent risk factor for ACS.[10]

Association of ACS with MOF

We have examined the effect of ACS on outcome in patients requiring DCS following trauma.[24] In this study, ACS was defined as an IAP > 20 mm Hg in association with renal,

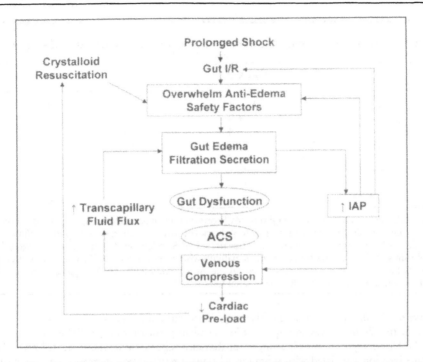

Figure 2. The salt water vicious cycle. Crystalloid resuscitation initially improves hemodynamics by increasing cardiac preload; however, prolonged shock results in gut ischemia/reperfusion (I/R) which injures the interstitial matrix of the bowel. Further crystalloid administration overwhelms anti-edema safety factors leading to bowel wall edema and net fluid loss into the bowel lumen, "filtration secretion". The resulting increase in intra-abdominal pressure (IAP) causes venous compression which decreases cardiac preload and increases interstitial pressure in the gut. This further increases transcapillary fluid flux into the bowel wall and lumen which, if unchecked, results in abdominal compartment syndrome (ACS).

pulmonary or cardiac dysfunction. Of 77 patients who underwent DCS, 36% developed ACS. Patients who developed ACS were not significantly different from patients who did not develop ACS in terms of patient demographics, Injury Severity Score, emergency department vital signs, intensive care unit admission indices or 24 hour fluid balance. Despite these similarities, the development of ACS was associated with a significantly worsened outcome (Table 1). Importantly, the incidence of multiple organ failure, which was the most common cause of death in both groups, was significantly higher in patients who developed ACS (32% vs. 8%, p<0.01). The mortality of patients with ACS who developed MOF was 85%, and the overall mortality of patients with ACS was 43% compared to 12% in patients who did not develop ACS. We hypothesized that ACS is a second insult to the patients' inflammatory response, which has been sufficiently primed by the inciting trauma and subsequent DCS.

A recent report by Balogh et al[19] prospectively evaluated high risk trauma patients for the development of ACS. They used multiple logistic regression analysis to identify early, independent predictors of ACS (Table 2) and used these variables to develop an early prediction model for ACS. Despite vigorous monitoring for ACS and timely decompression, the mortality of patients who developed ACS remained much higher than that of patients without ACS (primary ACS = 64%, secondary ACS = 53%, no ACS = 17%). Similar to previous reports,[10,23,24] MOF was the most common cause of death and occurred in just over 50% of patients who developed ACS but in only 12% of patients without ACS. Logistic regression analysis again confirmed that ACS was an independent predictor of both MOF (odds ratio = 9.2; 95% confidence intervals, 3.8-22.8; p < 0.0001) and mortality (odds ratio = 8.4; 95% confidence

Table 1. Effect of ACS on outcome in patients undergoing damage control surgery

	ACS	No ACS	P Value
Ventilator days	22 ± 3	15 ± 2	0.046
ICU days	26 ± 3	18 ± 2	0.03
Hospital days	40 ± 3	26 ± 2	0.002
Number of complications/patient	3.2 ± 0.4	1.9 ± 0.2	0.005
ARDS	39%	18%	0.04
ARF	32%	12%	0.03
MOF	32%	8%	0.006
Mortality	43%	12%	0.002

Effect of abdominal compartment syndrome (ACS) on outcome in patients undergoing damage control surgery. Despite similar demographics, injury/shock severity and emergency department/intensive care unit (ICU) parameters, patients who developed ACS following damage control surgery had significantly worsened outcome compared to patients who did not develop ACS. Specifically, multiple organ failure (MOF) was four times higher in the ACS compared to no ACS group. ARDS: adult respiratory distress syndrome; ARF: acute renal failure.

intervals, 3.5-20.6; p < 0.0001). In this study, the amount of crystalloid volume infused (>7.5 L) in the first 24 hours was an important independent predictor of ACS (Table 2).

The University of Texas-Houston group has also reported that the increased crystalloid volume infusion associated with supranormal trauma resuscitation increases the incidence of ACS.[26] In this study, a computerized protocol was used to guide the early resuscitation of trauma patients presumed to be at high risk for MOF. They compared the hemodynamic responses and outcomes of patients resuscitated to a supranormal oxygen delivery goal ($DO_2I>600$ mL/min/m^2) versus a normal goal ($DO_2I>500$ mL/min/m^2). The two groups of patients had equivalent demographics and injury/shock severity at presentation. Both groups exhibited similar responses in terms of cardiac index and mixed venous oxygen saturation, and optimization of base deficit and serum lactate levels occurred during a similar time frame. In order to achieve the supranormal DO_2I goal, patients received nearly twice as much crystalloid in the first 24 hours of ICU care (13 ± 2 vs. 7 ± 1 L, p<0.05) compared to the normal resuscitation group. The increased crystalloid infusion in the supranormal resuscitation group was associated with an increased incidence of intra-abdominal hypertension (42% vs. 20%, p<0.05), ACS (16% vs. 8%, p<0.05), MOF (22% vs. 9%, p,0.05) and death (27% vs. 11%, p<0.05) (Table 3). An interesting finding was that the gastric regional mucosal carbon dioxide minus end tidal carbon dioxide ($GAPCO_2$) was significantly higher in the supranormal resuscitation group indicating that these patients had worsened intestinal perfusion (Table 3).

The authors hypothesized that this suboptimal intestinal perfusion accounted for the worsened outcome in the supranormal resuscitation group. Although crystalloid infusion may have a beneficial effect on cardiac index by improving preload, it can also have a detrimental effect by increasing bowel edema. Crystalloid loading decreases colloid oncotic pressure and increases hydrostatic pressure in the capillary vascular bed. When administered to patients with postreperfusion capillary leak, these effects compound to result in a futile crystalloid preloading cycle (Fig. 2), where further crystalloid infusion worsens the bowel edema and increases the intra-abdominal pressure which then leads to further organ dysfunction and ACS.

These clinical studies have collectively demonstrated an association between ACS and MOF; however, it is difficult to prove a causal relationship. Both animal and human studies have shown that ACS results in remote organ dysfunction and that decompression of ACS rapidly improves organ function. However, despite an almost uniform initial improvement in organ

Table 2. Independent predictors of ACS

	Independent Predictor	Odds Ratio	95% Confidence Interval	SENS	SPEC	PPV	NPV
All ACS	GAP CO$_2$ ≥ 16	>999.9	22.1->999.9	96	76	42	99
	Crystalloid ≥ 7.5L	166.2	4.76->999.9	77	70	32	94
	UO ≤ 150 mL	89.8	4.49->999.9	80	76	38	96
	Hb ≤ 8 g/dL	252.5	9.89->999.9	46	92	50	90
	CI ≤ 2.6 L/min/m^2	12.5	1.02-153.64	65	77	34	92
1° ACS	Temp ≤ 34°C	22.9	1.39-378.25	55	94	43	96
	GAP$_{CO2}$ ≥ 16	54.3	2.15->999.9	91	76	22	99
	Hb ≤ 8 g/dL	206.1	7.41->999.9	73	92	40	98
	BD ≥ 12 mEq/L	3.5	1.37-839.50	46	98	56	96
2° ACS	GAP$_{CO2}$ ≥ 16	>999.9	<0.001->999.9	100	76	30	100
	Crystalloid ≥ 7.5L	38.7	3.19-469.55	87	70	24	98
	UO ≤ 150 mL	64.1	5.48-749.68	87	76	27	98

Independent predictors of abdominal compartment syndrome (ACS). Multiple logistic regression analysis was used in a prospective examination of severely injured patients to identify early independent predictors of ACS. GAP CO2: gastric regional mucosal carbon dioxide minus end tidal carbon dioxide; UO: urine output; Hb: hemoglobin; CI: cardiac index; temp: temperature, BD: base deficit. Reprinted with permission from Balogh et al, J Trauma 2003.

function following decompression of ACS, 50% of these patients develop MOF and die. This observation suggests that ACS may act as a second-insult in a two-event model of MOF.

The two-event model of MOF postulates that the trauma patient's innate immune system is primed by the inciting event (hemorrhagic shock and/or direct tissue injury) for an exaggerated response to a subsequent systemic inflammatory stimulus[27] (Fig. 3). We have previously shown that neutrophils obtained from injured patients at risk for MOF exhibit augmented superoxide production in response to the agonist FMLP compared to normal volunteers.[28] In subsequent studies,[29] we have shown that the postinjury neutrophil priming also exhibits augmented elastase release for membrane degradation, increased CD11b/CD18 expression for adhesion, increased release of IL-8 to attract more neutrophils and delayed apoptosis prolonging cytotoxicity. This augmented response pattern occurs during a defined time frame following trauma; and thus, the timing of the second-event may be important in terms of whether or not it triggers a systemic hyperinflammatory state resulting in MOF. Our study of human neutrophil priming following severe trauma suggests that the time period of maximal priming is 3-16 hours post-injury. Importantly, this corresponds to the time period when ACS most commonly occurs.[28,30]

ACS as a Cause of MOF

We have developed a small animal model of ACS to further examine its potential role as a second-insult in a two-event model of MOF.[31] In this study, rats were hemorrhaged to a MAP of 30 mm Hg (76% of blood volume) for 45 min and then resuscitated (HS/R) with the shed blood and crystalloid until MAP was returned to baseline. Similar to data in humans,[28,30] neutrophil CD11b expression was increased 3-fold at 8h following HS/R but not increased at 2h or 18h. The rats were then subjected to ACS at various times following HS/R. ACS was induced by insufflating air to a pressure of 25 mm Hg for 60 min followed by decompression. Animals were kept normothermic and crystalloid was administered to maintain MAP at greater than 75 mm Hg. At 24 hours after HS/R and ACS, animals were sacrificed to determine lung neutrophil sequestration (elastase concentration), lung injury (Evans blue dye in broncho-alveolar

Table 3. *Effect of supranormal oxygen delivery directed trauma resuscitation*

	GAP CO$_2$ (mm Hg)	IAH (mm Hg)	ACS (mm Hg)	MOF	Mortality
DO$_2$>600 (mL/min/m^2)	16 ± 2*	42%*	16%*	22%*	27%*
DO$_2$>500 (mL/min/m^2)	7 ± 1	20%	8%	9%	11%

Effect of supranormal oxygen delivery (DO$_2$I) directed trauma resuscitation. A computerized protocol was used to direct resuscitation of severely injured patients at high risk for multiple organ failure (MOF). Patients resuscitated to a supranormal oxygen delivery goal (DO$_2$I>600 mL/min/m^2) required significantly more crystalloid volume than patients resuscitated to a normal goal (DO$_2$I>500 mL/min/ m^2). This was associated was worsened mesenteric perfusion and an increased incidence of intra-abdominal hypertension (IAH), abdominal compartment syndrome (ACS), multiple organ failure and mortality. (* P < 0.05 vs. DO$_2$I>500 mL/min/m^2). Data from Balogh et al. Arch Surg 2003, with permission.

lavage fluid) and liver injury (ALT concentration). Lung PMN sequestration was maximal when ACS was induced at 8h after HS which is the time point when neutrophils were maximally primed. Accordingly, lung injury occurred when ACS was induced at 8h after HS/R but not when ACS was induced at 2h or 18h following HS/R (Fig. 4). Similarly, liver injury was maximal when ACS was induced 8h after HS. 24h mortality was 0% after HS alone and HS + 18h-ACS, 16% when ACS was induced 2h after HS and 33% in when ACS was induced at 8h following HS/R.

In this study, HS/R alone did not result in histologic or biochemical evidence of lung or liver injury; however, the addition of ACS during the "vulnerable period" of maximal neutrophil priming resulted in MOF and increased mortality. This study corroborates that ACS can indeed act as a second-event to trigger MOF. Furthermore, the observation that the deleterious effect of ACS occurred only when neutrophils were primed strongly implicates the neutrophil as a key mediator in ACS-induced MOF.

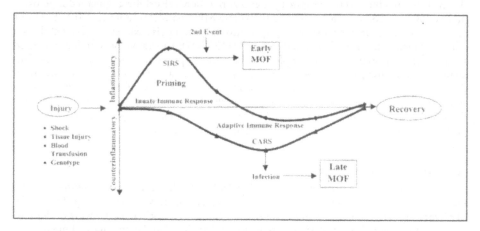

Figure 3. Two-event model of multiple organ failure. Mechanical injury and associated hemorrhagic shock trigger a cascade of proinflammatory reactions, manifesting clinically as the systemic inflammatory response syndrome (SIRS), that primes the innate immune system such that a secondary insult during this vulnerable window provokes an unbridled inflammatory response culminating in multiple organ failure (MOF). Simultaneously, the injury initiates events resulting in a depressed adaptive immune response, compensatory anti-inflammatory response syndrome (CARS), that renders the patient at risk for overwhelming infection which can result in delayed MOF.

It is well established that cytokines are important mediators of the hyperinflammatory state that precipitates post-injury MOF. To further examine the role of ACS in the etiology of MOF, we examined the effect of ACS on systemic cytokine levels and lung injury in a small animal model.[32] In this study, rats were subjected to ACS by increasing IAP to 20 mm Hg for 60 min or 90 min followed by decompression. MAP was reduced by 40-45% during ACS and returned to baseline with decompression. We determined the effect of this moderate degree of elevated IAP on systemic levels of the pro-inflammatory cytokines tumor necrosis factor-α (TNF), interleukin (IL)-6 and IL-1β. All three cytokines increased either during ACS or following decompression. Lung myeloperoxidase (MPO) activity was also examined to determine the effect of ACS on lung neutrophil sequestration. Lung MPO was significantly increased in the animals that underwent 60 min of ACS followed by decompression. Histopathologic evidence of lung injury (lung neutrophil/macrophage accumulation, alveolar edema) was also observed in these animals.

An interesting finding in this study is that both the increase in TNF and lung MPO did not occur until the ACS was decompressed (Fig. 5). This suggests that TNF promotes sequestration of neutrophils in the lung, perhaps via induction of adhesion molecules both on the neutrophil and pulmonary endothelium. Both animal and human studies have demonstrated that mesenteric blood flow is dramatically reduced during ACS and improves following decompression.[8,9,33,34] Consequently, we hypothesize that the development of ACS followed by decompression allows a "bolus" of inflammatory agents to enter the lungs and systemic circulation, which is analogous to what occurs following an ischemia/reperfusion insult to the bowel.[35] Further, we propose that mesenteric lymph is the conduct for these gut-derived proinflammatory agents (Fig. 6).

Some investigators[36-38] but not all[39] have reported that the decreased mesenteric perfusion during ACS results in bacterial translocation from the gastrointestinal tract. The lack of consistency in these observations is likely due to inter-species variability. Importantly, the relevance of bacterial translocation in the pathogenesis of MOF in humans is controversial in that clinical studies have demonstrated that bacterial translocation occurs infrequently following hemorrhagic shock and does not correlate with the development of MOF.[40] Despite the failure of clinical studies to confirm a causative role for bacterial translocation in MOF, the gut is still felt by many to play an important role in the pathogenesis of MOF.[41] We have demonstrated that neutrophil-mediated lung injury following hemorrhagic shock is at least in part secondary to toxic lipid moieties present in mesenteric lymph.[42] This observation is supported by studies demonstrating that mesenteric lymph, obtained from rats subjected to hemorrhagic shock, primes neutrophils for cytotoxicity,[43,44] activates pulmonary endothelium[45] and provokes lung injury.[44] Furthermore, ligation of the mesenteric lymph duct prior to HS protects rats from HS-induced lung injury.[44,46] We have also documented similar findings in a swine model.[47] Similar to mesenteric blood flow, mesenteric lymph flow is also substantially reduced during increased IAP and promptly returns and even increases following decompression.[3,48,49]

The recent work of Sener and colleagues[50-52] provides further evidence supporting a role for ACS as a second insult in a two-event model of MOF. These investigators report that the development of ACS and subsequent decompression results in oxidative injury to remote organs due to mesenteric ischemia/reperfusion. They subjected rats to 1 hour of increased IAP (20 mm Hg) followed by 1 hour of decompression and then examined the intestines, liver, lung and kidneys for evidence of oxidative injury. They found that tissue levels of malondialdehyde (MDA, an index of lipid peroxidation) and MPO were elevated and glutathione (GSH, an endogenous anti-oxidant) was decreased in all organs studied following ACS and decompression. Organ injury was examined by measuring serum AST, BUN and creatinine levels which were all elevated following ACS and decompression. An important observation was that these changes were not observed with ACS alone but only after ACS was decompressed. This further supports the hypothesis that decompression of ACS allows a "bolus" of inflammatory mediators to enter the systemic circulation.

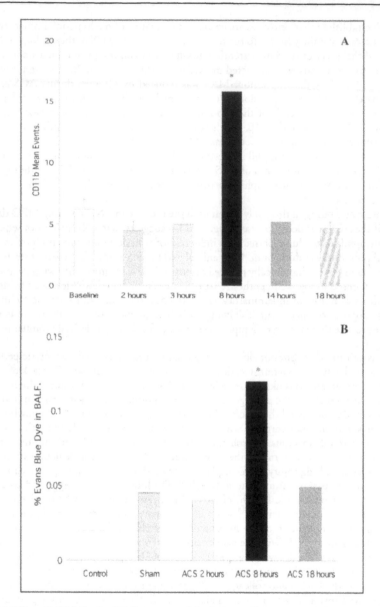

Figure 4. Effect of ACS following HS/R on neutrophil priming and lung injury. A) Flow cytometry was used to examine the time course of neutrophil CD11b expression in rats following hemorrhagic shock and resuscitation (HS/R). Neutrophils were primed (increased CD11b expression) at 8 hours following HS/R (* P <0.05 vs. baseline). B) Rats were then subjected to the abdominal compartment syndrome (ACS) at various times following HS/R and lung injury was assayed by measuring Evans blue dye concentration in brochoalveolar fluid. HS/R alone or followed by ACS at 2 hours or 18 hours did not result in lung injury; however, ACS induced lung injury when it occurred at 8 hours following HS/R (+ P <0.05 vs. sham).

These investigators went on to demonstrate that the administration of either melatonin[52] (a free-radical scavenger and anti-oxidant) or octreotide[50,51] (a mesenteric vasoconstrictor) immediately prior to decompression of ACS abrogated the increase in MDA and MPO, preserved

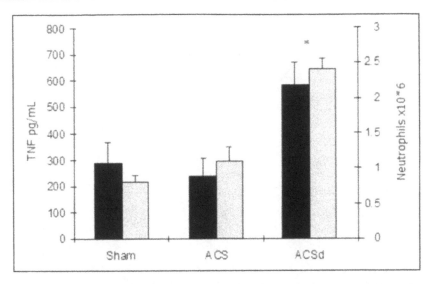

Figure 5. Enzyme-linked immunosorbent assay was used to measure serum tumor necrosis factor (TNF)-α levels in rats (black bars) during (60 min at 20 mm Hg) and after decompression of the abdominal compartment syndrome (ACS). The serum TNF level did not increase until after ACS was decompressed (* P <0.05 vs. sham and ACS). Myeloperoxidase (MPO) activity was examined to determine the effect of ACS and decompression on lung neutrophil accumulation (grey bars). Accumulation of neutrophils in the lung paralleled TNF levels; no increase in lung neutrophils occurred until ACS was decompressed (* P <0.05 vs. sham and ACS).

tissue GSH levels and prevented the rise in serum AST, BUN and creatinine. These data suggest that the mechanism of remote organ injury following ACS and decompression is due to ischemia and reperfusion and that interventions aimed at limiting the oxidative injury inherent to reperfusion can limit organ injury. We have previously reported that blocking xanthine oxidase prior to splanchnic ischemia/reperfusion (temporary occlusion of superior mesenteric artery) attenuates oxidative injury in the liver and lung;[53] however, Sener and colleagues are the first to report similar findings in ACS and decompression. The exact mechanism of octreotide in protecting animals from the reperfusion injury is unclear; however, the authors speculate that the one-time dose of octreotide may result in a more gradual return of mesenteric perfusion that is better tolerated.

Summary

It has been argued that MOF associated with ACS is a reflection of the global hemodynamic compromise precipitated by the initial insult and that ACS is merely the terminal manifestation of MOF. Indeed, it is difficult to discern a direct adverse effect of ACS on remote organ function as ACS rarely occurs in the absence of hemodynamic instability. However, in our in vivo work of HS/R followed by ACS, crystalloid infusion was used to maintain a MAP of 75 mm Hg during ACS. Despite this intervention, ACS lead to both lung and liver injury as well as increased 24 hour mortality when instituted at 8 hours following HS/R. Furthermore, Diebel et al have demonstrated that increased IAP results in decreased mesenteric, hepatic, and portal venous blood flow even when MAP[8,36,54] and cardiac output[8,54] are maintained at normal values with fluid resuscitation. Ivatury and colleagues[9] have also shown in a clinical study that increased IAP results in mesenteric ischemia (gastric mucosal acidosis) in the absence of the classic signs of ACS (cardiac, pulmonary and renal dysfunction). These findings further support the hypothesis that ACS is not merely a late manifestation but rather a cause of MOF.

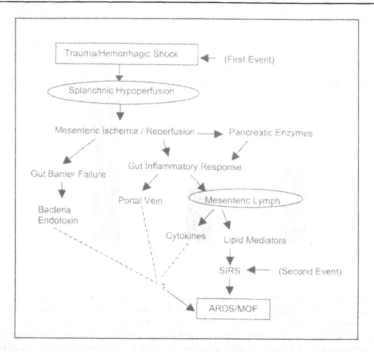

Figure 6. Mesenteric lymph is the conduit for gut-derived mediators of systemic hyperinflammation. Hemorrhagic shock provokes mesenteric ischemia/reperfusion which results in failure of the gut barrier and damage to the interstitial matrix of the bowel. Whether gut barrier failure results in bacterial translocation in humans is controversial. Comparison of portal vein and systemic cytokine levels have failed to show a consistent increase in portal vein levels following hemorrhagic shock. Toxic lipid moieties present in mesenteric lymph obtained from animals following hemorrhagic shock induce a systemic inflammatory response (SIRS). We hypothesize that the abdominal compartment syndrome, by causing further mesenteric ischemia/reperfusion, serves as a second-event to exacerbate the systemic inflammatory response ultimately leading to multiple organ failure.

The advent of damage control surgery has undoubtedly increased the survival of severely injured patients; however, in doing so it has also transformed the abdominal compartment syndrome from an interesting laboratory phenomenon to a common, devastating clinical entity. Clinical studies have demonstrated a strong association between ACS and MOF, and clinically relevant animal models have provided convincing evidence that ACS plays a causal role. Current investigation has identified several early independent variables that predict which patients develop ACS. The University of Texas-Houston group has demonstrated that the volume of crystalloid used during the initial 24 hours of trauma resuscitation is an independent predictor for the development of ACS.[19,26] Therefore, further investigation of alternative strategies in trauma resuscitation such as the use of colloids, blood substitutes and hypertonic saline may prove to decrease the incidence of ACS by reducing the crystalloid volume requirement. Moreover, the potential benefits of blood substitutes such as polymerized hemoglobin[55] and hypertonic saline[56] for the resuscitation of patients at high risk for the development of ACS may extend beyond a simple reduction in the volume of crystalloid infusion.

We have reported that severely injured patients resuscitated with polymerized hemoglobin have lower systemic cytokine levels[57] and do not undergo neutrophil priming[58] compared to similar patients resuscitated with red blood cell transfusions. Thus, polymerized hemoglobin as a resuscitation fluid may decrease the inflammatory response and resultant MOF associated with early transfusion of red blood cells. Similarly, we[59,60] and others[61,62] have demonstrated that hypertonic saline inhibits neutrophil priming both in vitro and in vivo.

While decreasing crystalloid administration during resuscitation may reduce the incidence of ACS, it will not entirely prevent it. Early recognition and prompt abdominal decompression remain the mainstay of treatment for ACS. Monitoring of gastric mucosal pH may identify a subset of patients with increased IAP and mesenteric ischemia without the classic signs of ACS.[9] These patients might benefit from earlier abdominal decompression. Animal work has suggested that anti-oxidant therapies initiated prior to decompression of ACS may also limit remote organ injury. Specific identity of proinflammatory agents in mesenteric lymph may provide additional therapeutic strategies. While we now recognize the adverse physiologic effects of intra-abdominal hypertension in the injured patient, we have only begun to elucidate the fundamental mechanisms that will be key to ultimately reduce the life-threatening consequences of the abdominal compartment syndrome.

Commentary

Manu L. N. G. Malbrain

With the advent of new studies, the correlation between intra-abdominal pressure (IAP), intramucosal pH (pHi) and increased gut permeability (as demonstrated by bacterial translocation) seems to become stronger and stronger every day. The association between increased gut permeability and the subsequent development of multiple organ failure (MOF) and death has also been recently demonstrated. With regard to the abdominal compartment syndrome (ACS) the question still remains whether the intra-abdominal hypertension (IAH) is the cause of or an epi-phenomenon in the emergence of MOF. What is the chicken and what is the egg? Nonbelievers will indeed point towards increased IAP as a mere side effect of the resuscitative strategies in trauma, septic or burn patients, whereas believers will point towards the direct negative effects of IAH on organ perfusion increasing intestinal and capillary permeability, and requiring further and ongoing resuscitation that will eventually lead to a vicious cycle with ongoing IAH and ACS. This chapter will take the reader through some historical perspectives, followed by the association between and the cause and effect of ACS and MOF.

References

1. Kron IL, Harman PK, Nolan SP. The measurement of intra-abdominal pressure as a criterion for abdominal reexploration. Ann Surg 1984; 199:28-30.
2. Moore FA, Moore EE, Seagraves A. Nonresectional management of major hepatic trauma. An evolving concept. Am J Surg 1985; 150:725-9.
3. Balogh Z, McKinley BA, Cox Jr CS et al. Abdominal compartment syndrome: The cause or effect of postinjury multiple organ failure. Shock 2003; 20:483-92.
4. Burch JM, Moore EE, Moore FA et al. The abdominal compartment syndrome. Surg Clin North Am 1996; 76:833-42.
5. Johna S. Can we use the bladder to estimate intra-abdominal pressure? J Trauma 2001; 51:1218.
6. Fusco MA, Martin RS, Chang MC. Estimation of intra-abdominal pressure by bladder pressure measurement: Validity and methodology. J Trauma 2001; 50:297-302.
7. Sanchez NC, Tenofsky PL, Dort JM et al. What is normal intra-abdominal pressure? Am Surg 2001; 67:243-8.
8. Diebel LN, Wilson RF, Dulchavsky SA et al. Effect of increased intra-abdominal pressure on hepatic arterial, portal venous, and hepatic microcirculatory blood flow. J Trauma 1992; 33:279-82, discussion 282-3.
9. Ivatury RR, Porter JM, Simon RJ et al. Intra-abdominal hypertension after life-threatening penetrating abdominal trauma: Prophylaxis, incidence, and clinical relevance to gastric mucosal pH and abdominal compartment syndrome. J Trauma 1998; 44:1016-21, discussion 1021-3.
10. Meldrum DR, Moore FA, Moore EE et al. Prospective characterization and selective management of the abdominal compartment syndrome. Am J Surg 1997; 174:667-72, discussion 672-3.
11. Friedlander MH, Simon RJ, Ivatury R et al. Effect of hemorrhage on superior mesenteric artery flow during increased intra-abdominal pressures. J Trauma 1998; 45:433-89.
12. Oda J, Ivatury RR, Blocher CR et al. Amplified cytokine response and lung injury by sequential hemorrhagic shock and abdominal compartment syndrome in a laboratory model of ischemia-reperfusion. J Trauma 2002; 52:625-31, discussion 632.

13. Simon RJ, Friedlander MH, Ivatury RR et al. Hemorrhage lowers the threshold for intra-abdominal hypertension-induced pulmonary dysfunction. J Trauma 1997; 42:398-403, discussion 404-5.
14. Varela JE, Cohn SM, Giannotti GD et al. Near-infrared spectroscopy reflects changes in mesenteric and systemic perfusion during abdominal compartment syndrome. Surgery 2001; 129:363-70.
15. Peppriell JE, Bacon DR. Acute abdominal compartment syndrome with pulseless electrical activity during colonoscopy with conscious sedation. J Clin Anesth 2000; 12:216-9.
16. Gorecki PJ, Kessler E, Schein M. Abdominal compartment syndrome from intractable constipation. J Am Coll Surg 2000; 190:371.
17. Gecelter G, Fahoum B, Gardezi S et al. Abdominal compartment syndrome in severe acute pancreatitis: An indication for a decompressing laparotomy? Dig Surg 2002; 19:402-4, discussion 404-5.
18. Balogh Z, McKinley BA, Cocanour CS et al. Secondary abdominal compartment syndrome is an elusive early complication of traumatic shock resuscitation. Am J Surg 2002; 184:538-43, discussion 543-4.
19. Balogh Z, McKinley BA, Holcomb JB et al. Both primary and secondary abdominal compartment syndrome can be predicted early and are harbingers of multiple organ failure. J Trauma 2003; 54:848-59, discussion 859-61.
20. Biffl WL, Moore EE, Burch JM et al. Secondary abdominal compartment syndrome is a highly lethal event. Am J Surg 2001; 182:645-8.
21. Ivy ME, Possenti PP, Kepros J et al. Abdominal compartment syndrome in patients with burns. J Burn Care Rehabil 1999; 20:351-3.
22. Tremblay LN, Feliciano DV, Rozycki GS. Secondary extremity compartment syndrome. J Trauma 2002; 53:833-7.
23. Hong JJ, Cohn SM, Perez JM et al. Prospective study of the incidence and outcome of intra-abdominal hypertension and the abdominal compartment syndrome. Br J Surg 2002; 89:591-6.
24. Raeburn CD, Moore EE, Biffl WL et al. The abdominal compartment syndrome is a morbid complication of postinjury damage control surgery. Am J Surg 2001; 182:542-6.
25. Offner PJ, de Souza AL, Moore EE et al. Avoidance of abdominal compartment syndrome in damage-control laparotomy after trauma. Arch Surg 2001; 136:676-81.
26. Balogh Z, McKinley BA, Cocanour CS et al. Supranormal trauma resuscitation causes more cases of abdominal compartment syndrome. Arch Surg 2003; 138:637-42, discussion 642-3.
27. Moore FA, Moore EE, Read RA. Postinjury multiple organ failure: Role of extrathoracic injury and sepsis in adult respiratory distress syndrome. New Horiz 1993; 1:538-49.
28. Botha AJ, Moore FA, Moore EE et al. Postinjury neutrophil priming and activation: An early vulnerable window. Surgery 1995; 118:358-64, discussion 364-5.
29. Biffl WL, Moore EE, Zallen G et al. Neutrophils are primed for cytotoxicity and resist apoptosis in injured patients at risk for multiple organ failure. Surgery 1999; 126:198-202.
30. Zallen G, Moore EE, Johnson JL et al. Circulating postinjury neutrophils are primed for the release of proinflammatory cytokines. J Trauma 1999; 46:42-8.
31. Rezende-Neto JB, Moore EE, Masuno T et al. The abdominal compartment syndrome as a second insult during systemic neutrophil priming provokes multiple organ injury. Shock 2003; 20:303-8.
32. Rezende-Neto JB, Moore EE, Melo de Andrade MV et al. Systemic inflammatory response secondary to abdominal compartment syndrome: Stage for multiple organ failure. J Trauma 2002; 53:1121-8.
33. Bongard F, Pianim N, Dubecz S et al. Adverse consequences of increased intra-abdominal pressure on bowel tissue oxygen. J Trauma 1995; 39:519-24, discussion 524-5.
34. Caldwell CB, Ricotta JJ. Changes in visceral blood flow with elevated intraabdominal pressure. J Surg Res 1987; 43:14-20.
35. Bathe OF, Chow AW, Phang PT. Splanchnic origin of cytokines in a porcine model of mesenteric ischemia-reperfusion. Surgery 1998; 123:79-88.
36. Diebel LN, Dulchavsky SA, Brown WJ. Splanchnic ischemia and bacterial translocation in the abdominal compartment syndrome. J Trauma 1997; 43:852-5.
37. Eleftheriadis E, Kotzampassi K, Papanotas K et al. Gut ischemia, oxidative stress, and bacterial translocation in elevated abdominal pressure in rats. World J Surg 1996; 20:11-6.
38. Gargiulo 3rd NJ, Simon RJ, Leon W et al. Hemorrhage exacerbates bacterial translocation at low levels of intra-abdominal pressure. Arch Surg 1998; 133:1351-5.
39. Doty JM, Oda J, Ivatury RR et al. The effects of hemodynamic shock and increased intra-abdominal pressure on bacterial translocation. J Trauma 2002; 52:13-7.
40. Moore FA, Moore EE, Poggetti R et al. Gut bacterial translocation via the portal vein: A clinical perspective with major torso trauma. J Trauma 1991; 31:629-36, discussion 636-8.
41. Hassoun HT, Kone BC, Mercer DW et al. Post-injury multiple organ failure: The role of the gut. Shock 2001; 15:1-10.

42. Moore EE. Mesenteric lymph: The critical bridge between dysfunctional gut and multiple organ failure. Shock 1998; 10:415-6.
43. Gonzalez RJ, Moore EE, Ciesla DJ et al. Mesenteric lymph is responsible for post-hemorrhagic shock systemic neutrophil priming. J Trauma 2001; 51:1069-72.
44. Zallen G, Moore EE, Johnson JL et al. Posthemorrhagic shock mesenteric lymph primes circulating neutrophils and provokes lung injury. J Surg Res 1999; 83:83-8.
45. Gonzalez RJ, Moore EE, Ciesla DJ et al. Post-hemorrhagic shock mesenteric lymph activates human pulmonary microvascular endothelium for in vitro neutrophil-mediated injury: The role of intercellular adhesion molecule-1. J Trauma 2003; 54:219-23.
46. Deitch EA, Adams C, Lu Q et al. A time course study of the protective effect of mesenteric lymph duct ligation on hemorrhagic shock-induced pulmonary injury and the toxic effects of lymph from shocked rats on endothelial cell monolayer permeability. Surgery 2001; 129:39-47.
47. Sarin E, Moore EE, Ciesla DJ et al. Mesenteric lymph duct ligation protects against hemmorhagic shock induced lung injury. J Trauma 2004, In Press.
48. Drake RE, Gabel JC. Effect of outflow pressure on intestinal lymph flow in unanesthetized sheep. Am J Physiol 1991; 260:R668-71.
49. Drake RE, Teague RA, Gabel JC. Lymphatic drainage reduces intestinal edema and fluid loss. Lymphology 1998; 31:68-73.
50. Kacmaz A, Polat A, User Y et al. Octreotide: A new approach to the management of acute abdominal hypertension. Peptides 2003; 24:1381-6.
51. Kacmaz A, Polat A, User Y et al. Octreotide improves reperfusion-induced oxidative injury in acute abdominal hypertension in rats. J Gastrointest Surg 2004; 8:113-9.
52. Sener G, Kacmaz A, User Y et al. Melatonin ameliorates oxidative organ damage induced by acute intra-abdominal compartment syndrome in rats. J Pineal Res 2003; 35:163-8.
53. Poggetti RS, Moore FA, Moore EE et al. Simultaneous liver and lung injury following gut ischemia is mediated by xanthine oxidase. J Trauma 1992; 32:723-7, discussion 727-8.
54. Diebel LN, Dulchavsky SA, Wilson RF. Effect of increased intra-abdominal pressure on mesenteric arterial and intestinal mucosal blood flow. J Trauma 1992; 33:45-8, discussion 48-9.
55. Gould SA, Moore EE, Hoyt DB et al. The first randomized trial of human polymerized hemoglobin as a blood substitute in acute trauma and emergent surgery. J Am Coll Surg 1998; 187:113-20, discussion 120-2.
56. Shukla A, Hashiguchi N, Chen Y et al. Osmotic regulation of cell function and possible clinical applications. Shock 2004; 21:391-400.
57. Johnson JL, Moore EE, Gonzalez RJ et al. Alteration of the postinjury hyperinflammatory response by means of resuscitation with a red cell substitute. J Trauma 2003; 54:133-9, discussion 139-40.
58. Johnson JL, Moore EE, Offner PJ et al. Resuscitation with a blood substitute abrogates pathologic postinjury neutrophil cytotoxic function. J Trauma 2001; 50:449-55, discussion 456.
59. Gonzalez RJ, Moore EE, Ciesla DJ et al. Hyperosmolarity abrogates neutrophil cytotoxicity provoked by post-shock mesenteric lymph. Shock 2002; 18:29-32.
60. Zallen G, Moore EE, Tamura DY et al. Hypertonic saline resuscitation abrogates neutrophil priming by mesenteric lymph. J Trauma 2000; 48:45-8.
61. Angle N, Hoyt DB, Coimbra R et al. Hypertonic saline resuscitation diminishes lung injury by suppressing neutrophil activation after hemorrhagic shock. Shock 1998; 9:164-70.
62. Rizoli SB, Kapus A, Fan J et al. Immunomodulatory effects of hypertonic resuscitation on the development of lung inflammation following hemorrhagic shock. J Immunol 1998; 161:6288-96.

CHAPTER 13

Postinjury Secondary Abdominal Compartment Syndrome

Zsolt Balogh* and Frederick A. Moore

Definition and Historical Perspectives

Postinjury ACS is defined by the presence of intra-abdominal hypertension (IAH) with intra-abdominal pressure (IAP) greater than 25 mm Hg accompanied by organ dysfunction(s) such as cardiac, respiratory and renal.[1] ACS is referred to as "secondary" when there are no intraperitoneal injuries. To avoid misclassification, several issues need to be clarified. Patients whose abdominal parenchymal organ injuries are managed nonoperatively should not be categorized as secondary ACS since they have intraperitoneal injuries. Pelvic fracture related retroperitoneal hematomas without intraperitoneal injury are classified as secondary ACS. However, retroperitoneal vascular, renal, duodenal, pancreatic etc injuries requiring laparotomy with and without packing should not be classified secondary ACS group. Burrows et al reported the first trauma related secondary ACS case in his series of primary ACS cases in 1998.[2] The terminology ("secondary ACS") was attributed to Maxwell et al who described 6 cases and mentioned the potential connection with massive resuscitation.[3]

Epidemiology

Incidence

It is difficult to determine the true incidence of postinjury secondary ACS due to its elusive nature. Maxwell et al reported 13% incidence among trauma patients who required abdominal mesh closure.[3] Based on a prospective shock trauma database, Balogh et al reported the incidence of postinjury secondary ACS to be 0.09% of all trauma admissions, 0.7% of all trauma ICU admissions, 8% of shock trauma patients requiring aggressive resuscitation [with ISS>15, requiring more than 6 units of packed red blood cell (PRBC) transfusions during the first 12 hours and having initial base deficit (BD) greater than 6 mEq/L] and it represented 58% of all cases of postinjury ACS.[1]

Time to Develop Secondary ACS from Hospital Admission

The initial studies[3,4] describing secondary ACS reported long delays in surgical decompression (up to 108 hours from hospital admission). This may be related to the late recognition of the syndrome and/or to futile attempts to overcome IAH related cardiac dysfunction with fluid challenges, which was the recommended treatment at that time.[5] More recent data suggest that postinjury secondary ACS is a much earlier phenomenon. Among trauma patients requiring massive resuscitation ACS typically manifest itself within 12 to 14 hours after hospital admission.[1,6] It is important to concentrate diagnostic, preventive and predictive efforts during this

*Corresponding Author: Zsolt Balogh—Department of Traumatology, University of Szeged, Szeged, Hungary. Email: zsoltbalogh@yahoo.com

Abdominal Compartment Syndrome, edited by Rao R. Ivatury, Michael L. Cheatham, Manu L. N. G. Malbrain and Michael Sugrue. ©2006 Landes Bioscience.

Table 1. Demographics, injury characteristics and outcome of patients with and without ACS

	Primary ACS (n = 11)	Secondary ACS (n = 15)	Non ACS (n = 162)
Demographics:			
Age (years)	36 ± 5	45 ±4	39 ±1
Gender (male %)	73	80	76
Injury mechanism (blunt %)	82	86	85
Severity of shock:			
Initial ED BD (mEq/L)	11 ± 1	9 ± 2	9 ± 0.5
Lowest ED SBP (mm Hg)	79 ± 3[a]	82 ± 4[a]	93 ± 2
12 hrs PRBCs (Units)	14 ± 4[a]	11 ± 2[a]	8 ± 1
% of urgent interventions (IR/OR)	82	87	85
Injury severity and pattern:			
ISS	29 ± 3	28 ± 2	27 ± 1
ATI	18 ± 1[a,b]	4 ± 2[a,b]	10 ± 2
GCS	13 ± 1	13 ±1	13 ± 0.2
AIS head	1.3 ± 0.4	1.2 ± 0.2	1.6 ± 0.05
AIS face	0.8 ± 0.1	0.7 ± 0.1	1 ± 0.01
AIS chest	2.7 ± 0.3	2.4 ± 0.2	2.8 ± 0.05
AIS abdomen	3.9 ± 0.2[a]	0[a,b]	2.6 ± 0.05
AIS extremity	2.7 ± 0.1	4.1 ± 0.2[a,b]	2.8 ± 0.05
AIS external	1.1 ± 0.2	1.3 ± 0.3	1.3 ± 0.01
Times from ED admission:			
ED discharge (hours)	0.9 ± 0.1[a,b]	3 ± 0.3	2 ± 0.1
ICU admission (hours)	3.7 ± 0.5[a,b]	6.2 ± 0.6	7 ± 0.25
Outcome:			
Mechanical ventilation (days)	13 ± 3[a]	14 ± 3[a]	8 ± 2
ICU LOS (days)	14 ± 5	16 ± 3	12 ± 2
MOF (%)	55[a]	53[a]	12
Mortality (%)	64[a]	53[a]	17

ACS: abdominal compartment syndrome; ED BD: emergency department base deficit; ED SBP: emergency department systolic blood pressure; PRBCs: packed red blood cells; IR/OR: interventional radiology/operating room; ISS: injury severity score; ATI: abdominal trauma index; GCS: Glasgow coma scale; AIS: abbreviated injury scale; ICU: intensive care unit; MOF: multiple organ failure; ICU LOS: intensive care unit length of stay. Univariate comparisons: [a] $p<0.05$ ACS vs nonACS, [b] $p<0.05$ primary vs. secondary ACS

early timeframe parallel with resuscitation, hemorrhage control and completion of diagnostic studies.

Distinct Characteristics Compared to Primary ACS

All case reports describing postinjury secondary ACS mention massive resuscitation. A prospective evaluation of primary and secondary ACS patients compared to non ACS patients is depicted in Table 1. Patients who develop these syndromes have similar demographics, injury severity, injury mechanism and initial base deficit.[8] ACS patients differ from nonACS in their initial systolic blood pressure (SBP). The lower SBP reflected in more aggressive resuscitation in the ACS patients' preICU course. By definition secondary ACS patients have no intraperitoneal injury thus abdominal abbreviated injury scale (AIS) is zero compared to 3.9 ± 0.2 in primary ACS patients. Secondary ACS patients typically have multiple extremity injuries and severe pelvic fractures or penetrating chest injuries or extremity vascular injuries. Patients with major vascular injuries presenting in shock undergo massive resuscitation which is known to be related to secondary ACS. Trauma patients with multiple pelvic and long bone fractures without obvious abdominal or chest injuries undergo extended diagnostic evaluations which may include pelvic angiography which is reflected in their significantly longer preICU course.[7] 82% of primary ACS patients had hemorrhage control in the OR and only 9% of them was taken to interventional radiology (IR) while in secondary ACS patients hemorrhage control was attempted in the IR in 47% and in the OR in 40% of the cases. The longer preICU course (6.2 vs 3.7 hours) results in longer periods of less controlled resuscitation. Compared to primary ACS patients, secondary ACS patients received significantly more crystalloid infusion (32 vs. 20 Liters in first 24 hours) and had a much higher ratio of liters of crystalloid/unit of PRBCs (1.92 vs 0.55). Primary and secondary ACS patients are both decompressed within 14 hours of hospital admission, but because of longer pre ICU times secondary ACS is an earlier ICU phenomenon (~6 hours) than primary (~10 hours). Another reason for the differences in time is that primary ACS patients arrive from the OR after damage control and their temporary abdominal closure prevents the very early development of ACS. After decompression the time to definitive fascial closure is shorter in secondary ACS patients (3 ±0.8 vs. 9 ±2 days) and they are less likely to develop abdominal abscesses. The outcome of both syndromes is poor.

Mechanism

The pathologic mechanisms of postinjury 2° ACS should be searched for in the early injury response of patients arriving with exsanguinating hemorrhage (see Fig. 1). Traumatic shock and subsequent standard of care resuscitation leads to whole body ischemia/reperfusion injury due to effects of inflammatory cells and mediators. The hemodilution after massive crystalloid resuscitation together with the increased permeability and hydrostatic pressure are the key early driving forces for interstitial fluid accumulation. The edematous bowel (increasing peritoneal content) and the retroperitoneal hematoma (decreasing peritoneal volume) are both important elements of the intra-abdominal hypertension. The elevated intra-abdominal pressure impairs venous and lymphatic outflow from the gut, and thus worsens gut edema by increasing capillary filtration pressure. Therapeutic interventions initiated to reverse organ dysfunction (further fluid resuscitation, increased positive pressure ventilation) can be added factors to the already developed IAH. In secondary ACS patients pelvic fractures with significant blood loss and retroperitoneal hematoma are common findings, these as major sources of hemorrhagic shock and by decreasing peritoneal volume are contributors to IAH.

Outcome

The reported mortality of postinjury secondary ACS ranges between 38-67% (Table 2). Early studies reporting later recognition and decompression had higher mortality.[3,4] These studies concluded that earlier decompression improves outcome. More recent studies in which all patients were decompressed within 16 hours of hospital admission found no significant difference between survivors and nonsurvivors related to the time elapsed until decompression.[1,6] 55% of secondary ACS patients develop multiple organ failure (MOF) which is defined after the first 48 hours so as to not confuse early organ dysfunctions related to inadequate resuscitation and to the effects of IAH. Based on our studies postinjury ACS is a strong independent predictor of postinjury death and MOF.[8] With the available basic science and clinical research postinjury ACS appears to be a modifiable link between hemorrhagic shock and MOF.[8-12]

Prediction

Risk Factors

Seminal case series during the late 1990s published risk factors based on expert opinion without strong statistical background.[2-4] These included severe trauma, massive resuscitation,

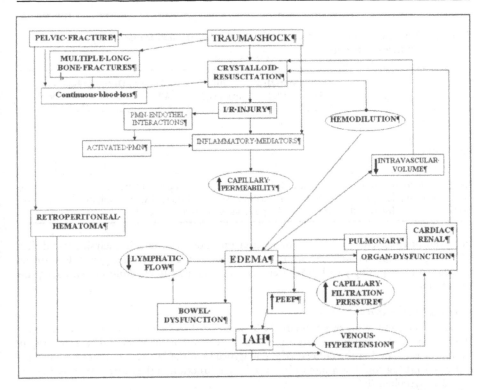

Figure 1. The proposed mechanism of postinjury secondary abdominal compartment syndrome. I/R= ischemia/reperfusion; PMN= neutrophil leukocytes; PEEP= positive end-expiratory pressure; IAH= inta-abdominal hypertension.

hemorrhagic shock and coexisting cirrhosis. These reports also empirically set up crystalloid cut-points above which IAP monitoring is recommended.[3] Based on 6 secondary ACS cases Maxwell recommended 10 liters of crystalloid or 10 units of PRBCs. Ivy et al (10 postburn secondary ACS patients) suggested >0.25 L/kg crystalloid resuscitation as the trigger.[13] Biffl et al (8 trauma, 6 general surgical secondary ACS patients) have found both cut-offs ineffective and recommended 6 liters or more crystalloid or 6 units of PRBCs in a 6 hours period among patients with base deficit >10 mEq/L especially if they are on vasopressors.[6]

Independent Predictors

Given the early occurrence of postinjury secondary ACS we focused our prediction models on the first 6 hours from hospital admission.[8] We developed two prediction models: Emergency Department model (0-3 hours, all patients are over their initial diagnostic work-up, completed their initial laboratory results and discharged from the emergency department) and Intensive Care Unit model (0-6 hours, all patients admitted to ICU and their first physiologic measurements and laboratory parameters on the resuscitation protocol are available). Although primary and secondary ACS patients develop the same symptoms and predecompression physiology, their injury pattern, resuscitation and hospital times are different. We thus hypothesized that their predictors would be different. The variables, which were entered into the multivariate prediction models included demographics, shock severity, injury severity, interventions, hospital times, crystalloid and blood volumes, vital signs, initial pulmonary artery catheter readings on the ICU, respiratory parameters, gastric tonometry data, blood gas results, laboratory and coagulation results. From these variables, the ones listed in Table 3 turned out to be

Table 2. **Summary of the reported postinjury secondary abdominal compartment syndrome patients**

| | No. | ISS | Mortality (%) | UBP (mm Hg) | Time to Decompression (h) | | |
					All	Survived	Died
Maxwell 1999[7]	6	25	67	33	18	3	25
Kopelman 2000[8]	6	17[†]	67	30.5	108	48	192
Biffl 2001[9]	8[‡]	20	38	30	10	12	5
Balogh 2002	11	28	54	34	12	14	10

No: the number of postinjury secondary abdominal compartment patients; ISS: injury severity score; UBP: urinary bladder pressure at the time of decompression; h: hours; NA: not available; [†]: calculated from the injuries listed in the paper; [‡]: 7 of 8 patients were decompressed.

independent risk factors for ACS. The receiver operator characteristic analysis showed that ACS in general can be predicted with 0.88 accuracy at the time of emergency department discharge and surprisingly with 0.99 accuracy one hour after ICU admission with adequate monitoring.

Clinical Indicators

The clinical indicators of secondary ACS are elevated IAP, elevated peak airway pressure (PAP), poor cardiac index and low urine output. It is important to note that urine output should be evaluated in the context of the magnitude of the resuscitation. The urine output of ~ 1 mL/kg could be a sign of impaired renal function in the setting of resuscitation of ~ 2 L crystalloids/h.[8]

Table 3A. **Emergency department model: independent predictors**

	Independent Predictor	Odds Ratio	95% Conf.	SENS	SPEC	PPV	NPV
2° ACS	Crystalloid = 3L	15.8	1.74-143.85	93	74	25	99
	No urgent surgery	0.3	0.073-0.94	40	21	4	79
	PRBC = 3 units	5.6	1.03-30.83	87	78	27	98

Table 3B. **Intensive care unit model: independent predictors**

	Independent Predictor	Odds Ratio	95% Conf.	SENS	SPEC	PPV	NPV
2° ACS	$GAP_{CO2} = 16$	>999.9	<0.001->999.9	100	76	30	100
	Crystalloid = 7.5 L	38.7	3.19-469.55	87	70	24	98
	UO = 150 mL	64.1	5.48-749.68	87	76	27	98

2° ACS: secondary abdominal compartment syndrome; 95% Conf.:95% confidence intervals; SENS: sensitivity; SPEC: specificity; PPV: positive predictive value; NPV: negative predictive value; No urgent surgery: no surgery before ICU admission; PRBC: packed red blood cells.

Diagnosis of Secondary ACS

Postinjury secondary ACS is an elusive potentially lethal syndrome where early recognition is the key to achieve optimal outcome and to terminate causative factors such as indiscriminant crystalloid resuscitation of uncontrolled bleeding. Since physical examination is inaccurate to determine IAP,[14,15] measurement of IAP is required to define the syndrome. The most widely accepted method is the intravesical technique via the urinary catheter.[16] Given the early occurrence of the postinjury secondary ACS, intermittent IAP measurements (even 4 hourly) are not adequate. In cases with ongoing massive resuscitation the continuous IAP monitoring via the standard three-way catheter is warranted (Balogh-Sugrue technique).[17]

Treatment

Decompression

To date the standard treatment of secondary ACS is surgical decompression and the application of a temporary abdominal closure. The previously recommended (crystalloid) fluid challenges in patients with impending ACS generally do not improve cardiac output[18] and are harmful.[19] Earlier decompression was associated with better outcome when late decompression happened after 24 hours from the initial signs of the IAH.[3,4] In series where all secondary ACS patients decompressed within 24 hours there was no difference between earlier and later decompression.[6,8] Critically ill secondary ACS patients on aggressive ventilator settings are not good candidates for operating room trips. If suspicion of life threatening intra-abdominal bleeding is low, these patients can be safely decompressed in the ICU with aseptic surgical prep and draping.[8] The regular 48-72 hours temporary abdominal closure changes may also be performed by the bedside ICU procedure team. This helps to adhere to the planned steps of closure without competing with other cases in the operating room. During decompression the second most common finding after bowel edema is ascites. This might warrant percutaneous drainage to relieve the IAH. This method is well described in the management of post-burn secondary ACS but not in postinjury scenarios.[13,20]

Response to Decompression

Unless the decompression is performed in the terminal phase, uniformly good physiologic responses to decompression are reported in the literature.[5,21] Unfortunately, a good physiologic response to decompression does not insure a better outcome. In our experience only the improvement in cardiac output and urine output discriminated survivors from nonsurvivors (Fig. 2).[8]

Prediction, Prevention and Surveillance

The prevention of postinjury secondary ACS revolves around the treatment of the traumatic shock. Timely hemorrhage control and fine-tuned resuscitation are paramount. Resuscitation of uncontrolled bleeding or supranormal resuscitation goals especially with large volume of crystalloids are potential instigators of the syndrome and should be avoided.[19] The previously recommended hypervolemic resuscitation of impending ACS improves the preload[5] but not cardiac output and contributes to increased gut edema thus precipitating the full blown syndrome.[18] Given the elusive nature and early occurrence of the syndrome adequate monitoring, such as continuous IAP measurement and gastric tonometry together with the organ specific monitoring of the heart, kidneys and lungs are essential for the timely recognition. Among the typical secondary ACS scenarios the source of bleeding could be obvious (penetrating chest and major extremity vascular injuries) or rather obscure such as multiple blunt extremity and pelvic fractures. Ideally all significant bleeding should be addressed in order to prevent uncontrolled resuscitation.

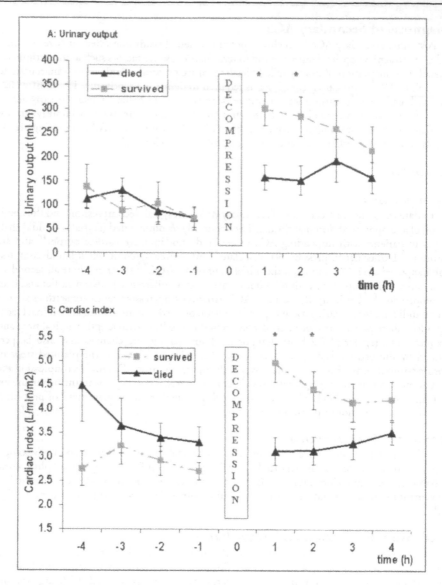

Figure 2. Response to abdominal decompression ACS: abdominal compartment syndrome. *p<0.05 survivors versus nonsurvivors.

Future Directions

Postinjury secondary ACS is a serious early complication of current resuscitation strategies for patients arriving with exsanguinating bleeding. With timely thorough hemorrhage control, fine-tuned resuscitation with crystalloid thresholds and monitoring we believe most cases of secondary ACS can be prevented. Basic laboratory studies should allow development of alternative resuscitation strategies to limit gut inflammation and edema formation that characterize this disease process. Given the strong statistical association between the amount of crystalloid infusion and secondary ACS, the early use of hypertonic saline and colloids in high risk patients is attractive. Patients who have the ED model's independent risk factors require frequent

IAP monitoring, (preferably continuous IAP) and gastric tonometry monitoring, (a highly valuable ICU predicting tool). Given the difficulty demonstrating improved outcome with early decompression, the concept of early presumptive decompression of high risk patients with IAH must be prospectively assessed with careful documentation of the inherent threats of open abdomen management.

Commentary

Andrew Kirkpatrick

See "Commentary" section at the end of Chapter 14.

References

1. Balogh Z, McKinley BA, Cocanour CS et al. Secondary abdominal compartment syndrome: An elusive complication of traumatic shock resuscitation. Am J Surg 2002; 184:538-544.
2. Burrows R, Edington J, Robbs JV. A wolf in wolf's clothing: The abdominal compartment syndrome. South Afr Med J 1995; 85:46-48.
3. Maxwell RA, Fabian TC, Croce MA et al. Secondary abdominal compartment syndrome: An underappreciated manifestation of severe hemorrhagic shock. J Trauma 1999; 47:995-999.
4. Kopelman T, Harris C, Miller R et al. Abdominal compartment syndrome in patients with isolated extraperitoneal injuries. J Trauma 2000; 49:744-749.
5. Burch JM, Moore EE, Moore FA et al. The abdominal compartment syndrome. Surg Clin North Am 1996; 76:833-842.
6. Biffl WL, Moore EE, Burch JM et al. Secondary abdominal compartment syndrome is a highly lethal event. Am J Surgery 2001; 182:645-648.
7. Johnson JW, Gracias VH, Schwab CW et al. Evolution in damage control for exsanguinating penetrating abdominal injury. J Trauma 2001; 51:261-9.
8. Balogh Z, McKinley BA, Holcomb JB et al. Both primary and secondary abdominal compartment syndrome (ACS) can be predicted early and are harbingers of multiple organ failure. J Trauma 2003; 54:848-861.
9. Rezende-Neto JB, Moore EE, Masuno T et al. The abdominal compartment syndrome as a second insult during systemic neutrophil priming provokes multiple organ injury. Shock 2003; 20:303-8.
10. Oda J, Ivatury RR, Blocher CR et al. Amplified cytokine response and lung injury by sequential hemorrhagic shock and abdominal compartment syndrome in a laboratory model of ischemia-reperfusion. J Trauma 2002; 52:625-631.
11. Rezende-Neto J, Moore EE, Melo de Andrade MV et al. Systemic inflammatory response secondary to abdominal compartment syndrome: Stage for multiple organ failure. J Trauma 2002; 53:1121-1128.
12. Balogh Z, McKinley BA, Cox Jr CS et al. Abdominal compartment syndrome: The cause or effect of postinjury multiple organ failure. Shock 2003; 20:483-492.
13. Ivy ME, Atweh NA, Palmer J et al. Intra-abdominal hypertension and abdominal compartment syndrome in burn patients. J Trauma 2000; 49:387-391.
14. Kirkpatrick AW, Brenneman FD, McLean RF et al. Is clinical examination an accurate indicator of raised intra-abdominal pressure in critically injured patients? Can J Surg 2000; 43:207-11.
15. Sugrue M, Bauman A, Jones F et al. Clinical examination is an inaccurate predictor of intraabdominal pressure. World J Surg 2002; 26:1428-31.
16. Krone IL, Harman PK, Nolan SP. The measurement of intra-abdominal pressure as a criterion for exploration. Ann Surg 1984; 199:28-30.
17. Balogh Z, Jones F, D'Amours SK et al. Continuous intra-abdominal pressure measurement: A new gold standard. Am J Surg 2004; 188:679-84.
18. Balogh Z, McKinley BA, Cocanour CS et al. Patients with impending abdominal compartment syndrome do not respond to early volume loading. Am J Surg 2003; 186:602-7
19. Balogh Z, McKinley BA, Cocanour CS et al. Supranormal trauma resuscitation causes more cases of abdominal compartment syndrome. Arch Surg 2003; 138:637-42.
20. Corcos AC, Sherman HF. Percutaneous treatment of secondary abdominal compartment syndrome. J Trauma 2001; 51:1062-1064.
21. Ivatury RR, Porter JM, Simon RJ et al. Intra-abdominal hypertension after life-threatening penetrating abdominal trauma: Prophylaxis, incidence, and clinical relevance of gastric mucosal pH and abdominal compartment syndrome. J Trauma 1998; 44:1016-1023.
22. Sugrue M, D'Amours S. The problems with positive end expiratory pressure (PEEP) in association with abdominal compartment syndrome (ACS). J Trauma 2001; 51:419-20.

CHAPTER 14

Secondary Abdominal Compartment Syndrome in Burns

Michael E. Ivy*

Abstract

Patients with large burns (50% or greater or with associated inhalation injury) are at risk of developing intra-abdominal hypertension (IAH). Patients with burns that are very large, greater than 70% TBSA, are at risk of developing abdominal compartment syndrome (ACS), particularly if they have a concurrent inhalation injury. The development of IAH and ACS is related to the volume of crystalloid fluid infused during the burn resuscitation and does not require abdominal injury or operation or even the presence of abdominal wall burn eschar. Burn patients are also at risk for developing IAH and ACS during subsequent septic episodes.

A variety of management options exist for IAH including observation, sedation, pharmacologic paralysis, abdominal wall escharotomy, and percutaneous catheter decompression of the peritoneal cavity. Options for the management of ACS include the standard decompressive laparotomy and in some cases percutaneous catheter decompression. Most patients who require intervention improve significantly at the time. However, these patients have very large burns, often severe inhalation injuries and frequently die later in their hospitalization from complications of their burns that are unrelated to their ACS.

Introduction

Recognition of the existence of abdominal compartment syndrome (ACS) has been one of the most important advances in the care of the surgical patient over the past 20 years. While some of the basics of the syndrome are reasonably well understood, many areas still need to be fully characterized. The importance of intra-abdominal hypertension (IAH) and ACS in burn patients has not been extensively studied to date and remains an area that warrants further investigation.

Kron and colleagues at the University of Virginia wrote the first article that recognized the importance of elevated intra-abdominal pressure (IAP) in postoperative patients in 1984.[1] The authors described 4 patients who developed the signs and symptoms we now recognize as part of the ACS without IAP measurements being performed. They subsequently measured the IAP in 10 patients with a normal postoperative course and 7 patients who developed an elevated IAP (30 to 77 mm Hg) along with oliguria despite adequate cardiac output. Four of the

*Michael E. Ivy—Associate Director of Surgery, Department of Surgery, Hartford Hospital, 80 Seymour Street, Hartford, Connecticut 06102, U.S.A.; Assistant Professor of Surgery, Department of Surgery, University of Connecticut School of Medicine, 263 Farmington Avenue, Farmington, Connecticut 06030, U.S.A. Email: Mivy@harthosp.org

Abdominal Compartment Syndrome, edited by Rao R. Ivatury, Michael L. Cheatham, Manu L. N. G. Malbrain and Michael Sugrue. ©2006 Landes Bioscience.

patients who developed elevated IAP underwent reexploration and abdominal decompression and they all improved; unfortunately two of the four later died of sepsis. The three who did not undergo decompressive laparotomy all died in the early postoperative period. All of the patients who developed ACS had undergone major abdominal surgery including abdominal aortic aneurysm repair, major oncologic surgery, portosystemic shunts, and repair of major vascular injury. The authors concluded that an IAP greater than 25 mm Hg was associated with a need for reoperation or death. Over the next ten years, the occurrence of ACS in patients with abdominal trauma became widely recognized by trauma surgeons and surgical intensivists. It was noted to be a complication occurring primarily in patients with abdominal trauma or surgery who were in profound shock and required massive fluid resuscitation.

Several years later, the initial reports of ACS in patients without abdominal surgery or injury began appearing. Burrows reported a patient with abdominal compartment syndrome who had not sustained abdominal trauma.[2] Maxwell and colleagues reported the Memphis experience in 1999.[3] Of 1216 consecutive trauma admissions to their ICU, 46 (4%) underwent temporary abdominal closure because of edema in the abdominal cavity. Six of the patients had not sustained significant abdominal trauma, but had been in shock and had required massive fluid resuscitation. They reported that survivors were, on average, decompressed 20 hours earlier than nonsurvivors, which led them to suggest that early decompression of ACS might improve patient outcomes. They recommended measuring bladder pressures in all patients who receive 10 liters of crystalloid resuscitation or who are transfused with 10 or more units of packed red blood cells.

The Burn Literature

In 1994, Greenhalgh and Warden reported four cases of elevated IAP in their pediatric burn unit, which prompted the prospective trial reported in the same article.[4] They prospectively measured IAP in all children (30 patients) admitted with burns large enough to require resuscitation and placement of a foley catheter. They noted that 11 of 30 patients (36%) had at least one reading of more than 30 mm Hg, their working definition of intra-abdominal hypertension (IAH). The average total body surface area (TBSA) burn size for the patients with IAH was 67%; in contrast, the patients who did not develop IAH had a mean TBSA burn size of 50%. The mortality of the group with IAH was 55%, which was significantly higher than the 16% mortality in the group without IAH. Nearly half of the patients (5/11) in the prospective series who developed the IAH did so during their initial burn resuscitation, the rest developed IAH during subsequent bouts of sepsis. Two patients who developed IAH during their initial resuscitation had resolution of their IAH after performance of abdominal escharotomies. The other three patients had return of their IAP to normal levels spontaneously.

Interestingly, most of the elevated IAP measurements reported in the Greenhalgh and Warden series occurred during septic episodes.[4] Three of these patients in the prospective series were treated with chemical paralysis with resolution of their IAH. Two patients were observed and their IAP returned to < 30 mm Hg spontaneously. One patient required percutaneous catheter drainage of intraperitoneal fluid and one patient required laparotomy and bowel resection for ischemic bowel.

The case reports included in the article were not part of the prospective series, and are somewhat confusing because two of the patients developed ACS as a result of a tension pneumoperitoneum rather than the typical ACS caused entirely by accumulation of intra-abdominal fluid.[4] The first case was a 3 year old with an 81% TBSA burn. He developed shock and severe respiratory failure with marked CO_2 retention during his initial burn resuscitation. He improved dramatically with laparotomy and eventually survived to discharge. The second patient developed pneumoperitoneum presumably due to barotrauma, which eventually caused IAH. She initially responded to percutaneous catheter decompression, but later required bilateral chest tubes and abdominal and chest wall escharotomies. Despite these efforts, she eventually died from worsening respiratory failure. The third case also developed a pneumoperitoneum related to severe pulmonary disease, but had a minimal improvement in response to the initial percutaneous catheter but did respond to a second catheter placed days later. Nevertheless, he died several days later from progressive pulmonary failure. Their fourth case developed abdominal compartment syndrome during a bout of candida sepsis. She did not improve with percutaneous but did improve with laparotomy and open decompression, but a month later died from recurrent sepsis.

This article established that abdominal compartment syndrome occurs in patients with very large burns during their initial burn resuscitation and also occurs during subsequent episodes of sepsis. The article provides evidence that burn patients who develop IAH have a high risk of mortality, and it suggests that part of the risk is due to the development of ACS. Aggressive operative decompression seemed to improve their patients' physiological status initially. Intra-abdominal hypertension was shown to occur more frequently than ACS and was often managed successfully with a variety of nonoperative techniques including observation, abdominal escharotomy, use of nondepolarizing paralytic agents and percutaneous catheter decompression of the peritoneal cavity.

In 1998 we reported 3 cases of ACS in adult burn patients occurring during their burn resuscitation.[5] All of the patients had sustained greater than 70% TBSA burns and two had associated inhalation injuries. Their intra-abdominal pressures at the time of diagnosis were 49, 50 and 36 mm Hg. All three of the patients subsequently died. We noted that all three patients had received more than 20 liters of crystalloid prior to developing ACS. One patient developed recurrent ACS after the initial decompression.

After the case report, we conducted a prospective study of ACS in adult burn patients to evaluate the frequency of intra-abdominal hypertension and ACS in adults with burns.[6] We prospectively recorded intra-abdominal pressures (IAP) in patients with burn shock admitted to our burn unit. Our cutoff for IAH was 25 mm Hg, which developed in 7 of 10 patients. We reserved the diagnosis of abdominal compartment syndrome for patients with IAH as well as decreased pulmonary compliance or oliguria despite adequate filling pressures and cardiac output. Only two patients, (20%) developed ACS. One patient with ACS had an 80% TBSA burn that was essentially all full-thickness. He developed an IAP of 35 mm Hg after receiving 48 liters of fluid resuscitation. He was decompressed with a laparotomy and a temporary abdominal wall closure. He responded well to decompression and was eventually discharged to home. The other patient with ACS had a similar 80% TBSA burn. He was diagnosed with ACS after 38 liters of fluid resuscitation and similarly responded well to decompression. He eventually died from recurrent sepsis on post-burn day 86.

A linear regression analysis demonstrated that IAP correlates with the volume of fluid infused, and that this one variable explained 33% of the variance in intra-abdominal pressure.[6] We calculated that a resuscitation of 250 mL/kg body weight results in an IAP of 25 mm Hg in burn patients. While this number should be useful in burn patients, it should be extrapolated to other patient populations with great caution. Most other patient populations in shock will develop peripheral edema, but not to the extent that burn patients develop edema under their burn eschar. Additionally, patients who have sustained abdominal trauma or undergone abdominal surgery may have a significant hematoma in their retroperitoneum that will directly contribute to elevating their IAP. Consequently, most patients who have undergone abdominal surgery will develop an IAP of 25 mm Hg with a significantly smaller volume of fluid than 250 mL/kg.

Given the significant drawbacks associated with laparotomy in burn patients, we used an algorithm to assist us with the management of uncomplicated IAH (Fig. 1).[6] If a patient's IAP was greater than 25 mm Hg, we initially gave them additional sedation, as it is clear that patient agitation can dramatically elevate IAP. If the IAP remained elevated, we chemically paralyzed them. If their pressure was still high and we clinically thought they had an adequate intra-vascular volume we tried to diurese them with furosemide. Most patients with IAH did not require operative intervention, so it seems clear that patients with mildly elevated pressures can avoid decompression. If their pressures remained mildly elevated without renal or pulmonary sequelae we observed them until their pressures returned to normal.

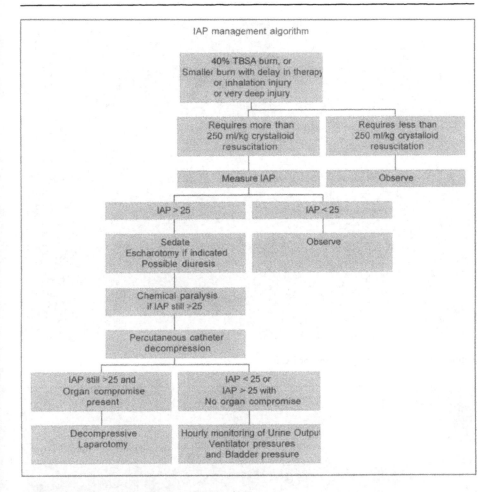

Figure 1. Algorithm for management of IAH in burns.

Despite all of these efforts to avoid laparotomy, some patients still require a decompressive laparotomy (Figs. 2 and 3). Combining our two articles, the patients who required laparotomy had a mean TBSA burn of 84%, with a range of 74 to 98%.[5,6] Thus, the need for laparotomy to treat ACS in our hands is limited to patients with very extensive burns and a massive fluid requirement. Both patients in our prospective study had received in excess of 40 liters during their initial 48 hours of treatment.

These articles established that intra-abdominal hypertension is a common occurrence in adult patients with large burns (> 50% TBSA) and that abdominal compartment syndrome is seen in patients with very large burns (> 70% TBSA and often with associated inhalation injury).

Corcos and Sherman in 2001 reported their experience with percutaneous decompression (PD) of burn patients with ACS.[7] They described three patients with large TBSA burns who developed IAH. The first patient was initially drained with an 18 gauge angiocatheter, which resolved his IAH. However due to a recurring need for drainage due to intermittent IAH, a peritoneal dialysis catheter was placed and was very successful in draining excess fluid from his peritoneal cavity and in reducing his IAP. The other two patients described were decompressed initially with peritoneal dialysis catheters. All three patients had immediate resolution of their

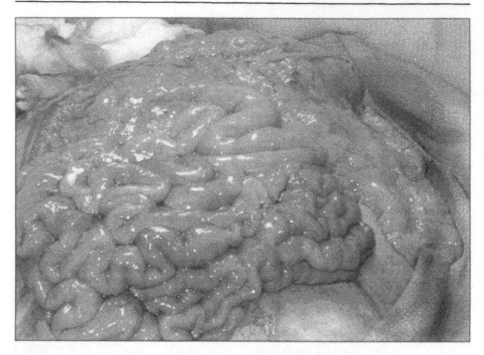

Figure 2. Decompression laparotomy for IAH.

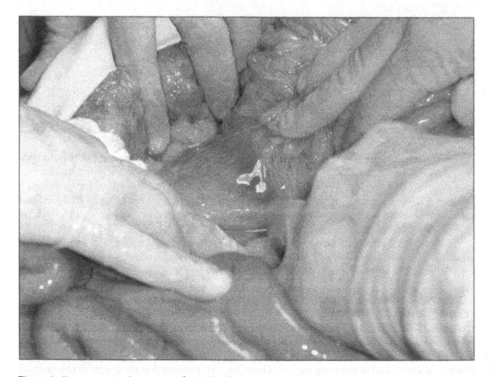

Figure 3. Decompressive laparotomy for IAH. Note mesenteric edema.

IAH after placement of the dialysis catheters. The adult patient in the series had removal of 1800 mL in the first hour after placement. They noted a survival rate of 33% and the mean TBSA burn was 60%.

More recent articles on ACS in burn patients include a report from UC Davis by Hobson and colleagues.[8] They conducted a retrospective review of over 1,000 patients admitted between 1998 and 2000. They diagnosed 10 patients with ACS. They initially manage IAH with nonoperative methods including NG decompression, bowel care, sedation, pharmacologic paralysis and abdominal wall escharotomy when indicated. The authors identified 10 patients, six children and four adults, who developed an IAP > 30 mm Hg with signs of physiologic compromise consistent with ACS that required a procedure to decrease their IAP. The mean TBSA burn of patients who developed ACS in this series was 70%. The authors performed decompressive laparotomy on eight patients. Two of their patients were decompressed by placement of a large-bore hemodialysis catheter. They noted two distinct times that burn patients were at risk for ACS; 6 patients developed ACS during their initial burn resuscitation and 4 developed it during a later septic episode. The IV fluid requirements of the patients who developed ACS during their initial resuscitation were 237 mL/kg in the first 12 hours of resuscitation. While the numbers are very small, they noted a 50% survival rate in children and a 25% survival rate in adults.

Latenser and colleagues at Cook County prospectively studied the utility of diagnostic peritoneal lavage (DPL) catheter decompression of the abdominal cavity in patients with large (> 40% TBSA) burn patients.[9] While 22 patients met criteria for inclusion in the study, nine were excluded because they were "comfort care only". This left 13 patients in the study, of which 9 (69%) developed IAH. They attempted to treat all occurrences of IAH, defined as an IAP > 25 mm Hg, with percutaneous decompression. Consequently, of the nine patients with IAH, five patients improved with DPL catheter decompression and four required laparotomy despite percutaneous DPL decompression. The patients requiring laparotomy had sustained very large burns with a mean TBSA burn of > 80% in addition to inhalation injury. All of their patients requiring laparotomy after failed DPL decompression died, while the mortality rate for the group that responded to DPL decompression was 40%.

Tsoutsos and colleagues recently have studied the role of torso eschar formation in the development of IAH.[10] They studied patients with > 35% TBSA burns involving the torso. They performed a standardized set of escharotomy incisions and documented a resultant decrease in abdominal pressures. Clearly a nearly circumferential abdominal wall eschar will elevate IAP as patients are resuscitated. Consequently, escharotomy should decrease the elevation in IAP related to the restrictive eschar. This may not be as valuable in patients with IAH not related to abdominal wall eschar.

Fluid Resuscitation Issues

There are several factors that are known to increase the volume of fluid required to successfully resuscitate burn patients. These factors include increasing extent of TBSA burn, greater depth of the burns, the presence of inhalation injury, a delay in the initiation of resuscitation, and the indiscriminate use of fluid boluses during the burn resuscitation. There are also concerns that we are in an era that is simply more aggressive with the volume of fluid given during resuscitation. In an editorial on over aggressive fluid resuscitation that accompanied an article on ACS in burn patients, Dr. Pruitt linked the occurrence of ACS in burn patients to the general increase in the volume of crystalloid used to resuscitate patients in the current era.[11] Cartotto et al studied the volume of fluid given to burn patients being resuscitated in their institution.[12] They retrospectively reviewed the data on burn patients admitted within 6 hours of injury with > 15% TBSA burns and they intentionally excluded patients with inhalation injury, concomitant trauma and electrical injury as well as compassionate care only patients. They looked at 31 patients with a mean TBSA burn of 27%. They received a mean of 13 liters in the first 24 hours, which far exceeded the Parkland formula estimate. The mean hourly urine output in their patients was 1.2 mL/kg/hr. Notably, while the IV fluid rate decreased by 34%

Table 1. Characteristics of burn patients with IAH or ACS

Series	# Pts	TBSA Burn	Inhalation	IAP	Early	Late
Greenhalgh case series[4]	4	47%	N/A	N/A	1	3
Greenhalgh prospective[4]	11	67%	N/A	> 30	5	7
Ivy[5]	3	87%	1/3	45	3	0
Latenser[9]	9	59%	7/9	34	9	0
Hobson[8]	10	71%	N/A	40	6	4
Corcos[7]	3	60%	2/3	52	3	0
Ivy[6]	7	47%	28%	37	7	0
Tsoutsos[10]	10	57%	60%	23	10	0

pts: number of patients in series that developed IAH and or ACS; TBSA Burn: mean percentage of total body surface area burned in patients with IAH or ACS; Inhalation: presence of inhalation injury; IAP: mean IAP; Early: development of IAH or ACS within 72 hours of burn injury; Late: development of IAH or ACS during sepsis episode later in hospitalization; N/A: not available; IAP: intra-abdominal pressure in mm Hg

in 16 patients at 8 hours, it actually increased by 47% in the remaining 15 patients. The patients that were successfully managed with a lower rate of IV fluid resuscitation had less severe injuries with a mean area of full-thickness burn of only 3%, while the remaining patients had full-thickness burns covering 14% of their TBSA. The groups were otherwise very similar. The mean 24 hour fluid resuscitation volume in the group where the rate was decreased was 5.6 mL/kg/%TBSA and was 7.7 mL/kg/%TBSA in the other group. The implication of this study is that a 70 kg patient with a 50% TBSA burn with more than 15% of the TBSA burn being full-thickness will receive approximately 27 liters over the first 24 hours even in the absence of an inhalation injury.

A survey conducted by Engrav et al reviewed data on 50 patients and reported that 58% received more than the 4.3 mL/kg/% TBSA compared with the 12% incidence reported by Baxter.[13] These studies corroborate the suspicion of Dr Pruitt and others that we are significantly more aggressive with our fluid resuscitation than in years past. This increase in the volume of fluid resuscitation probably does contribute to the relatively high incidence of IAH and ACS in burn patients.

Balogh and colleagues from Houston have demonstrated that an aggressive fluid resuscitation increases the risk of IAH and ACS in other populations of critically injured patients as well.[14] They compared the outcomes from a standardized trauma resuscitation protocol using a target of $DO_2I > 500$ mL/min/m^2 instead of a target of >600 mL/min/m^2. The authors found that the lower target required significantly less fluid and halved the incidence of IAH (20% vs. 42%) and the incidence of ACS (8% vs. 16%). A separate study by that group retrospectively analyzed data from 188 patients who developed ACS over a 44 month period.[15] Their multiple logistic regression analysis linked the development of secondary ACS to the volume of crystalloid infused, an average of over 30 liters in the first 24 hours in that population, a number that is in relatively close agreement with the multiple burn series mentioned above.

Summary

Large volume fluid resuscitations required for patients with major burns place them at significant risk of developing intra-abdominal hypertension. Patients who develop IAH or ACS generally have TBSA burns of approximately 50% or more and often have a concurrent inhalation injury[4-10] (Table 1). The subgroup of patients that develop ACS have a mean TBSA burn of 75% (Table 2), and more than half of the burn patients who develop ACS have

Table 2. Characteristics of patients with burns and ACS

Series	# pts	TBSA Burn	Inhalation	IAP	Survival
Latenser[9]	4	83%	4/4	34	0/4
Hobson[8]	10	71%	N/A	40	4/10
Corcos[7]	3	60%	2/3	52	1/3
Ivy[5]	3	87%	1/3	45	0/3
Ivy[6]	2	80%	1/2	40	1/2
Total	22	75%	8/12	41	6/22 (27%)

pts: number of patients; TBSA: % Total body surface area burned; Inhalation: presence of inhalation injury; IAP: mean intra-abdominal pressure of patients in that series; Survival: survival to discharge from hospital; IAP: intra-abdominal pressure in mm Hg

sustained an inhalation injury.[5-9] In the general trauma population, less aggressive use of crystalloid resuscitation utilizing protocols has been shown to drop the risk of IAH and ACS by 50%.[14] Hopefully, the incidence of IAH and ACS in patients with large burns can also be decreased by a resuscitation that utilizes a very tightly titrated fluid resuscitation protocol in order to maintain a urine output greater than 0.5 mL/kg/h but less than 1 mL/kg/h. This hypothesis can only be evaluated by a prospective trial performed collaboratively by multiple burn centers.

An aggressive approach to IAH and ACS emphasizing early diagnosis and treatment may be of benefit to some burn patients, but this has not been clearly demonstrated by a randomized clinical trial. A variety of techniques have been used to decrease IAP in patients with IAH including sedation, use of paralytic agents, diuresis, abdominal wall escharotomy and percutaneous catheter drainage of intraperitoneal fluid.[4-10] While several studies have documented an increased risk of mortality in patients who develop IAH, most patients with IAH do not go onto develop full-blown ACS.[4-10] Some patients with ACS seem to benefit from percutaneous catheter decompression, but often patients with ACS will require decompressive laparotomy.[4,9] Burn patients with ACS universally benefit from decompression during the initial postoperative period.[4-6,8,9] Unfortunately, most burn patients who require decompressive laparotomy for ACS do not survive to discharge.[4-6,8,9] The overall rate of survival to discharge in the burn literature is only 29% (5/17) but this is substantially better than the expected survival rate of 0% (Table 3).

Table 3. Outcome of laparotomy for ACS

Series	LAP	Survival
Greenhalgh case series[4]	2	1
Ivy[5]	1	0
Ivy[6]	2	1
Latenser[9]	4	0
Hobson[8]	8	3
Total	17	5 (29%)

LAP: number of patients undergoing decompressive laparotomy for ACS; Survival: number of patients surviving to discharge

Commentary

Andrew Kirkpatrick

Tremendous progress has been made in the last century in treating the most seriously burned and injured. This progress has often been necessitated by both human conflict and recognition of previously unknown complications unmasked by prior advances. The first world war in the first half of this century led to the discovery that shock resulted from intravascular volume deficits and that colloid and blood should be administered preoperatively.[1-3] In the second half of the last century, the frequent complication of early renal failure was all but eliminated through the administration of aggressive crystalloid fluid resuscitation in burns and serious injury. Volumes administered sometimes seemed enormous (especially burns with inhalation injury).[4] Unfortunately, this practice was followed by the acute respiratory distress syndrome as a major cause of morbidity and mortality in the third quarter.[5] Dedicated trauma care in the last quarter has led to efficient trauma systems providing early hemorrhage control, often entailing damage control surgery to reduce exsanguinating deaths. This has also been followed by multi-disciplinary care to alleviate single system organ failure, with multi-system organ failure becoming a leading cause of in-hospital trauma death.[6] The last few years of the 20th century have now seen a description of another post-injury complication that is also likely related to treatment of hemorrhagic shock and severe burns. This is the secondary abdominal compartment syndrome (SACS). It is likely that this syndrome, closely associated and potentially bridging shock and multi-system organ failure, is related to our current resuscitation practices. How we resuscitate shock (either traumatic or burn-related), as well as how quickly the underlying insult is addressed, may be important in the pathophysiology of the SACS.

Dr Balogh et al, from the University of Texas have published more than anyone else on the SACS, and the world literature on this subject is thus greatly influenced by their thinking. They have described the syndrome as an elusive, but very early complication of severe trauma and resuscitation.[7] While pressure thresholds and definitions are controversial, they recognize intra-abdominal pressures greater than 25 mm Hg accompanied by cardiac, respiratory, and renal function as being diagnostic. They refer to SACS when there are no intra*peritoneal* injuries, and specifically consider pelvic fracture related intraperitoneal hematomas without intra-peritoneal injury to be secondary,[7,8] in consensus with other authors.[9,10] This is an important point as this particular injury comprises 73% of the SACS group. As other authors have considered SACS to apply only to those without injury or disease in the *abdominopelvicregion*,[11] this distinction should be noted. As almost half of this patient group required urgent angiographic embolization for hemorrhage control in the Texas experience, they exemplify the presumed risk factors for secondary ACS; namely delay in hemorrhage control temporized by with massive crystalloid resuscitation in a prolonged pre-intensive care unit course.[8] Secondary ACS has been recognized to occur earlier in the hospital admission and unfortunately earlier decompression does not appear to improve outcome in this select minority of injured patients.[7,11]

Dr. Ivy has concisely reviewed the limited literature describing the epidemiology and therapeutic implications of ACS occurring in setting of major burn injury, a literature which also bears tribute to his personal work. He has also introduced the concept that clinicians may be potentially harming their patients through over-vigorous fluid resuscitation. Sicker patients with larger burns are now surviving, and these patients, with body surface area burns > 70% and often with an associated inhalation injury are at great risk of developing the abdominal compartment syndrome.[12,13] While combined abdominal injury is always a concern, the vast majority of ACS seen in the setting of major burns is not associated with primary abdominal pathology or even abdominal wall eschar.[12-14] ACS in this setting has been termed secondary, and it typically occurs either in response to our therapy, that being massive crystalloid infusion, or as a result of a septic complication.[14] Factors relating to the required fluid volume are burn surface area, depth of the burn, presence of inhalation injury, which are fixed, and the indiscriminate use of fluid boluses and timing of resuscitation which are clinician-dependant.

Linear regression predicts elevated intra-abdominal pressure to 25 mm Hg, a level that usually requires intervention,[15] after 250 mL/kg or a burn injury of 63% BSA according to the Parkland resuscitation formula. This syndrome has been variably described by various authors but essentially reflects clinically evident end-organ dysfunction on the basis of acute and sustained raised intra-abdominal pressures.[16,17] It also likely that raised intra-abdominal pressures also adversely affect the course of critically injured patients through impaired visceral perfusion, and potentiation of multi-organ dysfunction, even if the full-blown syndrome is not evident.[18,19] When the SACS does develop, it typically has a dramatic response to decompressive laparotomy, although the presence of SACS after burn injury has ominous implications for hospital discharge (29% survivors).

Fluid resuscitation has been defined as a treatment regimen involving fluid replacement intended to minimize the effects of (hemorrhagic) shock and to stabilize the hemodynamic response.[2] The goal being to maintain perfusion to each and every cell of the organism, especially to the penumbra of injury. In contradistinction to previous generations who may have under-resuscitated patients though, we appear to have entered an era of "fluid creep",[20,21] with crystalloid fluids administered in excess of that required, so that perfusion and oxygen delivery are not increased, but are actually paradoxically decreased due to the detrimental systemic effects of SACS.[22,23]

Precedent has shown that identified limitations and problems that were previously obscured by conditions now successfully treated, quickly become topics for new research problems; hopefully the SACS will be as well.[4] Many anti-inflammatory approaches have been or are being investigated to ameliorate the primary insult, including hypertonic saline resuscitative strategies. Closed loop computer-driven resuscitation using fluid administration proportional to the measured value of a predefined end-point may achieve favorable outcomes with less over-all fluid administration and avoidance of over-resuscitation,[24] although choosing the appropriate end-point is still critical in avoiding over-resuscitation and the ACS.[22] Until novel approaches are proven though, due diligence will be required of all clinicians to avoid all possible delay in hemorrhage control and to optimize the fluid resuscitation to the best of their abilities, implying a continued presence at the foot of the bed.

Commentary References

1. Cannon WB. Traumatic shock. New York: D Appelton Co, 1923.
2. Introduction. Pope A, French G, Longnecker DE, eds. Fluid Resuscitation: State of the Science for Treating Combat Casualties and Civilian Injuries. Washington, DC: National Academy Press, 1999:9-18.
3. Pruitt BA, Pruitt JH, Davis JH. History. Moore EE, Feliciano DV, Mattox KL, eds. Trauma. 5th ed. New York: McGraw-Hill, 2004:3-19.
4. Pruitt BA. The development of the International Society for Burn Injuries and progress in burn care: the whole is greater than the sum of its parts. Burns 1999; 25:683-96.
5. Moore FA, McKinley BA, Moore EE. The next generation in shock resuscitation. Lancet 2004; 363:1988-96.
6. Sauaia A, Moore FA, Moore EE et al. Epidemiology of trauma death: a reassessment. J Trauma 1995; 38:185-93.
7. Balogh Z, McKinley BA, Cocanour CS et al. Secondary abdominal compartment syndrome is an elusive early complication of traumatic shock resuscitation. Am J Surg 2002; 184:538-44.
8. Balogh Z, McKinley BA, Holcomb JB et al. Both primary and secondary abdominal compartment syndrome can be predicted early and are harbringers of multiple organ failure. J Trauma 2003; 54:848-61.
9. Maxwell RA, Fabian T, Croce M et al. Secondary abdominal compartment syndrome: an underappreciated manifestation of severe hemorrhagic shock. J Trauma 1999; 47:995-9.
10. Kopelman T, Harris C, Miller R et al. Abdominal compartment syndrome in patients with isolated extraperitoneal injuries. J Trauma 2000; 49:744-9.
11. Biffl WL, Moore EE, Burch JM et al. Secondary abdominal compartment syndrome is a highly lethal event. Am J Surg 2001; 182:645-8.
12. Ivy ME, Possenti PP, Kepros J et al. Abdominal compartment syndrome in patients with burns. J Burn Care Rehabil 1999; 20:351-3.

13. Ivy ME, Atweh NA, Palmer J et al. Intra-abdominal hypertension and abdominal compartment syndrome in burn patients. J Trauma 2000; 49:387-91.
14. Hobson KG, Young KM, Ciraulo A et al. Release of abdominal compatment syndrome improves survival in patients with burn injury. J Trauma 2002; 53:1129-34.
15. Nathens AB, Brenneman FD, Boulanger BR. The abdominal compartment syndrome. Can J Surg 1997; 40:254-8.
16. Schein M, Wittman DH, Aprahamian CC et al. The abdominal compartment syndrome: the physiolgical and clinical consequences of elevated intra-abdominal pressure. J Am Coll Surg 1995; 180:745-52.
17. Kirkpatrick AW, Brenneman FD, McLean RF et al. Is clinical examination an accurate indicator of raised intra-abdominal pressure in critically injured patients. Can J Surg 2000; 43:207-11.
18. Ivatury RR, Porter JM, Simon RJ et al. Intra-abdominal hypertension after life-threatening penetrating abdominal trauma: prophylaxis, incidence, and clinical relevance to gastric mucosal pH and abdominal compartment syndrome. J Trauma 1998; 44:1016-23.
19. Diebel LN, Dulchavsky SA, Brown WJ. Splanchnic ischemia and bacterial translocation in the abdominal compartment syndrome. J Trauma 1997; 43:852-5.
20. Engrav LH, Colescott PL, Kemalyan N et al. A biopsy of the use of the Baxter formula to resuscitate burns or do we do it like Charlie did it? J Burn Care Rehabil 2000; 21:91-5.
21. Pruitt BA. Protection from excessive resuscitation: "Pushing the pendulum back". J Trauma 2000; 49:567-8.
22. Balogh Z, McKinley BA, Cocanour CS et al. Supranormal trauma resuscitation causes more cases of abdominal compartment syndrome. Arch Surg 2003; 138:637-43.
23. Balogh Z, MooreFA, McKinley BA. Supranormal trauma resuscitation and abdominal compartment syndrome: In reply [letter]. Arch Surg 2004; 139:226-7.
24. Chaisson NF, Kirschner RA, Deyo DJ et al. Near-infrared spectroscopy-guided closed-loop resuscitation of hemorrhage. J Trauma 2003; 54:S183-S192.

References

1. Kron IL, Harman PK, Nolan SP. The measurement of intra-abdominal pressure as a criterion for abdominal reexploration. Ann Surg 1984; 199:28-30.
2. Burrows R, Edington J, Robbs JV. A wolf in wolf's clothing-the abdominal compartment syndrome. S Afr Med J 1995; 85:46-48.
3. Maxwell RA, Fabian TC, Croce MA et al. Secondary abdominal compartment syndrome: An underappreciated manifestation of severe hemorrhagic shock. J Trauma 1999; 47:995-999.
4. Greenhalgh DG, Warden GD. The importance of intra-abdominal pressures measurements in burned children. J Trauma 1994; 36:685-690.
5. Ivy ME, Possenti PP, Kepros J et al. Abdominal compartment syndrome in patients with burns. J Burn Care Rehabil 1999; 20:351-353.
6. Ivy ME, Atweh NA, Palmer J et al. Intra-abdominal hypertension and abdominal compartment syndrome in burn patients. J Trauma 2000; 49:387-391.
7. Corcos AC, Sherman HF. Percutaneous treatment of secondary abdominal compartment syndrome. J Trauma 2001; 51:1062-1064.
8. Hobson KG, Young KM, Ciraulo A et al. Release of abdominal compartment syndrome improves survival in patients with burn injury. J Trauma 2002; 53:1129-1134.
9. Latenser BA, Kowal-Vern A, Kimball D et al. A pilot study comparing percutaneous decompression with decompressive laparotomy for acute abdominal compartment syndrome in thermal injury. J Burn Care Rehabil 2002; 23:190-195.
10. Tsoutsos D, Rodopoulou S, Keramidas E et al. Early escharotomy as a measure to reduce intra-abdominal hypertension in full-thickness burns of the thoracic and abdominal area. World J Surg 2003; 27:1323-1328.
11. Pruitt BA. Protection from excessive resuscitation:"Pushing the pendulum back". J Trauma 2000; 49:567-568.
12. Cartotto RC, Innes M, Musgrave MA et al. How well does the parkland formula estimate actual fluid resuscitation volumes? J Burn Care Rehabil 2002; 23:258-265.
13. Engrav LH, Colescott PL, Kemalyan N et al. A biopsy of the use of the Baxter formula to resuscitate burns or do we do it like Charlie did it? J Burn Care Rehabil 2000; 21:91-95.
14. Balogh Z, McKinley BA, Cocanour CS et al. Supra-normal trauma resuscitation causes more cases of abdominal compartment syndrome. Arch Surg 2003; 138:637-643.
15. Balogh Z, Mckinley BA, Holcomb JB et al. Both primary and secondary abdominal compartment syndrome can be predicted early and are harbingers of multiple organ failure. J Trauma 2003; 54:848-861.

CHAPTER 15

Morbid Obesity and Chronic Intra-Abdominal Hypertension

Giselle G. Hamad* and Andrew B. Peitzman

Abstract

Morbid obesity has achieved epidemic proportions in the United States. A vast number of comorbid conditions are associated with morbid obesity, including metabolic syndrome, which consists of central obesity, insulin resistance, hypertension, and hyperlipidemia. An association between obesity and intra-abdominal hypertension has been demonstrated, which explains the predisposition to pseudotumor cerebri, hypertension, pulmonary disorders, stress urinary incontinence, gastroesophageal reflux, and incisional hernia in morbid obesity. Weight loss results in a reduction in intra-abdominal pressure and resolution of comorbidities.

Morbid Obesity and Weight-Related Comorbidities

Morbid obesity has achieved epidemic proportions in the United States. An estimated 66 percent of American adults are overweight and 31 percent are obese.[1] In the United States, obesity-attributable medical expenditures reached $75 billion per year in 2003.[2] Obesity induces a multitude of chronic illnesses affecting virtually every organ system (Table 1). These weight-related comorbidities account for the heightened mortality among the morbidly obese.

Two types of obesity have been described: peripheral, or gynecoid, and central, or android. Central obesity has been linked to an elevation in intra-abdominal pressure (IAP), creating a chronic intra-abdominal compartment syndrome.[3] Central obesity is one of the hallmarks of the metabolic syndrome, whose major components include insulin resistance, hypertension, and hyperlipidemia. Intra-abdominal fat is independently associated with all five criteria that comprise the metabolic syndrome.[4] In a study of over 14,000 subjects, Janssen and colleagues found that waist circumference, which reflects central obesity, was a significant predictor of comorbidity while body mass index (BMI), which is defined as body weight divided by the height squared (kg/m^2), was not.[5] The components of metabolic syndrome predispose a patient to cardiovascular disease. The etiology of metabolic syndrome remains unclear and the pathophysiologic mechanisms underlying the comorbid illnesses are controversial.

A number of studies have shown that urinary bladder pressure, which is a surrogate for IAP, is elevated in obesity.[6-9] Sugerman et al demonstrated the relationship between obesity and IAP and examined their association with weight-related comorbidities.[6] The investigators measured urinary bladder pressures in 84 morbidly obese patients prior to performing gastric bypass and compared them to measurements in five nonobese subjects. Sagittal abdominal diameters were recorded by measurement of the apex of the abdominal girth with the patient supine. Obese patients had significantly higher bladder pressures than the nonobese. Urinary bladder pressure

*Corresponding Author: Giselle G. Hamad—Department of Surgery, University of Pittsburgh Medical Center, Pittsburgh, Pennsylvania, U.S.A. Email: hamadg@upmc.edu.

Abdominal Compartment Syndrome, edited by Rao R. Ivatury, Michael L. Cheatham, Manu L. N. G. Malbrain and Michael Sugrue. ©2006 Landes Bioscience.

Table 1. Obesity-related comorbidities

Neurologic	**Hypercoagulable states**
Migraine headaches	Deep venous thrombosis
Pseudotumor cerebri	Venous stasis disease
Respiratory	Pulmonary embolism
Obstructive sleep apnea	**Genitourinary**
Asthma	Stress incontinence
Obesity hypoventilation syndrome	Renal insufficiency
Cardiovascular	**Reproductive**
Hypertension	Amenorrhea
Hyperlipidemia	Dysmenorrhea
Congestive heart failure	Polycystic ovary syndrome
Coronary artery disease	**Connective tissue**
Left ventricular hypertrophy	Degenerative joint disease
Metabolic syndrome	Gout
Varicose veins	Incisional hernia
Venous stasis disease	**Psychiatric**
Gastrointestinal	Depression
Gastroesophageal reflux	Anxiety
Cholelithiasis	**Malignancies**
Nonalcoholic steatohepatitis	Colorectal
Endocrine	Endometrial
Type II diabetes mellitus	
Hypothyroidism	

correlated significantly with sagittal abdominal diameter ($r = 0.6$, $r^2 = 0.36$, $p < 0.0001$). In obese patients with intra-abdominal pressure-related comorbidities, including obesity hypoventilation, gastroesophageal reflux, venous stasis, stress incontinence, and incisional hernia, bladder pressures were significantly elevated compared to obese patients without pressure-related comorbidity (19 ± 0.8 vs 15 ± 1.3, $P < 0.05$). The data suggest that central obesity elevates IAP, which then induces the comorbidities related to IAH.

Additional studies have demonstrated the relationship between elevated IAP and obesity. A significant correlation between IAP and BMI has been demonstrated.[7,8] IAP, measured transvaginally, transrectally, or transvesically, correlated strongly with BMI, with a correlation coefficient of 0.76 ($p<0.0001$) with transrectal measurements and a correlation coefficient of 0.71 ($p<0.0001$) for bladder pressures. McIntosh et al measured IAP during pressure flow studies in 100 consecutive men and demonstrated a correlation coefficient of 0.52 between BMI and IAP.[9]

Weight loss results in a reduction in IAP. In a prospective study of obese patients undergoing gastric bypass, significant reductions in sagittal abdominal diameter (32 ± 1 to 20 ± 2 cm, $p<0.001$) and bladder pressure (17 ± 2 to 10 ± 1 cm H_2O, $p < 0.001$) were observed one year following gastric bypass.[10] Patients lost $69 \pm 4\%$ of excess body weight. A decrease in the number of weight-related comorbidities per patient was also observed after gastric bypass (2.9 ± 0.4 to 1 ± 0.2). These data illustrate the relationship between obesity, abdominal pressure, and weight-related comorbidities.

Hemodynamic Alterations in Obesity

The deleterious effect of intra-abdominal hypertension (IAH) on the cardiovascular system has been evaluated in large animal models. Intra-abdominal hypertension creates reproducible disturbances in the hemodynamic system. An increase in IAP produces hypertension in a canine model.[11] An intraperitoneal balloon was progressively inflated weekly over four weeks to a pressure of 25 mm Hg above baseline, maintained at that pressure for two weeks, and then deflated over two weeks. With balloon inflation, systolic and diastolic blood pressure increased compared to controls, and returned to baseline with decompression. In a swine model using an intra-abdominal balloon, elevations in IAP caused a reduction in cardiac index and systemic vascular resistance index and an increase in central venous pressure and cardiac filling pressures.[12]

In central obesity, intra-abdominal hypertension results from an excess of intra-abdominal fat, which has been found to promote hypertension. In a study of obese women by Kanai et al, the amount of intra-abdominal fat was determined by calculating the ratio of the area of intra-abdominal visceral fat to subcutaneous fat by CT scan.[13] The authors found a correlation between intra-abdominal fat accumulation and hypertension. Compared to obese normotensive subjects, the obese hypertensive subjects had a significantly higher proportion of intra-abdominal visceral fat. The authors suggested that intra-abdominal fat accumulation may contribute to the pathogenesis of hypertension in obesity. Hayashi et al studied 563 Japanese-American subjects and measured intra-abdominal fat area by CT.[14] Intra-abdominal fat area was found to be a significant predictor of hypertension.

Data suggest that weight loss improves blood pressure by decreasing intra-abdominal fat. Kanai et al reported on the effect of diet-induced weight loss on blood pressure.[15] Obese hypertensive women were placed on a 12-week hypocaloric diet. The ratio of visceral to subcutaneous fat was significantly reduced with weight loss. Mean blood pressure fell from 112 +/- 9 to 101 +/- 12 mm Hg (p < .001). The change in mean blood pressure after weight loss did not correlate with the change in body weight or BMI, but did correlate with the reduction in visceral fat or ratio of visceral fat to subcutaneous fat. The authors concluded that in obese hypertensive subjects, a decrease in intra-abdominal fat, rather than simply body weight, reduced blood pressure.

The derangements in hemodynamic parameters in the morbidly obese are more pronounced with laparoscopic surgery as a result of the carbon dioxide pneumoperitoneum. A significant elevation in systolic blood pressure accompanied pneumoperitoneum in obese patients relative to controls, whose systolic pressure did not rise significantly.[16] Left ventricular wall stress was higher in obese patients at baseline and significantly increased with pneumoperitoneum. Therefore, obese patients are more susceptible to a compromise in myocardial oxygenation with pneumoperitoneum.

Whether the pathogenesis of hypertension is related to the mechanical pressure of visceral fat on the cardiovascular system leading to an activation of the renin-angiotension-aldosterone system, or it is the result of proinflammatory mediators released by the adipose tissue is unclear.

Pseudotumor Cerebri

Elevated IAP is associated with pseudotumor cerebri (PTC), or idiopathic intracranial hypertension, a condition that is most frequently seen in obese females. Clinical manifestations include debilitating headache, dizziness, nausea, vomiting, pulsatile tinnitus, and papilledema. Intracranial hypertension is present without evidence of hydrocephalus, mass lesion, structural, or vascular lesion. Sugerman and colleagues hypothesized that PTC is secondary to IAH.[17] They found that bladder pressure and sagittal abdominal diameter were significantly elevated in patients with elevated intracranial pressure compared with nonobese controls. The transesophageal pleural pressure, central venous pressure, mean pulmonary artery pressure, and pulmonary artery occlusion pressure were markedly higher than in obese patients without PTC. The data suggest that central obesity raises IAP with resultant increases in pleural pressure and cardiac filling pressure. Venous return from the brain is therefore impaired, leading to the increased intracranial pressure associated with PTC.

An additional study by Sugerman et al further supports the hypothesis that IAH causes PTC in obesity.[18] Seven morbidly obese women with PTC were treated with a negative

Table 2. Initial effect of ABSHELL™ on symptoms of pseudotumor cerebri in morbid obesity

	Headache	Tinnitus
Before ABSHELL™	6.8 ± 0.8	4.2 ± 0.5
5 min after ABSHELL™	4.2 ± 0.8*	1.8 ± 0.5**
1 h after ABSHELL™	2.2 ± 0.8†	1.7 ± 0.5†

* $p < 0.05$; ** $p < 0.02$; †$p < 0.01$. Reprinted with permission from: Sugerman HJ et al. Int J Obes Relat Metab Disord 2001; 25:486-90. ©2001 Nature Publishing Group.

abdominal pressure device designed to alleviate the effects of IAH. The device consisted of a fiberglass shell that was applied to the abdomen, vacuum tubing, and two vacuum pumps. Headache and pulsatile tinnitus resolved within five minutes (Table 2); relief lasted as long as the patients wore the devices. Symptoms gradually returned once the devices were removed. Patients who wore the device as they slept awoke without headache or tinnitus.

For more durable control of symptoms, surgically induced weight loss may reduce symptoms of PTC. Sugerman and colleagues demonstrated a reduction in headache and tinnitus in 18 out of 19 patients with PTC within four months of bariatric surgery.[19]

Intra-abdominal hypertension causes disturbances in cranial hemodynamics. Bloomfield et al reported significant increases in intracranial pressure and pleural pressure in a swine model of abdominal hypertension, while cerebral perfusion pressure decreased significantly.[12] The data suggest that an elevation in intra-abdominal pressure causes a rise in intrathoracic pressure, which induces an increase in intracranial pressure and a decrease in cerebral perfusion pressure.

Pulmonary Abnormalities and Obesity

The morbidly obese suffer from a higher risk of pulmonary complications. This may be explained by the adverse effects of IAH on the mechanical properties of the respiratory system and pulmonary hemodynamics in association with obesity. In the previously described intra-abdominal balloon experiment in swine by Bloomfield et al, significant increases in pleural pressure and pulmonary artery occlusion pressure were associated with IAH.[12] These findings are in concordance with the abnormalities seen clinically in morbidly obese patients with obesity hypoventilation syndrome (OHS), which is defined as an arterial oxygen tension ≤ 55 mg Hg and/or arterial carbon dioxide tension ≥ 47 mm Hg.

Obesity has adverse mechanical effects on the pulmonary system. Pelosi et al measured bladder pressure and pulmonary mechanics in 8 nonsmoking obese subjects[20] and compared these with data from a study of 16 nonobese subjects by D'Angelo et al.[21]

Functional residual capacity was lower in the obese subjects, while IAP was higher. Airway resistance and the alveolar-arterial oxygenation gradient were higher in the obese subjects. The data suggest that the unopposed elevation in abdominal pressure contributes to the reduction in lung volume and hypoxemia seen in obesity.

Intra-abdominal hypertension causes disturbances in pulmonary hemodynamics. Sugerman performed pulmonary artery catheterization in 46 morbidly obese patients, 26 with and 20 without OHS.[22] Mean pulmonary artery pressure and mean pulmonary artery occlusion pressures were significantly higher in the OHS patients compared to those without OHS. Despite mechanical ventilation and correction of hypoxemia and hypercarbia, which would reduce the effect of hypoxic pulmonary artery vasoconstriction, cardiac filling pressures remained elevated. Following surgically induced weight loss, pressures normalized.

Gastroesophageal Reflux

Gastroesophageal reflux disease (GERD) and hiatal hernia are prevalent among the morbidly obese. Factors that promote GERD include a short intra-abdominal lower esophageal sphincter (LES), reduced LES pressure, elevated intra-abdominal pressure, abnormal esophageal clearance, and impaired gastric emptying. The IAH associated with obesity alters the pressure gradient between the stomach and the gastroesophageal junction,[23,24] leading to displacement of the LES above the diaphragm. A sliding hiatal hernia results, which promotes GERD.[25] In a study of 345 patients with morbid obesity who underwent evaluation for GERD prior to undergoing gastric bypass, hiatal hernia was present in 52.6% and reflux esophagitis was diagnosed in 31.4%.[26] Twenty-four pH monitoring was abnormal in 73%. Ruhl and Everhart published the results of the National Health and Nutrition Examination Survey (NHANES I), in which approximately 12,000 patients were followed for a median of 18.5 years.[27] The data suggested that hospitalization for reflux was associated with a 5kg/m^2 increment in BMI.

Obese patients undergoing traditional antireflux operations have a higher recurrence rate than normal weight or overweight patients. Patients with documented GERD who had failed medical management were classified into groups based on their BMI: normal, overweight, and obese.[28] Following laparoscopic Nissen fundoplication or Belsey Mark IV, 31.3% of the obese patients had recurrent reflux, while only 4.5% of the normal group (p < 0.0001 vs obese) and 8% of the overweight group (p = 0.001 vs obese) recurred. The authors suggested that the increase in IAP in obesity contributes to the failure of antireflux surgery by loosening the crural repair and fundoplication.

In light of the failure of traditional antireflux surgery in obesity, Roux-en-Y gastric bypass has been recommended for treatment of GERD in this population.[29,30] In contrast, vertical banded gastroplasty is not only unsuccessful at controlling GERD,[31,32] but also exacerbates reflux.[33]

Genitourinary System

The pathogenesis of hypertension and renal dysfunction in obesity may be explained by IAH. In a porcine model, elevated IAP caused an elevation in renal venous pressure with a concomitant reduction of urine output, an increase in plasma renin and an increase in aldosterone.[34] Therefore, sodium and water are retained and vasoconstriction occurs. Furthermore, glomerulopathy and proteinuria may result from the increase in renal venous pressure.

Obesity is associated with stress urinary incontinence (SUI).[35] Abrupt increases in IAP from coughing or sneezing lead to involuntary leakage of urine. Bai et al compared 98 women with SUI and 102 control women without SUI.[36] The BMI in the SUI group was significantly higher compared to controls, but there was no correlation between BMI and urodynamic parameters. In Noblett's study of BMI and IAP, 48% had genuine stress urinary incontinence and 13% had severe prolapse.[37] The authors suggested that the etiology of stress urinary incontinence might be explained by the chronic elevation of IAP in obesity, which exerts stress on the pelvic floor.

Massive weight loss may result in an improvement in lower urinary tract function. Bump et al demonstrated a significant improvement in both objective and subjective measures of stress urinary incontinence following surgically-induced weight loss.[38] Only three of 12 patients with SUI prior to bariatric surgery complained of incontinence postoperatively. A significant reduction was seen in vesical pressure, the change in bladder pressure with coughing, the number of incontinence episodes, and the need for absorptive pads.

Incisional Hernias

Chronic intra-abdominal hypertension has been postulated to be a cause of incisional hernia in obesity. Sugerman et al published the results of a study comparing the incidence of incisional hernia following open gastric bypass for morbid obesity versus total colectomy with ileal pouch-anal anastomosis (IPAA) for ulcerative colitis.[39] Among the gastric bypass patients, 20% subsequently developed incisional hernia, compared to 4% of the IPAA patients (p<0.001). Five out of 7 IPAA patients with incisional hernia were obese (BMI = 30 kg/m^2). In contrast, only 1% of the nonobese IPAA patients developed an incisional hernia. In a study of 50 patients undergoing laparoscopic ventral hernia repair, Raftopoulos et al stratified the patients according to BMI.[40] The surface area of the hernia correlated with BMI.

Following incisional herniorrhaphy, obesity is an important predictor of hernia recurrence.[39,41,42] Sauerland et al found a recurrence rate increase of 1.10% per unit of BMI in a study of 160 patients undergoing open incisional herniorrhaphy.[43]

In Sugerman's hernia study, the rate of recurrent hernia was 41% among the gastric bypass patients.[39]

Conclusion

The comorbid illnesses of obesity adversely affect virtually every organ system. The pathogenesis of many of these conditions may be related to intra-abdominal hypertension. Weight loss results in resolution of the majority of these comorbidities.

Commentary

Michael L. Cheatham

Intra-abdominal hypertension (IAH) has been increasingly documented in recent years to occur in a wide-variety of critically ill patient populations. In this chapter, Drs. Hamad and Peitzman identify and describe a clinical scenario in which IAH may develop in a *nonacute*, chronic setting: the morbidly obese patient. As with IAH in the critically ill, elevated intra-abdominal pressure (IAP) in the morbidly obese patient can have far reaching effects on cerebral, cardiac, pulmonary, and renal physiology and function. As illustrated by the authors, disease processes common to the morbidly obese such as pseudotumor cerebri, obesity hypoventilation syndrome, gastroesophageal reflux, and stress urinary incontinence are now recognized as being caused, in large part, by the IAH incurred with an elevated body mass index (BMI). Further, the propensity to poor fascial healing and increased incisional hernia rates among the obese has been established as being due to IAH-induced reductions in abdominal wall and rectus sheath blood flow, pathophysiologic changes that, once recognized, can lead to outcome-altering therapeutic interventions.

This chapter raises three crucial points. First, the IAH-related complications of morbid obesity as outlined above generally respond to weight loss with resolution of patient symptoms and improved quality of life. Second, the morbidly obese are potentially at increased risk for developing symptomatic IAH and even abdominal compartment syndrome (ACS) as a result of preexisting baseline IAH and organ dysfunction. The morbidly obese patient may well develop IAH / ACS either earlier in a given disease process or following a physiologic insult of reduced severity compared to their nonobese counterpart. Clinicians should have a low threshold for measuring IAP in the obese patient in order to detect what could well be termed "silent IAH". Failure to recognize such elevations in IAP may lead to potentially preventable organ dysfunction and failure in such patients. Third, and perhaps most importantly, just as IAH/ ACS was originally erroneously considered to be a disease process afflicting only the traumatically injured, IAH/ACS can no longer be regarded as solely a disease of the critically ill. As illustrated in the case of the morbidly obese, IAH clearly occurs outside of the intensive care unit and even the hospital setting. As a result, the list of accepted risk factors for IAH must be reconsidered. Just as our understanding of the critically ill patient populations vulnerable to elevated IAP has evolved, so must our recognition of nontraditional patient populations such as the morbidly obese. These nonacute patients with silent IAH may well stand to gain as much from our diagnosis of their elevated IAP as have our critically ill patients.

References

1. Hedley AA, Ogden CL, Johnson CL et al. Prevalence of overweight and obesity among US children, adolescents, and adults, 1999-2002. JAMA 2004; 291:2847-2850.
2. Finkelstein EA, Fiebelkorn IC, Wang G. State-level estimates of annual medical expenditures attributable to obesity. Obes Res 2004; 12:18-24.
3. Sugerman HJ. Effects of increased intra-abdominal pressure in severe obesity. Surg Clin North Am 2001; 81:1063-1075.
4. Carr DB, Utzschneider KM, Hull RL et al. Intra-abdominal fat is a major determinant of the National Cholesterol Education Program Adult Treatment Panel III criteria for the metabolic syndrome. Diabetes 2004; 53:2087-2094.
5. Janssen I, Katzmarzyk PT, Ross R. Waist circumference and not body mass index explains obesity-related health risk. Am J Clin Nutr 2004; 79:379-384.
6. Sugerman H, Windsor A, Bessos M et al. Intra-abdominal pressure, sagittal abdominal diameter and obesity comorbidity. J Intern Med 1997; 241:71-79.
7. Sanchez NC, Tenofsky PL, Dort JM et al. What is normal intra-abdominal pressure? Am Surg 2001; 67:243-248.
8. Noblett KL, Jensen JK, Ostergard DR. The relationship of body mass index to intra-abdominal pressure as measured by multichannel cystometry. Int Urogynecol J Pelvic Floor Dysfunct 1997; 8:323-326.
9. McIntosh S, Drinnan M, Griffiths C et al. Relationship of abdominal pressure and body mass index in men with LUTS. Neurourol Urodyn 2003; 22:602-605.
10. Sugerman H, Windsor A, Bessos M et al. Effects of surgically induced weight loss on urinary bladder pressure, sagittal abdominal diameter and obesity comorbidity. Int J Obes Relat Metab Disord 1998; 22:230-235.
11. Bloomfield GL, Sugerman HJ, Blocher CR et al. Chronically increased intra-abdominal pressure produces systemic hypertension in dogs. Int J Obes Relat Metab Disord 2000; 24:819-824.
12. Bloomfield GL, Ridings PC, Blocher CR et al. A proposed relationship between increased intra-abdominal, intrathoracic, and intracranial pressure. Crit Care Med 1997; 25:496-503.
13. Kanai H, Matsuzawa Y, Kotani K et al. Close correlation of intra-abdominal fat accumulation to hypertension in obese women. Hypertension 1990; 16:484-490.
14. Hayashi T, Boyko EJ, Leonetti DL et al. Visceral adiposity and the prevalence of hypertension in Japanese-Americans. Circulation 2003; 108:1718-1723.
15. Kanai H, Tokunaga K, Fujioka S et al. Decrease in intra-abdominal visceral fat may reduce blood pressure in obese hypertensive women. Hypertension 1996; 27:125-129.
16. Prior DL, Sprung J, Thomas JD et al. Echocardiographic and hemodynamic evaluation of cardiovascular performance during laparoscopy of morbidly obese patients. Obes Surg 2003; 13:761-767.
17. Sugerman HJ, DeMaria EJ, Felton III WL et al. Increased intra-abdominal pressure and cardiac filling pressures in obesity-associated pseudotumor cerebri. Neurology 1997; 49:507-511.
18. Sugerman HJ, Felton 3rd WL, Sismanis A et al. Continuous negative abdominal pressure device to treat pseudotumor cerebri. Int J Obes Relat Metab Disord 2001; 25:486-490.
19. Sugerman HJ, Felton 3rd WL, Sismanis A et al. Gastric surgery for pseudotumor cerebri associated with severe obesity. Ann Surg 1999; 229:634-640; discussion 640-642.
20. Pelosi P, Croci M, Ravagnan I et al. Respiratory system mechanics in sedated, paralyzed, morbidly obese patients. J Appl Physiol 1997; 82:811-818.
21. D'Angelo E, Calderini E, Torri G et al. Respiratory mechanics in anesthetized paralyzed humans: Effects of flow, volume, and time. J Appl Physiol 1989; 67:2556-2564.
22. Sugerman HJ, Baron PL, Fairman RP et al. Hemodynamic dysfunction in obesity hypoventilation syndrome and the effects of treatment with surgically induced weight loss. Ann Surg 1988; 207:604-613.
23. Zacchi P, Mearin F, Humbert P et al. Effect of obesity on gastroesophageal resistance to flow in man. Dig Dis Sci 1991; 36:1473-1480.
24. Mercer CD, Wren SF, DaCosta LR et al. Lower esophageal sphincter pressure and gastroesophageal pressure gradients in excessively obese patients. J Med 1987; 18:135-146.
25. Barak N, Ehrenpreis ED, Harrison JR et al. Gastro-oesophageal reflux disease in obesity: Pathophysiological and therapeutic considerations. Obes Rev 2002; 3:9-15.
26. Suter M, Dorta G, Giusti V et al. Gastro-esophageal Reflux and esophageal motility disorders in morbidly obese patients. Obes Surg 2004; 14:959-966.
27. Ruhl CE, Everhart CE. Overweight, but not high dietary fat intake, increases risk of gastroesophageal reflux disease hospitalization: The NHANES i epidemiologic followup study. First national health and nutrition examination survey. Ann Epidemiol 1999; 9:424-435.

28. Perez AR, Moncure AC, Rattner DW. Obesity adversely affects the outcome of antireflux operations. Surg Endosc 2001; 15:986-989.
29. Jones Jr KB. Roux-en-Y gastric bypass: An effective antireflux procedure in the less than morbidly obese. Obes Surg 1998; 8:35-38.
30. Smith SC, Edwards CB, Goodman GN. Symptomatic and clinical improvement in morbidly obese patients with gastroesophageal reflux disease following Roux-en-Y gastric bypass. Obes Surg 1997; 7:479-484.
31. Di Francesco V, Baggio E, Mastromauro M et al. Obesity and gastro-esophageal acid reflux: Physiopathological mechanisms and Role of gastric bariatric surgery. Obes Surg 2004; 14:1095-1102.
32. Ortega J, Escudero MD, Mora F et al. Outcome of esophageal function and 24-hour esophageal pH monitoring after vertical banded gastroplasty and roux-en-Y gastric bypass. Obes Surg 2004; 14:1086-1094.
33. Schauer P, Hamad G, Ikramuddin S. Surgical management of gastroesophageal reflux disease in obese patients. Semin Laparosc Surg 2001; 8:256-264.
34. Bloomfield GL, Blocher CR, Fakhry IF et al. Elevated intra-abdominal pressure increases plasma renin activity and aldosterone levels. J Trauma Inf Crit Care 1997; 42:997-1005.
35. Dwyer PL, Lee ETC, Hay DM. Obesity and urinary incontinence in women. Br J Obstet Gynecol 1998; 95:91-96.
36. Bai SW, Kang JY, Rha KH et al. Relationship of urodynamic parameters and obesity in women with stress urinary incontinence. J Reprod Med 2002; 47:559-563.
37. Noblett KL, Jensen JK, Ostergard DR. The relationship of body mass index to intra-abdominal pressure as measured by multichannel cystometry. Int Urogynecol J Pelvic Floor Dysfunct 1997; 8:323-326.
38. Bump RC, Sugerman H, Fantl JA et al. Obesity and lower urinary tract function in women: Effect of surgically induced weight loss. Am J Obstet Gynecol 1992; 166:392-399.
39. Sugerman HJ, Kellum Jr JM, Reines HD. Greater risk of incisional hernia with morbidly obese than steroid-dependent patients and low recurrence with prefascial polypropylene mesh. Am J Surg 1996; 171:80-84.
40. Raftopoulos I, Vanuno D, Khorsand J et al. Outcome of laparoscopic ventral hernia repair in correlation with obesity, type of hernia, and hernia size. J Laparoendosc Adv Surg Tech 2002; 12:425-429.
41. Anthony T, Bergen PC, Kim LT et al. Factors affecting recurrence following incisional herniorrhaphy. World J Surg 2000; 24:95-101.
42. Langer C, Schaper A, Liersch T. Prognosis factors in incisional hernia surgery: 25 years of experience. Hernia 2004, [Epub ahead of print].
43. Sauerland S, Korenkov M, Kleinen T et al. Obesity is a risk factor for recurrence after incisional hernia repair. Hernia 2004; 8:42-46.

CHAPTER 16

Miscellaneous Conditions and Intra-Abdominal Hypertension

Ari Leppäniemi, Andrew Kirkpatrick, Anastazia Salazar, Davis Elliot, Savvas Nicolaou and Martin Björck

Intra-abdominal hypertension and the abdominal compartment syndrome are increasingly recognized in non-traumatic conditions in the critically ill patient. This chapter deals with three of those situations: acute pancreatitis, renal transplant and abdominal aortic aneurysm (the condition in which some of the very early cases of ACS were described).

Part A: Severe Acute Pancreatitis

Ari Leppäniemi*

Abstract

Under-diagnosed and untreated abdominal compartment syndrome (ACS) is a potential contributing factor to the development of early organ failure seen in patients with severe acute pancreatitis, and warrants routine measurement of intra-abdominal pressure in all patients treated for severe pancreatitis. The current estimate of the prevalence of intra-abdominal hypertension (IAH) in severe acute pancreatitis is about 40%, with about 10% overall progressing to ACS associated with increased hospital mortality rates. In the majority of cases, the development of IAH is rapid and mainly due to the combined effects of aggressive fluid resuscitation and the inflammatory process in the retroperitoneum leading to the development of visceral edema and pancreatic ascites within days or even hours from admission, although in some cases a delayed form of ACS has been associated with the emergence of infected peripancreatic necrosis. Percutaneous drainage of large amounts of pancreatic ascites may decrease the intra-abdominal pressure considerably and is the first line treatment if appropriate. In most cases, however, surgical decompression through a vertical midline incision without exploring the pancreas further is the most effective and safest procedure. Decompression performed 2-3 weeks after the onset of the disease can be combined with necrosectomy. Primary fascial closure of the abdominal wall following abdominal decompression can be attempted, but in most cases the prolonged inflammatory process in the abdomen and the risk of recurrent ACS favors the use of gradual closure or delayed reconstruction of the abdominal wall.

*Ari Leppäniemi—Department of Surgery, Meilahti Hospital, University of Helsinki, Finland. Email: ari.leppaniemi@hus.fi

Abdominal Compartment Syndrome, edited by Rao R. Ivatury, Michael L. Cheatham, Manu L. N. G. Malbrain and Michael Sugrue. ©2006 Landes Bioscience.

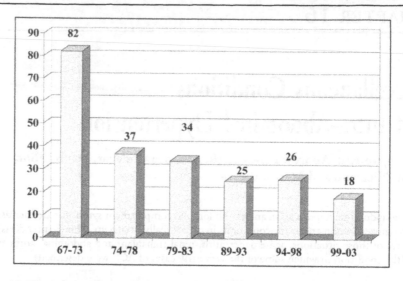

Figure A1. Hospital mortality rate (%) in patients treated for severe acute pancreatitis at the Meilahti Hospital, University of Helsinki, Finland in 1967-2003. (Halonen and Leppäniemi 2004, unpublished data.)

Introduction

The mortality rates for severe acute pancreatitis have shown steady decrease from 50-58% in 1978-1982 to 12-18% in 1993-1997.[1] Aggressive fluid resuscitation in the early phase, prophylactic antibiotic treatment, more accurate indications and timing of surgical intervention, advances in the monitoring and management of organ dysfunctions, increased use of enteral nutrition and early endoscopic sphincterotomy in patients with common bile duct stone-induced pancreatitis have been major factors contributing to improved survival. In addition to factors characterizing patient reserves, such as age or previous cardiovascular medication, the development of multiple organ dysfunction or failure is the major determinant of poor outcome.[2] Recent data from the Meilahti Hospital, University of Helsinki, Finland show that although the mortality rates in patients with severe acute pancreatitis and multiple organ failure (MOF) have improved considerably in the last 15 years (Fig. A1), the cumulative mortality charts (Fig. A2) show that there is little improvement in the early (within 14 days) mortality rate (Halonen and Leppäniemi, unpublished data, 2004). There is increasing clinical evidence that a major part of the deaths in the early phase of the disease, previously thought to be caused by an overwhelming acute inflammatory reaction leading to "early MOF", is associated with undiagnosed and untreated abdominal compartment syndrome (ACS).[3]

The presence of intra-abdominal hypertension (IAH) can also be used as a predictor of the severity of acute pancreatitis. In a study of 45 patients with acute pancreatitis using intra-abdominal pressure (IAP) levels higher than 16 mm Hg as a cut off point, there was a correlation with the increased IAP and severity of pancreatitis, mortality, peripancreatic infection rate, and need for surgical intervention.[4]

In 2002, the International Association of Pancreatology developed evidence-based guidelines for the surgical management of acute pancreatitis.[5] Of the 11 recommendations, 10 were grade B and one was grade A. The essence of the recommendations involved the indications for drainage and/or necrosectomy more than 2 weeks after the onset of the disease in patients with fine-needle aspiration biopsy proven infected necrosis in patients with sepsis syndrome. Although there was no mention of abdominal compartment syndrome, recommendation six encompasses all indications indicating early surgery, such as major hemorrhage, intestinal necrosis or perforation: "Early surgery within 14 days after onset of disease is not recommended in patients with necrotizing pancreatitis unless there are specific indications (recommendation

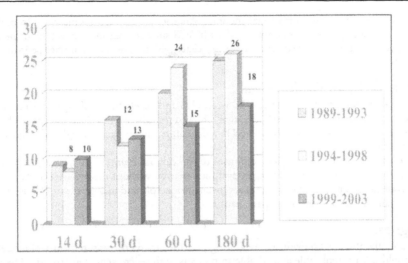

Figure A2. Cumulative mortality rate (%) at 14, 30, 60 and 180 days post-admission in patients treated for severe acute pancreatitis at the Meilahti Hospital, University of Helsinki, Finland in 1989-2003. (Halonen and Leppäniemi 2004, unpublished data.)

grade B)". Rapidly accumulating evidence supports the inclusion of early monitoring and treatment of abdominal compartment syndrome in the management algorithms of patients with severe acute pancreatitis.

Prevalence

The true prevalence of IAH in patients with severe acute pancreatitis is not known. In a study of 41 patients with severe acute pancreatitis, 44% of the patients had IAP levels higher than 12 mm Hg, and 4 patients (10%) had IAP levels higher than 25 mm Hg with severe organ dysfunction and undergoing abdominal decompression.[6] In another study of 37 patients treated in the ICU for severe acute pancreatitis, 10 patients (27%) had IAP levels higher than 25 mm Hg, with an overall frequency of 10/120 (8%) if all patients with severe acute pancreatitis treated in other units (surgical ICU, high dependency unit) were included.[7] A study of 297 patients treated for severe acute pancreatitis in China showed that the overall incidence of IAH (defined as IAP higher than 15 mm Hg) was 36%.[3]

Thus, it can be estimated that the overall prevalence of IAH in patients with severe acute pancreatitis is about 40%, and the frequency of ACS requiring surgical decompression about 10%.

Due to the more aggressive fluid resuscitation policy in the early phase of severe acute pancreatitis, and the decreasing proportion of patients requiring necrosectomy, it is likely that the incidences of both early and late onset ACS, respectively, are higher today than in the past.

Severe acute pancreatitis is one of the most common diseases associated with IAH in the ICU-environment. Among 18 patients treated in a surgical ICU in New York with documented ACS, the underlying condition was severe acute pancreatitis in 3 patients (17%).[8]

Time of the Development of ACS

The development of ACS in patients with severe acute pancreatitis occurs most commonly in the early course of the disease and is probably caused by the combined effects of the aggressive fluid resuscitation and inflammatory process in the retroperitoneum leading to the development of visceral edema and pancreatic ascites.[9,10] Among 297 patients treated in China, those who developed organ dysfunction within 72 hours after the onset of symptoms had a 78% incidence of IAH (>15 mm Hg) compared to 23% in patients with severe acute pancreatitis without early organ dysfunction.[3] Another study from China with 23 patients with severe acute pancreatitis complicated with ACS showed that in 6 cases (26%) the ACS developed at a later stage of the disease and was associated with the presence of infected peripancreatic necrosis.[11]

ACS and Organ Dysfunction

Several studies show the association between IAH and development of organ dysfunction in severe acute pancreatitis.[6,7,12] De Waele et al[6] showed that there was a 94% incidence of respiratory failure, 94% cardiovascular and 89% renal failure rate in patients with IAP higher than 12 mm Hg.

Because the pathophysiological responses to the cellular events early in the course of the disease have the ability to induce organ dysfunction even without the presence of ACS, more studies are needed to characterize the exact mechanisms, role and magnitude of ACS leading to early organ failure in severe acute pancreatitis.

Diagnosis

Currently, there is no evidence suggesting that the standard measurement techniques for IAP could not be applicable and reliable in patients with severe acute pancreatitis. The most commonly used technique is the bladder pressure measurement through a Foley catheter.[7,12]

Due to the nature of the development of ACS in patients with severe acute pancreatitis with other potential causes for organ dysfunctions, the routine measurement of IAP in all patients with severe acute pancreatitis is warranted.

Treatment

The treatment of ACS in patients with severe acute pancreatitis is based on the recognition of the principal cause of the IAH. If an ultrasound examination confirms the presence of large volumes of pancreatic ascites, the first line treatment would be the percutaneous drainage of the intraperitoneal exudate which can lead to a significant drop in IAP (Lisi et al 2004, abstract). In most cases, however, visceral edema is the principal factor contributing to the development of ACS, and decompressive laparotomy and temporary abdominal closure with a Bogota bag or equivalent is the most effective way of decreasing the IAP.[3,6,9,11,13]

During the last 10 years, the proportion of patients with infected necrosis or other indications for surgical necrosectomy in severe acute pancreatitis has dropped from 30% to about 15% at the Meilahti Hospital, University of Helsinki, Finland (Fig. A3) (Halonen and Leppäniemi, unpublished data 2004). When needed, the necrosectomy is performed through a bilateral subcostal incision. Considering the small proportion of patients eventually requiring necrosectomy and the (intuitively) inadequate decompression provided by a transverse incision, it is probably better to perform the early decompressive laparotomy through a long, vertical midline incision. Because of the unripe nature of the (sterile) necrosis and the risk of introducing infection to the peripancreatic space, there are no indications to explore the pancreas further or open the gastrocolic ligament at this stage but limit the operation to decompression and drainage of ascites only (Gonzales Santamaria et al 2004, abstract). Moreover, premature exploration can lead to fatal retroperitoneal bleeding.[6]

If the decompression is done more than 2-3 weeks after the onset of the disease and there is evidence of extensive necrosis on a CT scan or established infection of the peripancreatic necrosis, it is the policy of the author to perform a necrosectomy in conjunction with the decompressive laparotomy.

The management of the open abdomen following decompression in severe acute pancreatitis is a challenge. There are some isolated reports of primary closure of the abdominal wall within a week from decompression, and that should be the aim in all cases.[9,11] However, unlike in most cases with abdominal trauma, especially following damage control surgery with intra-abdominal packing, or other conditions where the initial event leading to ACS can be

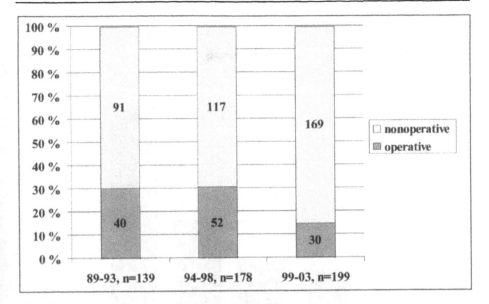

Figure A3. Proportion of patients undergoing operative or nonoperative treatment for severe acute pancreatitis at the Meilahti Hospital, University of Helsinki, Finland in 1989-2003. (Halonen and Leppäniemi 2004, unpublished data.)

rapidly reversed with aggressive management of the fluid balance, the inflammatory process in severe acute pancreatitis usually continues for several weeks prohibiting the early closure of the abdominal wall due to the risk of developing recurrent ACS. The best currently available technique and the one used by the author is the utilization of the vacuum-assisted closure technique aiming for gradual closure of the abdominal wall. If not possible, skin grafting over the bowel and late abdominal wall plastic surgical reconstruction 9-12 months later is the safest option, although associated with considerable morbidity and discomfort to the patient.

Mortality

In a study comparing patients with or without ACS (IAP>25 mm Hg) treated in the ICU for severe acute pancreatitis, the hospital mortality rate for patients with ACS was 50% compared with 15% in patients without ACS.[7] There was a clear correlation between the maximum IAP value within the first 2 weeks and the mortality rate (Fig. A4). In a logistic regression analysis, however, only the maximal SOFA (Sequential Organ Failure Assessment) score was an independent risk factor for hospital mortality.

Although some of the reported survival rates of patients undergoing abdominal decompression for ACS in severe acute pancreatitis are low, 1/3,[13] 1/4,[6] respectively, early decompression is associated with better outcome than delayed decompression or none at all. Among 23 patients with ACS and severe acute pancreatitis, 18 patients underwent decompression with a mortality rate of 3/18 (17%), whereas 4/5 (80%) of the nonoperatively treated patients died. Moreover, there were no deaths among patients who underwent emergency abdominal decompression within 5 hours after the confirmation of the presence of ACS.[11]

References

1. Bank S, Singh P, Pooran N et al. Evaluation of factors that have reduced mortality from acute pancreatitis over the past 20 years. J Clin Gastroenterol 2002; 35:50-60.
2. Halonen K, Leppäniemi A, Puolakkainen P et al. Severe acute pancreatitis – prognostic factors in 270 consecutive patients. Pancreas 2000; 21:266-271.

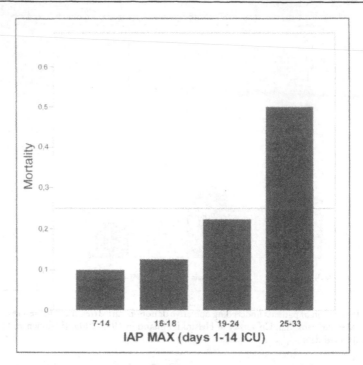

Figure A4. Maximum IAP value and mortality rate in 37 patients treated in the ICU for severe acute pancreatitis.[7]

3. Tao H-Q, Zhang J-X, Zou S-C. Clinical characteristics and management of patients with early acute severe pancreatitis: Experience from a medical center in China. World J Gastroenterol 2004; 10:919-921.
4. Navarro S, Hidalgo JM, Bejarano N et al. Intra-abdominal pressure (IAP) as an indicator of severity in acute pancreatitis. Inaugural World Congress Abdominal Compartment Syndrome, Noosa, Australia, 2004.
5. Uhl W, Warshaw A, Imrie C et al. IAP Guidelines for the surgical management of acute pancreatitis. Pancreatology 2002; 2:565-573.
6. De Waele J, Hoste E, Blot S et al. Intraabdominal hypertension and severe acute pancreatitis. Inaugural World Congress Abdominal Compartment Syndrome, Noosa, Australia, 2004.
7. Keskinen P, Leppäniemi A, Pettilä V et al. Intra-abdominal pressure in acute necrotizing pancreatitis. Inaugural World Congress Abdominal Compartment Syndrome, Noosa, Australia, 2004.
8. Mcnelis J, Soffer S, Marini CP et al. Abdominal compartment syndrome in the surgical intensive care unit. Am Surg 2002; 68:18-23.
9. Gonzales Santamaria JR, Perez Garcia R, Navarro Ramirez J et al. Handling of the abdominal compartmental syndrome in a patient with serious acute pancreatitis of alcohol etiology: report of a case. Inaugural World Congress Abdominal Compartment Syndrome, Noosa, Australia 2004.
10. Lisi M, Sne B, Baillie F et al. Computed tomography guided percutaneous drainage of the abdominal cavity is a safe alternative to celiotomy in patients with abdominal compartment syndrome due to severe phlegmonous pancreatitis: a case report. Inaugural World Congress Abdominal Compartment Syndrome, Noosa, Australia, 2004.
11. Ao J, Wang C, Chen L et al. Diagnosis and management of severe acute pancreatitis complicated with abdominal compartment syndrome. Journal of Huazhong University of Science and Technology. Medical Sciences 2003; 23:399-402.
12. Snippe K, Plaudis H, Pupelis G. Increased intra-abdominal pressure, is it of any consequence in severe acute pancreatitis? Inaugural World Congress Abdominal Compartment Syndrome, Noosa, Australia, 2004.
13. Gecelter G, Fahoum B, Gardezi S et al. Abdominal compartment syndrome in severe acute pancreatitis: an indication for a decompressive laparotomy? Dig Surg 2002; 19:402-4.

Part B: The Renal Allograft Compartment Syndrome in Perspective: An Organ Specific Compartment Syndrome with Illustrative Pathophysiology

Andrew Kirkpatrick,* Anastazia Salazar, Davis Elliot and Savvas Nicolaou

Abstract

Renal allograft compartment syndrome (RACS) is a specific entity described in the transplanted kidney due to closure of fascia with excessive force and may be predisposed to by large donors and small recipients, and adult kidneys transplanted in children. It is often relieved by opening the compartment. Diagnosis is by sonography of the allograft. This section will discuss this special variant of abdominal compartment syndrome.

Introduction

Abdominal compartment syndrome (ACS) affects every human organ system, but is recognized clinically as cardiac, respiratory, and renal dysfunction.[1-3] Although first described over 100 years ago, these adverse effects, to which the kidneys are particularly susceptible, have only recently been widely appreciated, in no small measure due to the works[4-7] of Michael Sugrue, one of the editors of this volume. IAH has been statistically associated with increased renal dysfunction, multi-system organ failure, and death.[5,7] Untreated, the ACS invariably leads to death in it's fully developed form,[8] and it is assumed that either prophylaxis or earlier intervention to reduce IAH might improve the previously dismal outcomes of the established ACS.[9-10] Routine IAP measurements are now easily obtained and should be performed in all critically ill patients.[3,11] Raised intra-abdominal pressure (IAP) may begin to affect renal function at only 10 mm Hg and markedly impairs renal function at pressures as low as 20 mm Hg.[12,13] Critically ill populations may have a 50% incidence of raised IAH (IAP > 12 mm Hg) though.[14] As renal function has not been shown to improve with decompression of the hypertensive peritoneal cavity,[6] other markers are needed to identify which patients with borderline IAH will go on to develop the ACS from those who will not, and thus could be spared the open abdomen.

The Renal Allograft Compartment Syndrome

A specific renal allograft compartment syndrome (RACS) has recently been recognized, adding another specific syndrome to our ever increasing knowledge of IAH and ACS.[15-18] It is suspected intra-operatively during renal transplant if the fascial closure requires excessive force, or results in graft turgor/color change or diminished renal artery pulsation. All cases diagnosed postoperatively have been on the basis of poor physiologic function (urine production) and abnormal Doppler US examinations rather than thorough measurement of intraperitoneal or compartment pressures. Without early recognition and reexploration, graft loss is felt inevitable.[17] It may be predisposed to by large donors and small recipients, and especially when an adult kidney is donated to a child.[15-18] In the absence of pressure measurements, the basis for classifying these cases as true compartment syndromes rather than cases of renal vein thrombosis or vascular kinking has been the absence of these findings on re-exploration. In addtion, vascular normalization has ensued upon opening of the compartment.[15,17] Further study will be needed to clarify whether the RACS is a localized ACS syndrome or a true generalized intra-abdominal phenomenon. If determined to be truly intra-abdominal the condition should be classified as either a specific secondary ACS syndrome, as the site of the raised pressure is

*Corresponding Author: Andrew Kirkpatrick—Departments of Surgery, Critical Care Medicine, Radiology, Vancouver Hospital & Health Sciences Centre, Vancouver, British Columbia, Canada. Email: andrew.kirkpatrick@calgaryhealthregion.ca

technically, extraperitoneal, but intra-abdominal.[19,20] The relationship of RACS to the Page kidney syndrome, wherein chronic extrinsic compression of the renal parenchyma causes unilateral renal ischemia and increased renin release with potential systemic hypertension,[21,22] is currently unknown.

All cases diagnosed postoperatively have been suspected on the basis of poor graft function, and confirmed through postoperative Doppler US demonstrating reversed diastolic flow in the interlobar and segmental arteries and absent venous flow.[15,16] Of four cases diagnosed postoperatively in London, Ontario three underwent early reexploration with graft preservation, while one with a delayed reexploration was lost. In cases of graft survival opening of the fascia led to immediately improved visual and Doppler findings. Treatment of the RACS may include the use of a prosthetic mesh, skin-closure only, or intraperitoneal relocalization of the graft.[17,18] At our centre, out of 458 renal transplants, there were 11 diagnoses of compartment syndromes (2.4% incidence). Six cases were diagnosed intra-operatively based on observation of allograft perfusion. There were five postoperative diagnoses based on graft dysfunction, corroborated by the vascular parameters of Doppler ultrasound. All cases were re-explored. Three extraperitoneal grafts were relocated to the peritoneal cavity, and in two cases the fascia was left open with skin-closure only. In all instances graft function was recovered and no grafts were lost. From this limited experience with a few cases of RACS important sonographic principles may be speculated upon. It has been simply stated that, "reversal of diastolic flow in the intrarenal arteries or absence of flow in the renal veins may be a sign of intra-abdominal compartment syndrome."[15] Might this simple bed-side technology offer a non-invasive marker of kidneys at risk from the ACS?

Sonography of the Renal Allograft

There is great clinical experience with renal sonography in the transplant kidney, with sonographic examination being a basic technique of examining graft health and function. Evaluation of the course of the main renal arteries in their entirety is usually not feasible, although the intrarenal vasculature can be detected in virtually all patients.[23] A tardus-parvus waveform with slowed systolic acceleration and low amplitude is sought as a distal sign of proximal obstruction. Renal vascular indices are commonly calculated after transplant. Although it does not discriminate the actual pathology, the finding of diminished, absent, or reversed diastolic flow in renal allografts almost always indicates pathologically increased renovascular impedance, except when severe hypotension is present.[24,25] The resistive index (RI) is typically calculated from the interlobar or arcuate arteries and thus does not sample true cortical perfusion in the cortical vessels (interlobular arteries and venules) beyond.[26,27] RI's in the range of 0.6–0.8 are considered normal, 0.8–0.9 are equivocal, and greater than 0.9 suggest increased vascular resistance[23] (Fig. B1). While the RI appears to be a useful tool to discern the overall vascular health of a renal graft, it is not necessarily sensitive to anomalies in cortical perfusion.[26] Causes of an increased RI include rejection, acute tubular necrosis, cyclosporine toxicity, ureteral obstruction, renal vascular complications, and external compression of the transplant.[23] The renal cortical index (RCI) is a measure of the relative decrease in individual resistive indices from the renal artery to the interlobar arteries, compared to the deriving arcuate arteries of the kidney in question, which may better reflect cortical function.[26] The RCI was even more predictive than the RI alone, in predicting acute dysfunction despite confusing biochemical and clinical pictures within 24 hours of grafting.[26]

Is There a Potential Utility in the Native Kidney Subjected to Raised Intra-Abdominal Pressure?

The pathophysiologic mechanisms responsible for impaired renal function with raised intraperitoneal pressure are a multifactorial combination of reduced renal blood flow, renal vein and parenchymal compression, elevated renal tubular pressure, and endocrine alterations.[28] As volume loading to increase cardiac output can often precipitate ACS and no primary therapies

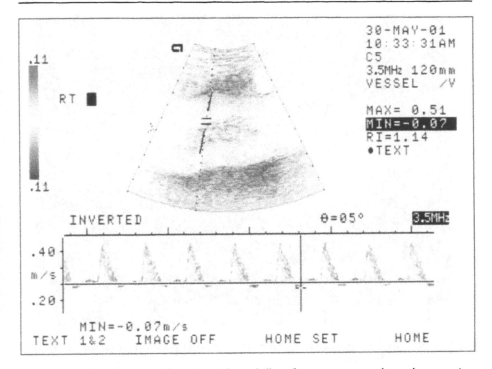

Figure B1. Renal duplex image of a patient with renal allograft compartment syndrome demonstrating reversal of diastolic flow in the interlobar renal arteries.

are available for reversing the endocrinological abnormalities of renal therapy, early diagnosis of renal perfusion abnormalities is required.[12,28-31] Resistive indices measured with Duplex sonography show very close correlation when intraperitoneal pressure is raised from 0 mm Hg to 30 mm Hg in anesthetized swine (Figs. B2-B4). To date though, there have been no clinical reports regarding the utility of renal sonography in evaluating native kidneys subjected to potentially pathologic IAP.

The Reality of the ACS Patient and Other Potential Technologies

The cohort of patients at risk for the abdominal compartment syndrome are typically massively resuscitated, edematous and may have intra-abdominally packs. They are too unstable to transport to ultrasound suites, constituting patients often unsuitable for sonographic study. These practicalities question whether other strategies or related technologies might be required to study the viability and reserve of the critically ill kidney. Color Power Doppler (CPD) is another ultrasound technology that is superior to regular Doppler in determining the presence or absence of flow at the expense of directional information, with improved ability to identify low-velocity and low-volume flow (or motion) and has less angle dependence.[32-37] CPD sonography displays data influenced by the number of red blood cells moving in the region of interest and can produce a perfusion index for each pixel.[38] Thus, CPD is felt to be superior to conventional color Doppler in the demonstration of the normal intrarenal vessels.[33,37,39] Trillaud and colleagues compared power Doppler measurement of interlobar artery resistive index measurement with CPD imaging of the cortical vasculature and described a correlation with early CPD and functional graft recovery at 12 months not seen with color Doppler sonography.[27] CPD with high-frequency probes improve the assessment of the cortical microvascularity

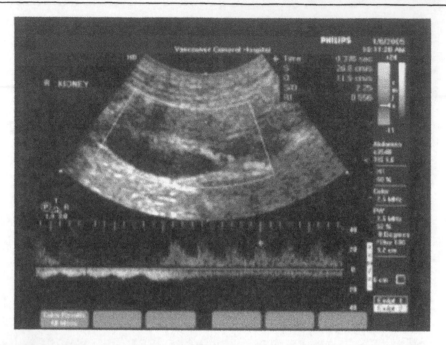

Figure B2. Renal duplex image of a porcine kidney subjected to an intra-abdominal pressure of 0 mm Hg. Resistive index = 0.556 (see text).

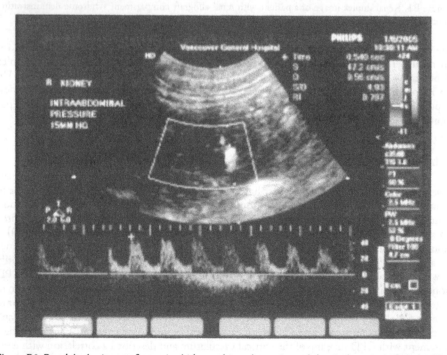

Figure B3. Renal duplex image of a porcine kidney subjected to an intra-abdominal pressure of 15 mm Hg. Resistive index = 0.797 (see text).

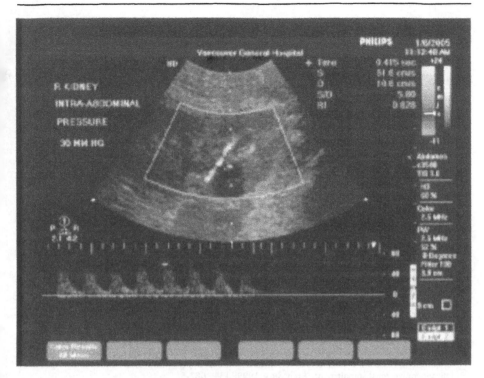

Figure B4. Renal duplex image of a porcine kidney subjected to an intra-abdominal pressure of 30 mm Hg. Resistive index = 0.828 (see text).

increasing the rate of detection for cortical defects.[37] Bude compared conventional color Doppler sonography to power Doppler and noted that CPD demonstrated blood flow to the cortex in 90% of normal kidneys, while conventional color Doppler sonography did not depict the cortical flow in any.[33] They speculated that flow depiction with CPD was the summation of all flow in the interlobular vessels that cannot be depicted with color Doppler sonography.[33] Takano performed power-Doppler sonography of human kidneys during voluntary Valsalva maneuvers which raised intra-abdominal pressures from 0 to 100 mm Hg, in 20 mm Hg increments.[40] Further, power-Doppler studies demonstrated increased mean flow intensities in human kidneys after IAP had been decreased from 22 to 16 mm Hg by removing 1650 mL of fluid by paracentesis.[40] Investigators from the US Army Institute of Surgical Research evaluated the color power Doppler image intensity signals generated from the renal cortex during varying degrees of renal vascular occlusion ranging from none to 100%.[38] They were evaluating the rationale that urine output is presumed to represent renal cortical perfusion when it is used as a resuscitation endpoint.[38] They found that cortical blood flow correlated well with CPD image intensity when measured both electronically and subjectively.[38] Newer evolving technologies include both 3-D and 4-D US and may further overcome the difficult abdominal wall. Obstetrical indications have often driven sonographic advances. One-hundred and six (106) healthy fetuses were evaluated with 3-D CPD to asses fetal renal blood flow, by looking at the vascularization index, flow index, and a vascularization-flow index.[41] The deceptively simple finding of normal fetal renal vasculature and its increase with advancement of gestational age offers the suggestion that this technology can be applied to other indications. Other advancements in ultrasound technology involve the use of contrast media. Gas-filled microbubbles that remain entirely within the vascular tree and that have rheology similar to red blood cells can be used as sonographic contrast media.[42,43] These can be introduced through a femoral venous line, demonstrating excellent correlation (r = 0.82, p < 0.82) over a 2.5 fold range of renal blood flow changes in swine.[42]

Future Directions

The early appreciation of RACS should be facilitated by a greater awareness of the condition, supplemented by the liberal use of renal sonography. Both intracompartmental and intra-peritoneal pressures should be measured and correlated with graft function. It is likely that correlation of the sonographic findings of the transplant kidney subjected to increased pressure will lead to both a physiologic and technical appreciation of the sonographic features of the native kidney subjected to similar adversity.

References

1. Schein M, Wittman DH, Aprahamian CC et al. The abdominal compartment syndrome: the physiolgical and clinical consequences of elevated intra-abdominal pressure. J Am Coll Surg 1995; 180:745-52.
2. Nathens AB, Brenneman FD, Boulanger BR. The abdominal compartment syndrome. Can J Surg 1997; 40:254-8.
3. Kirkpatrick AW, Brenneman FD, McLean RF et al. Is clinical examination an accurate indicator of raised intra-abdominal pressure in critically injured patients. Can J Surg 2000; 43:207-11.
4. Sugrue M, Buist MD, Hourihan F et al. Prospective study of intra-abdominal hypertension and renal function after laparotomy. 1995; 82:235-8.
5. Sugrue M, Jones F, Lee A et al. Intraabdominal pressure and gastric intramucosal pH: Is there an association? World J Surg 1996; 20:988-91.
6. Sugrue M, Jones F, Janjua KJ et al. Temporary abdominal closure: A prospective evaluation of its effects on renal and respiratoy physiology. J Trauma 1998; 45:914-21.
7. Sugrue M, Jones F, Deane SA et al. Intra-abdominal hypertension is an independant cause of postoperative renal impairment. Arch Surg 1999; 134:1082-5.
8. Burch JM, Moore EE, Moore FA et al. The abdominal compartment syndrome. Surg Clin N Amer 1996; 76:833-42.
9. Maxwell RA, Fabian T, Croce M et al. Secondary abdominal compartment syndrome: an underappreciated manifestation of severe hemorrhagic shock. J Trauma 1999; 47:995-9.
10. Ivatury RR, Porter JM, Simon RJ et al. Intra-abdominal hypertension after life-threatening penetrating abdominal trauma: prophylaxis, incidence, and clinical relevance to gastric mucosal pH and abdominal compartment syndrome. J Trauma 1998; 44:1016-23.
11. Malbrain ML. Intra-abdominal pressure in the intensive care unit: clinical tool or toy? Vincent JL, ed. Yearbook of Intensive Care and Emergency Medicine. Berlin: Springer-Verlag, 2002:792-814.
12. Harman PK, Kron IL, McLachlan HD et al. Elevated intra-abdominal pressure and renal function. 1982; 196:594-7.
13. Herman FG, Winton RF. The interaction of intrarenal and extrarenal pressures. J Physiol 1936; 87:77P-8P.
14. Malbrain MLNG, Chiumello D, Pelosi P et al. Prevalence of intra-abdominal hypertension in critically ill patients: a multicentre epidemiological study. Intensive Care Med 2004; 30:822-9.
15. Wiebe S, Kellenberger CJ, Khoury A et al. Early Doppler changes in a renal transplant patient secondary to abdominal compartment syndrome. Pediatr Radiol 2004; 34:432-4.
16. Humar A, Sharpe J, Hollomby D. Salvage of a renal allograft with renal vein occlusion secondary to extrinsic compression. Am J Kidney Dis 1996; 28:622-3.
17. Beasley KA, McAlister VC, Luke PPW. Mesh hood fascial closure in renal allograft compartment syndrome. Transplant Proc 2003; 35:2418-9.
18. Borowicz MR, Hanevold CD, Cofer JB et al. Extrinsic compression in the iliac fossa can cause renal vein oclusion in pediatric kidney recipients but graft loss can be prevented. Transplant Proc 1994; 26:119-20.
19. Balogh Z, McKinley BA, Cocanour CS et al. Secondary abdominal compartment syndrome is an elusive early complication of traumatic shock resuscitation. Am J Surg 2002; 184:538-44.

20. Balogh Z, McKinley BA, Holcomb JB et al. Both primary and secondary abdominal compartment syndrome can be predicted early and are harbingers of multiple organ failure. J Trauma 2003; 54:848-61.
21. McCune TR, Stone WJ, Breyer JA. Page kidney: Case report and review of the literature. Am J Dis Kidney 1991; 5:593-9.
22. Engel WJ, Page IH. Hypertension due to renal compression resulting from subcapsular hematoma. J Urol 1955; 73:735-9.
23. Thurston W, Wilson SR. The urinary tract. In: Rumack CM, Wilson SR, Charboneau JW, eds. Diagnostic Ultrasound. 2nd ed. St. Louis: Mosby, 1998:329-97.
24. Kelcz F, Pozniak MA, Pirsch JD et al. Pyramidal appearance and resistive index: Insensitive and nonspecific sonographic indicators of renal transplant rejection. AJR 1990; 155:531-5.
25. Taylor KJW, Marks WH. Use of Doppler imaging for evaluation of dysfunction in renal allografts. AJR 1990; 155:536-7.
26. Drudi FM, Pretagostini R, Padula S et al. Color Doppler ultrasound in renal transplant: role of resistive index versus renal cortical ratio in the evaluation of renal transplant disease. Nephron Clin Pract 2004; 98:c67-c72.
27. Trillaud H, Merville P, Le Linh PT et al. Color Doppler sonography in early renal transplantation: Resistive index measurements versus power Doppler sonography. AJR 1998; 171:1611-5.
28. Reddy VG. Prevention of postoperative renal failure. J Postgrad Med 2002; 48:64-70.
29. Balogh Z, McKinley BA, Cocanour CS et al. Supranormal trauma resuscitation causes more cases of abdominal compartment syndrome. Arch Surg 2003; 138:637-43.
30. Balogh Z, McKinley BA, Cocanour CS et al. Patients with impending abdominal compartment syndrome do not respond to volume loading. 2003; 186:602-8.
31. Tang IY, Murray PT. Prevention of perioperative acute renal failure: what works? Best Practice & Research Clinical Anesthesiology 2004; 18:91-111.
32. Rubin JM, Bude RO, Carson PL et al. Power doppler US: A potentially useful alternative to mean-frequency based color Doppler US. Radiology 1994; 190:853-6.
33. Bude RO, Rubin JM, Adler RS. Power versus conventional color Doppler sonography: comparison in the depiction of normal intrarenal vasculature. Radiology 1994; 192:777-80.
34. Turetschek K, Kollman C, Dorffner R et al. Amplitude-coded color doppler: Clinical applications. Eur Radiol 1999; 9:115-21.
35. Lencioni R, Pinto F, Armillotta N et al. Assessment of tumour vascularity in hepatocellular carcinoma: Comparison of power doppler US and color doppler US. Radiology 1996; 201:3583-358.
36. Robbin ML. Invited commentary. Radiographics 1998; 18:1455-56.
37. Helenon O, Correas JM, Chabriais J et al. Renal vascular Doppler imaging: Clinical benefits of power mode. Radiographics 1998; 18:1441-54.
38. Kuwa T, Cancio LC, Sondeen JL et al. Evaluation of renal cortical perfusion by noninvasive power Doppler ultrasound during vascular occlusion and reperfusion. J Trauma 2004; 56:618-24.
39. Clautice-Engle T, Jeffrey RB. Renal hypoperfusion: Value of power Doppler imaging. AJR 1997; 168:1227-30.
40. Takano R, Ando Y, Taniguchi N et al. Power Doppler sonography of the kidney: Effect of Vasalva's maneuver. J Clin Ultrasound 2001; 29:384-8.
41. Chang CH, Yu CH, Ko HC et al. Quantitative three-dimensional power doppler sonography for assessment of the fetal renal blood flow in normal gestation. Ultrasound Med Biol 2003; 29:929-33.
42. Wei K, Le E, Bin JP et al. Quantification of renal blood flow with contrast-enhanced ultrasound. J Am Coll Cardiol 2001; 37:1135-40.
43. Keller MW, Egal SS, Kaul S et al. The behavior of sonicated albumin microbubbles in the microcirculation: a basis for their use during myocardial contrast echocardiography. Circ Res 1989; 65:458-66.

Part C: Abdominal Compartment Syndrome after Aortoiliac Surgery

Martin Björck*

Abstract

This section reviews the literature regarding the incidence and treatment of IAH/ACS after aortoiliac surgery. The complication has never been reported after elective endovascular abdominal aortic aneurysm (AAA -repair and is rare after open elective AAA-surgery. After open operation of a patient in shock due to a ruptured AAA (rAAA) the incidence of IAH grade III (IAP ≥ 20 mm Hg) occurs in 25-55%. The incidence of IAH/ACS seems to increase with improved survival, suggesting a survival benefit in the early recognition and treatment of IAH/ACS. This, however, remains to be shown in a prospective trial. No studies have been published on the incidence of IAH after endovascular repair of rAAA, but clinical ACS seems to occur in approximately 5%.

Special considerations when treating the AAA patient with increased IAP are: The high prevalence of coronary artery disease, the high risk of colonic ischaemia and the lethal consequences of prolonged hypovolaemia.

The reluctance among vascular surgeons to treat AAA-patients with an open abdomen is explained by the fear of graft infection and recurrent bleeding. The trade-off between the advantages and disadvantages of decompressing a vascular patient with an increased IAP are discussed. Large prospective trials are warranted to define the relevant thresholds.

Introduction

Several of the first reports on ACS, after the name of the condition had been proposed by Kron[1] and colleagues in 1984, included patients who developed ACS after aorto-iliac surgery. Most investigators, however, have reported experiences from mixed patient groups with a predominance of trauma victims. This chapter will focus on the problem among vascular surgery patients, and in particular the patient who was operated on for a ruptured abdominal aortic aneurysm (rAAA).

Incidence of ACS after Open Aortoiliac Surgery

How often does ACS occur after aortoiliac surgery? This straight-forward question is not easy to answer. Before the establishment of the Consensus Document of the ACS[2] in December 2004 there was no established definition of the ACS. Most investigators have measured IAP by baldder pressure, as suggested by Kron,[1] but there has been no uniform way of reporting IAP or organ dysfunction. The incidence differs depending on the clinical situation. It may be affected by the routine of resuscitation used, as well as by survival. The more patients survive operation for rAAA, the more postoperative complications may develop.

The first report trying to establish the incidence was published by Fietsam[3] et al in 1989. In a retrospective study covering the period from 1978 to 1988, four patients developed overt ACS and two were left with open abdomen at the end of rAAA repair, among 104 patients operated on for rAAA (5.8%). No patients in this study appeared to have been monitored with IAP measurement. No results of IAP were presented in the publication, not even among those who developed ACS.

Platell et al measured IAP at 2,4,8,14,20 and 26 hours after completion of abdominal aortic surgery in 42 patients[4] with an aim to observe the association between IAP and renal function. Only six patients were operated on emergently. The number of patients operated on for rAAA

*Martin Björck—Vascular Surgery, Uppsala University Hospital, SE-751 85 Uppsala, Sweden.
Email: martin@bjorck.pp.se

was not stated. Ten needed reoperation, either for bleeding or for bowel ischaemia, a surprisingly large proportion in this group of predominantly elective patients. Renal impairment was common, 53% when defined as a postoperative creatinine > 130 μmol/L. The authors were able to show with an ROC-curve that an IAP of 18 mm Hg predicted renal impairment with a positive predictive value of 85%, and a negative predictive value of 62%. Thus only 11 of 22 patients who developed renal impairment had IAP > 18 mm Hg. Though this study does not supply data on the incidence of IAP/ACS, it can be calculated from the data presented that 33% (14/42) patients, had an IAP > 18 mm Hg, after predominantly non-emergent, elective AAA surgery.

Akers et al reported on 23 patients operated on for a rAAA during two years.[5] Four were treated with delayed abdominal closure and two were decompressed due to ACS. Mortality among these six patients was 50%. No IAP measurements were reported.

Oelschlager et al, from The Harborview Medical Center in Seattle, performed a retrospective study on 38 patients treated for rAAA, of whom 39% died during surgery.[6] Among the 23 patients who survived initial repair, five had their abdomen closed with a silastic sheet and two by closure of the skin alone. Though no measurements of IAP were reported, 35% (8/23) showed clinical signs of IAH. The investigators also reported a trend towards a more favourable outcome among those treated with delayed abdominal closure.

We measured IAP in the bladder consistently every 6 hours during the postoperative ICU-stay among 25 patients operated on for AAA.[7] Four patients had isolated values above 18 mm Hg (25 cm H_2O) without any clinical consequences. Three had consistent, prolonged IAH above this level. The latter three patients had abdominal decompression. Two of them were also noted to have colonic gangrene. All three patients with prolonged IAH were in a group of five patients operated on for rAAA. There was no mortality.

In a report from the Mayo clinic, among 223 patients operated on for rAAA during a ten year period, 53 (24%) were treated with open abdomen.[8] In 43 of these patients mesh was used at the primary abdominal closure, but 10 patients (4.5%) underwent a decompressive laparotomy due to IAH. In this retrospective study, IAP measurements were not reported.

A report from Leicester, UK, analyzed 75 patients operated on for ruptured (22) or non-ruptured (53) AAA.[9] Unfortunately, IAP was only measured once every 24 h, and only when the patient was still on the ventilator. The consequence of this methodology was that in 13 patients (17%) the IAP was never measured. In 21 (28%) it was measured only once immediately postoperatively, since they were extubated within 24 h. Others had IAP measured only twice. Among the patients operated on for a rAAA, and who were on the ventilator at 24 h and thus had IAP measurement at least twice, 12/22 (55%) had an IAP-value ≥ 15 mm Hg. This methodology limits conclusions regarding the prevalence of IAH.

In a study of 70 consecutive patients operated on for AAA, 30 were operated on for rAAA. Seventeen were consistently monitored with IAP throughout the ICU-stay.[10] Among those, 8 (47%) never had an IAP > 20 mm Hg and they all survived. Six patients (35%) had IAP between 20-29 mm Hg. Only one of them had abdominal decompression. Two of the six patients died. Three patients (18%) had IAP > 30 mm Hg. All were decompressed and they all survived.

In a preliminary report from a prospective study on patients with rAAA, where all patients were monitored with IAP as well as with colonic tonometry, there was no mortality among 16 operated patients.[11] Eleven had IAP ≥ 20 mm Hg (39%) and four had ≥ 25 mm Hg (25%).

Data on the incidence of IAH/ACS after open surgery on the abdominal aorta is summarized in Table 1. The complication seems to be a prevalent problem after operation on patients with rAAA. Grade III of IAH[2] (IAP ≥ 20 mm Hg) develops in approximately 25-55% of these patients. The incidence of IAH/ACS is likely to increase with more aggressive resuscitation, improved survival as well as increased monitoring.

Incidence of IAH/ACS after Endovascular Repair of AAA

The situation is even less clear after endovascular repair (EVAR) of AAA. The complication is unheard of after EVAR of an intact AAA. In a recent report on complications among 311 patients treated with EVAR, ACS was not mentioned.[12] This, however, may change since EVAR is becoming increasingly applied on patients with rAAA. In this technique large amounts of retroperitoneal blood may be left in situ and there may be a risk to develop IAH/ACS. No prospective studies have yet been published on the incidence of the complication after EVAR.

In a discussion of a paper by Rasmussen et al, Frank Veith reported that among 22 patients treated with EVAR for rAAA, three (14%) developed ACS and required abdominal decompression.[8] The Hospital of Malmö, Sweden, is one of the pioneering clinics in EVAR. Among 40 patients treated with EVAR due to a ruptured AAA, two underwent laparotomy for ACS (5%).[13] Van Herzeele et al reported a successful decompression of a patient with ACS after operation of a rAAA with EVAR: 1.5 litres of blood was removed from the retroperitoneum through an 18 cm lumbotomy, and another 0.5 litre of blood was aspirated through the peritoneum.[14]

The Onset of IAH/ACS

An important issue is the time of onset of the condition. Fietsam reported on four patients with ACS after rAAA. Three of them developed oliguria within 7 hours and the fourth within 14 hours.[3] On the other hand, as previously underscored, most investigators have neither measured nor reported IAP in a way that makes it possible to analyze this issue. In my personal experience, based on IAP measurements in all patients operated on for rAAA since 1996, most patients do develop IAH/ACS during the first 24 hours. Others, however, will continue their capillary leakage for many days, and the IAP may continue to increase during a whole week. We reoperated on one patient on the fifth postoperative day due to colonic ischaemia, partly a result of prolonged IAH.[7] Again a large prospective study is needed to characterize the importance of the timing of onset of the complication.

Consequences of IAH after Aorto-Iliac Surgery

Generally speaking, the consequences of IAH for the patient operated on the abdominal aorta are similar to those experienced by any other patient. The IAH/ACS represents a "second hit" to the patient who was in preoperative shock, be it after trauma or after a ruptured AAA. There are, however, three considerations that are particularly important for the AAA-patient.

Although aneurysm disease is believed to be a distinct clinical entity, it has common risk factors with atherosclerotic diseases, in particular coronary artery disease (CAD).[15] Smoking is very common among AAA-patients. In a classical study from 1984, 1000 consecutive patients who were to be operated on for peripheral arterial disease underwent coronary arteriography, only 8% had normal findings and 1/3 had severe CAD.[16] Thus, the risk of ischaemic heart events is increased.

Secondly, colonic gangrene after AAA-surgery is a highly lethal complication which is associated with IAH.[7,10,11,17] The inferior mesenteric artery is either occluded due to the disease, or ligated at the time of surgery, transforming the left colon into a sentinel organ in the case of splanchnic hypoperfusion. In a combined cohort and case-control study of 2824 patients who underwent aortoiliac surgery, the strongest independent risk factor for colonic ischaemia was preoperative shock due to a ruptured AAA (relative risk = 5.8).[18] In Figure C1 the anatomical distribution of the colonic ischaemic lesions is illustrated.

Thirdly, these two tissue beds are at extreme risk for hypovolaemia in the early postoperative period in these patients. IAH results in a "falsely high" CVP that, if not recognized, may result in inadequate volume replacement, prolonged splanchnic hypoperfusion and secondary multiple organ failure. Tonometry of the left colon has been reported to be more sensitive and may be more helpful to detect such hypoperfusion than any other monitoring.[7,11,17,20]

Treating the Vascular Patient with an Open Abdomen

Vascular surgeons are reluctant to treat patients with an open abdomen for two major reasons. Firstly the patient with a rAAA who has a tense abdomen at the end of laparotomy very often has hypothermia, coagulopathy and diffuse bleeding. Closing the abdomen in this situation is traditionally thought to help in correcting hypothermia and also for creating counter-pressure which may be helpful to control diffuse bleeding. A better alternative is to close the skin temporarily with towel clips, leave the patient in the OR while warming the patient but measuring the IAP. When the patient has regained normal body temperature and has been optimized in coagulation factors, abdominal closure with mesh and skin coverage or an open abdomen approach may be elected.

The second rationale to avoid treatment with open abdomen after aortoiliac surgery is the fear for graft infection. Akers et al reported the development of an intra-abdominal abscess in one of six patients treated with open abdomen after rAAA.[5] That patient had been treated with an open abdomen for 51 days and positive bacterial cultures were obtained at the time of closure. The patient was discharged on day 87. Long-term follow-up as regards the development of graft-infection was not reported.

Oelschlager et al reported that they had no case of graft infection among four survivors that were treated with open abdomen after rAAA, and who were closed after 6-28 days.[6] This statement is weakened by the fact that no information was given on follow-up, and from other information in the paper follow-up time seems to be have been very short, an important consideration since graft infections often develop months or even years after AAA repair. Furthermore, there was no information on the four patients treated with open abdomen after rAAA who died in the postoperative period. Did they have signs of sepsis or graft infection? Did they undergo autopsy?

In the largest report on open abdomens after operation for rAAA, Rasmussen et al, in a retrospective study[8] reported no documented graft infection among 53 treated patients. However, only 15 of them survived the early postoperative period and no information is given on one-year survival, follow-up or autopsy-rate.

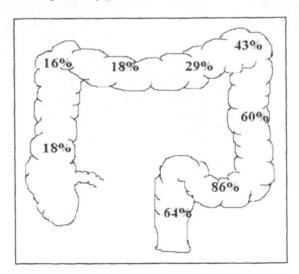

Figure C1. Distribution of ischaemic lesions in the colon and rectum among 63 patients who developed this complication after aortoiliac surgery.[19] Since more than one segment of the colon could be affected in the same patient, the sum exceeds 100%. In 95% of the patients some part of the left colon and/or rectum was affected.

In summary, the issue of the risk of graft-infection after treatment with open abdomen after AAA-repair remains unresolved and has to be addressed in a large prospective trial with long-term follow-up. The risk of this highly lethal complication has to be compared to the risk of not decompressing a patient with IAH. Should the patient who develops ACS after AAA surgery, and who has a synthetic graft prone for infection, be treated more aggressively in order to close the abdomen as fast as possible? On the other hand, such a strategy may result in either aggressive diuretic treatment or dialysis, the risk of hypovolemia and of further ischaemic injury to multiple organ systems. A choice between Skylla and Charybdis?

In order to be able to close the patient's abdomen quicker we have used a method of lateral incisions from the ribs to the iliac crest, permitting us to close the skin and leave the fascia open for late repair after 3-6 months, Figure C2. With this method it has been possible to close the abdomen after 3-5 days. In this situation, continued IAP monitoring is fundamental to be able to detect recurrent ACS.

It is probable that the vacuum assisted wound closure (WAWC) is an even better alternative in this situation. Suliburk et al reported on 35 trauma patients who were treated with open abdomen and WAWC. Among 29 survivors, 25 (86%) underwent primary fascial closure after a mean of 7 days (range 3-18).[21] Miller et al reported a similar experience on 53 trauma patients, where a primary fascial closure rate of 88% was achieved after a mean of 9.5 days among 45 survivors.[22] No report on AAA patients treated with WAWC has yet been published, but we have reasons to believe that this treatment modality may be superior in the AAA patient group also. A prospective trial is in progress.

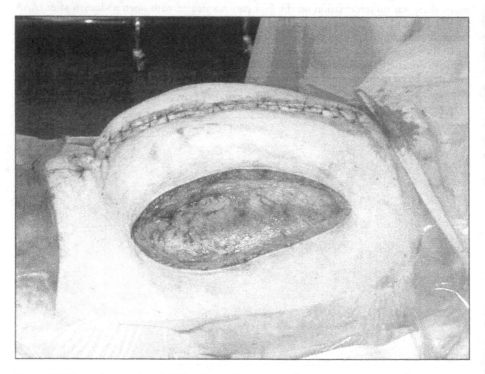

Figure C2. This patient developed ACS 18 hours after operation for a rAAA, and was treated with a Bogotá bag for three days. On the fifth day the abdominal skin could be closed by using diverting lateral skin incisions. IAP was monitored after skin closure, no recurrent ACS developed.

Controversial Issues

Open abdomen approach for the patient operated on for AAA does have costs in terms of increased risk of infection and bleeding as well as an increased morbidity in terms of reoperations, prolonged ICU-stay and incisional hernias. The benefits of early decompression in a situation of incipient ACS are also evident in terms of decreased risk of organ impairment and of colonic ischemia and probably an increased overall survival. The controversial issue is how to define the trade-off, when do the advantages outweigh the disadvantages? This difficult decision-making requires more information than a single IAP-measurement. We are moving towards continuous IAP-measurements[23] making it feasible to also measure continuous abdominal perfusion pressure (APP).[24] We must consider the area under the curve and realize that the ischaemic insult to the intra-abdominal organs, and particularly to the left colon, is a product of depth and duration.[17,20] Whereas a short ischaemic insult may be well tolerated, a prolonged intra-abdominal hypo-perfusion may prove lethal. Large prospective trials are warranted and are in progress to address these issues.

Commentary

Rao Ivatury

It is not surprising that IAH and ACS should be noted in critically ill, non-trauma patients. Conditions such as acute pancreatitis and abdominal aortic aneurysms contribute to raised intra-abdominal pressure by several mechanisms: space-occupying lesions in the retroperitoneal space, secondary distension of bowel from ileus, extra-vascular fluid sequestration (edema fluid, "pancreatic ascites" in pancreatitis and leaked blood in AAA) and aggressive fluid resuscitation. It is becoming increasingly evident that at least some of the morbidity of these pathologic conditions is related to undiagnosed intra-abdominal hypertension. The true incidence of this complication is yet to be determined, since monitoring of the IAP is begun only recently in these patients. Only by such data accrual and analysis we can balance the cost-benefit ratio of the difficult open abdomen management in these conditions.

Kirkpatrick and associates have given us a fascinating addition to our ever increasing knowledge of IAH and ACS. They report on a specific renal allograft compartment syndrome which, unless recognized and treated early, will lead to graft loss. In this syndrome, identified by diminished arterial and venous flow by color power Doppler, the changes are probably due to increased compartment pressure, since renal vein thrombosis or vascular kinking has been noticeably absent on reexploration. Further, vascular normalization occurs upon opening of the compartment. Further study will be needed to clarify whether the RACS is a localized ACS syndrome or a true generalized intra-abdominal phenomenon.

These considerations lend further support to the admonition by Malbrain: it is not wise not to think of intra-abdominal pressure in the care of the critically ill patient.

References

1. Kron IL, Harmon PK, Nolan SP. The measurement of intra-abdominal pressure as a criterion for abdominal re-exploration. Ann Surg 1984; 199:28-30.
2. On intra abdominal hypertension (IAH) and the abdominal compartment syndrome (ACS). Results from the international ACS concensus definitions conference. Available at: http://www.wsacs.org
3. Fietsam JR, Villalba M, Glover JL et al. Intra-abdominal compartment syndrome as a complication of ruptured abdominal aortic aneurysm repair. Am Surgeon 1989; 55:396-402.
4. Platell CF, Hall J, Clarke G et al. Intra-abdominal pressure and renal function after surgery to the abdominal aorta. Aust N Z J Surg 1990; 60:213-216.
5. Akers DL, Fowl RJ, Kempczinski RF. Temporary closure of the abdominal wall by use of silicone rubber sheets after operative repair of ruptured abdominal aneurysms. J Vasc Surg 1991; 14:48-52
6. Oelschlager BK, Boyle EM, Johansen K et al. Delayed abdominal closure in the management of ruptured abdominal aortic aneurysms. Am J Surg 1997; 172:411-415.

7. Björck M, Lindberg F. Abdominal compartment syndrome and colonic ischaemia after aortoiliac surgery, a prospective study. Abstract from the Swedish Surgical Week, published in Svensk Kirurgi (Swedish Surgery) 2000; 58:215.
8. Rasmussen TE, Hallett JW, Noel AA et al. Early abdominal closure with mesh reduces multiple organ failure after ruptured abdominal aortic aneurysm repair: guidelines from a 10 year case-control study. J Vasc Surg 2002: 35:246-253.
9. Papavassiliou V, Anderton M, Loftus IM et al. The physiological effects of intra-abdominal pressure following aneurysm repair. Eur J Vasc Endovasc Surg 2003; 26:293-298.
10. Djavani K, Björck M. Intra-abdominal hypertension (IAHT) and outcome after abdominal aortic surgery. Abstract from the First World Congress on Abdominal Compartment Syndrome, Australia, 2004.
11. Djavani K, Valtysson J, Björck M. Colonic ischemia and intra-abdominal hypertension following operation for ruptured Abdominal Aortic Aneurysm (rAAA). Abstract from the First World World Congress on the Abdominal Compartment Syndrome, Australia 2004.
12. Maldonado TS, Rockman CB, Riles E et al. Ischemic complications after endovascular abdominal aortic aneurysm repair. J Vasc Surg 2004; 40:703-710.
13. Malina M, Lindblad B. Personal communication, January 2005.
14. Van Herzeele I, De Waele JJ, Vermassen F. Translumbar extraperitoneal decompression for abdominal compartment syndrome after endovascular treatment of a ruptured AAA. J Endovasc Ther 2003; 10:933-5.
15. Wanhainen A, Bergqvist D, Boman K et al. Risk factors associated with abdominal aortic aneurysm: A population based study with historical and current data. J Vasc Surg 2005; 41:390-396.
16. Hertzer NR, Beven EG, Young JR et al. Coronary artery disease in peripheral vascular patients. A classification of 1000 coronary angiograms and results of surgical management. Ann Surg 1984; 199:223-233.
17. Björck M, Broman G, Lindberg F et al. pHi-monitoring of the sigmoid colon after aortoiliac surgery. A five-year prospective study. Eur J Vasc Endovasc Surg 2000; 20:273-280.
18. Björck M, Troëng T, Bergqvist D. Risk factors for intestinal ischaemia after aortoiliac surgery. A combined cohort and case-control study of 2824 operations. Eur J Vasc Endovasc Surg 1997; 13:531-539.
19. Björck M, Bergqvist D, Troëng T. Incidence and clinical presentation of bowel ischaemia after aortoiliac surgery - 2930 operations from a population-based registry in Sweden. Eur J Vasc Endovasc Surg 1996; 12:139-149.
20. Björck M, Hedberg B. Early detection of major complications after abdominal aortic surgery: predictive value of sigmoid colon and gastric intramucosal pH monitoring. Br J Surg 1994; 81:25-30.
21. Suliburk JW, Ware DN, Balogh Z et al. Vacuum-assisted wound closure achieves early fascial closure in open abdomens after severe trauma. J Trauma 2003; 55: 1155-1161.
22. Miller PR, Meredith JW, Johnson JC et al. Prospective evaluation of vacuum-assisted fascial closure after open abdomen: planned ventral hernia rate is substantially reduced. Ann Surg 2004; 239:608-616.
23. Balogh Z, Jones F, D'Amours S et al. Continuous intra-abdominal pressure measurement technique. Am J Surg 2004; 188:679-684.
24. Cheatham ML, White MW, Sagraves SG et al. Abdominal perfusion pressure: A superior parameter in the assessment of intra-abdominal hypertension. J Trauma 2000; 49:621-627.

CHAPTER 17

Abdominal Compartment Syndrome in the Pediatric Patient

M. Ann Kuhn* and David W. Tuggle

For all practical purposes, the original clinical model for the abdominal compartment syndrome (ACS) involved the repair of congenital abdominal wall defects such as omphalocele (Fig. 1) and gastroschisis (Fig. 2). Closure or coverage of these defects is always associated with an increase in intra-abdominal pressure.

Ambrose Pare first described omphalocele in 1634.[1] The first successful treatment of omphalocele was likely by Hey in 1803.[2] Most attempts at managing this neonatal problem were unsuccessful until Ahfeld promoted the topical treatment of the membrane to create an eschar in 1899.[3] Gross described the use of skin flaps to initially close giant omphaloceles in 1948, with good success.[4] Both topical treatment and skin flap closure were utilized to prevent the morbidity of ACS associated with closure of abdominal wall defects in newborns. Current therapy can include alternatives to immediate closure such as topical treatment with eschar promotion, and the use of preformed silastic silos with gradual reduction. Both of these management techniques tend to avoid much of the morbidity of immediate surgical closure in these infants.

Immediate gastroschisis repair will also create intra-abdominal hypertension in the newborn. Watkins reported the first survivor of surgical treatment in 1943.[5] Moore and Stokes established the present day criteria of classification, and noted that when the large intestinal mass is forced into the abdominal cavity, the resulting respiratory compromise may lead to death.[6] Izant described manually stretching the newborn abdominal wall to decrease the impact of immediate visceral reduction in 1966.[7] The prosthetic silo technique was described by Schuster in 1967 for omphalocele, and it was immediately employed for gastroschisis repair by most pediatric surgeons.[8] The survival of children with gastroschisis improved dramatically after the introduction of total parenteral nutrition. Many infants now undergo bedside visceral containment with a preformed silastic silo and delayed visceral reduction as a means to avoid the ACS in newborns with gastroschisis (Fig. 3).

As experience with difficult abdominal closures became more commonplace in children, the concept of estimating IAP directly or indirectly to guide abdominal closure gained popularity and was studied clinically and experimentally.[9-13] A different method to confirm the probability of safe closure of the newborn abdomen used continuous measurement of end tidal CO_2. With increasing abdominal pressure, $ETCO_2$ decreased, and was used as an intraoperative guide to successful visceral reduction and abdominal wall closure.[14] This was likely due to decreasing pulmonary blood flow as IAP increased.

As children with abdominal wall defects survived these common causes of intra-abdominal hypertension to a greater degree, the complications associated with an elevated IAP began to

*Corresponding Author: M. Ann Kuhn— Department of Surgery, Section of Pediatric Surgery, The University of Oklahoma College of Medicine, Oklahoma City, Oklahoma, U.S.A. Email: Ann-Kuhn@ouhsc.edu

Abdominal Compartment Syndrome, edited by Rao R. Ivatury, Michael L. Cheatham, Manu L. N. G. Malbrain and Michael Sugrue. ©2006 Landes Bioscience.

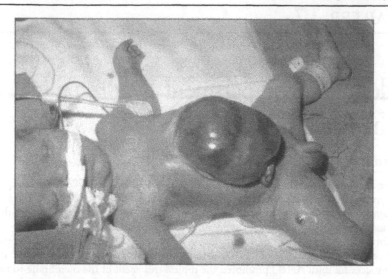

Figure 1. A newborn infant with omphalocele. Note the covering membrane.

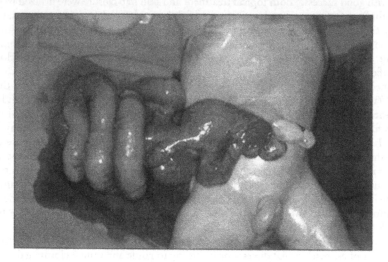

Figure 2. A newborn infant with gastroschisis.

appear. Chin showed in patients with an abdominal wall defect that there occurred an increased ascites leak, ventral hernia formation, edema, and oliguria in newborns with urinary bladder pressures greater than 20 mm Hg.[15] Yaster showed a 50% incidence of oliguria in eight cases after primary closure of abdominal wall defects in patients with gastric pressure of greater than 22 mm Hg.[16] Rizzo demonstrated a shorter stay and decreased hospital cost when abdominal wall closure was managed by urinary bladder pressure.[12] The role of elevated IAP in development of necrotizing enterocolitis after repair of abdominal wall defects has been suggested.[17,18] It has been demonstrated that infections and complications in children related to abdominal compartment syndrome occur less frequently in the staged closure group.[19]

The first experimental pediatric model for intra-abdominal hypertension found a 55% decrease in the cardiac index and cardiac output with a pressure of 20 mm Hg.[20] It is now commonly accepted that increased intra-abdominal pressure (IAP) causes a decrease in cardiac

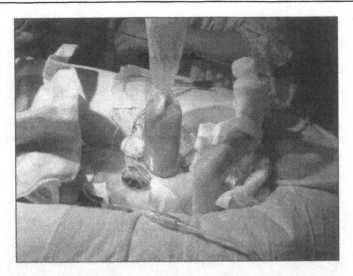

Figure 3. Newborn with gastroschisis. Note the preformed silastic silo placed at the bedside to reduce the likelihood of intra-abdominal hypertension.

output and hypoxia as a result of pulmonary ventilation restriction by increased peak respiratory pressures and hypercapnia. Visceral perfusion is decreased and oliguria occurs. It can also increase cerebrospinal pressure and intracranial pressure. The abdominal compartment syndrome in adults has been defined as abdominal distention with IAP >15 mm Hg which is accompanied by two of the following: oliguria or anuria, respiratory decompensation, hypotension or shock, or metabolic acidosis. In the pediatric population, the level of intra-abdominal pressure at which ACS occurs has not been clearly defined but a similar definition has been suggested by Beck.[21] In this same study, Beck and colleagues showed that ACS was infrequent when compared with adults but it did occur in critically ill children. Decompression of the abdomen resulted in improvement of physiologic parameters such as mean arterial pressure, PaO_2, PaO_2/FiO_2 ratio, urine output, $PaCO_2$, peak inspiratory pressure, PEEP, and base deficit but overall mortality was high. When compared to adults, children had more diverse primary diagnoses that included extra-abdominal conditions and central nervous system conditions. They concluded that the development of ACS in children seems to be related more to ischemia and reperfusion injury.

As in adults, children most often have abdominal pressure measured via an indwelling bladder catheter. An indirect assessment of intra-abdominal pressure can be determined by measurement of bladder pressure. Bladder capacity in children is estimated by a formula consisting of age in years = 2 x 30 mL. To measure bladder pressure in children a foley catheter is placed within the bladder. The empty bladder is then filled to 25-30% of bladder capacity. The sterile tubing of urinary drainage bag is cross-clamped just distal to the culture aspiration port. The end of the drainage bag tubing is connected to the foley. The clamp is released just enough to allow the tubing proximal to the clamp to flow fluid from the bladder then the clamp is reapplied. A 16 g needle is then used to y connect a manometer or pressure transducer through the culture aspiration port. The top of the symphysis pubis is used as a zero point. Increasing airway pressure is another method by which intra-abdominal hypertension can be suspected, especially during operative closure of the abdomen.[22] The presence of ACS in children can be suspected based upon CT findings.[23] These findings include narrowing of the IVC, direct renal compression or displacement, bowel wall thickening, and a rounded abdomen.

Several pediatric specific disease processes can lead to ACS. These processes can be categorized as neonatal vs. childhood and congenital vs. acquired. Neonatal congenital diseases

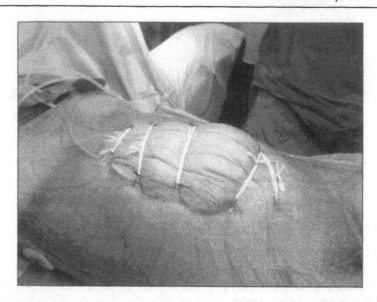

Figure 4. A patient who has undergone decompressive laparotomy for an abdominal compartment syndrome after liver injury. Note the umbilical tapes across the abdominal wound, maintaining fascial approximation.

include abdominal wall defects as mentioned previously such as omphalocele, gastroschisis, and diaphragmatic hernia, and ectopia cordis. Acquired neonatal processes include necrotizing enterocolitis, volvulus, and meconium perforation with cyst formation. In utero, increased IAP is tolerated due to the hormonal milieu of pregnancy. Pediatric processes that can cause ACS include trauma,[21] small bowel obstruction, renal tumors,[24] and burns.[25]

Therapy for ACS in children is decompression of the abdomen. Primary decompressive laparotomy is infrequently needed. More often, development of ACS occurs after a surgical procedure. Reopening the abdomen will result in an immediate decrease in IAP. In our hands the time needed to resolve the physiologic derangement found in children with ACS ranges from three days to three weeks. Even with prolonged use of an open abdomen technique, skin grafting should rarely be needed to close the abdomen in a child. A gradual staged closure of the open abdomen in children has been described to avoid skin grafting and to promote fascial approximation (Fig. 4).[26] Once the child is stabilized definitive therapy can be undertaken (Fig. 5).[27] There are also reports of pediatric ACS treated with paracentesis, thus avoiding decompressive laparotomy.[28,29]

Commentary

Rao R. Ivatury

Omphalocele and gastroschisis are the original clinical conditions that are closely associated with the phenomena of increased intra-abdominal pressure (IAP). We owe a debt of gratitude to the pediatric surgeons who were the first to deal with defects of abdominal wall and the consequences of their closure. Several series from the last decade document the manifestations of elevated IAP in children undergoing such repairs, the beneficial effects of monitoring IAP and the role of elevated IAP in the increased incidence of necrotizing enterocolitis. Kuhn and Tuggle succinctly summarize in this chapter these observations as well as the complete picture of abdominal compartment syndrome in the pediatric patient. They remind us of yet another area for paying attention to the IAP.

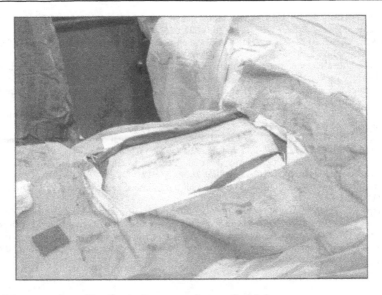

Figure 5. The patient from (Fig. 4), who has now undergone fascial closure.

References

1. Pare A. The workes of that famous chirigeon. In: Cotes T, Young R, eds. Book 24. London, 1977.
2. Hey W. Practical observations in surgery. Cadell T, Davies W. London, 1805.
3. Ahlfeld F. Der alkohol als desinficienz. Mschr Geburtsh 1899; G10:117.
4. Gross RE. A new method for surgical treatment of large omphaloceles. Surgery 1948; 24:277.
5. Watkins DE. Gastroschisis. Va Med 1943; 70:42.
6. Moore TC, Stokes GE. Gastroschisis. Surgery 1953; 33:112-120.
7. Izant RG, Brown F, Rothmann BF. Current embryology and treatment of gastroschisis and omphalocele. Arch Surg 1966; 93:49-53.
8. Shuster SR. A new method for the staged repair of large omphaloceles. Surg Gynecol Obstet 1967; 123:837-850.
9. Wesley JR, Drongowski R, Coran AG. Intragastric pressure measurement: A guide for reduction and closure of the silastic chimney and omphalocele and gastroschisis. J Pediatr Surg 1981; 16:264-270.
10. Lacey SR, Bruce J, Brooks SP et al. The relative merits of various methods of indirect measurement of intraabdominal pressure as a guide to closure of abdominal wall defects. J Pediatr Surg 1987; 22:1207-1211.
11. Lacey SR, Carris LA, Beyer 3rd AJ et al. Bladder pressure monitoring significantly enhances care of infants with abdominal wall defects: A prospective clinical study. J Pediatr Surg 1993; 28(10):1370-1374.
12. Rizzo A, Davis PC, Hamm CR et al. Intraoperative vesical pressure measurements as a guide in the closure of abdominal wall defects. Am Surg 1996; 62(3):192-196.
13. Yaster M, Buck JR, Dudgeon DL et al. Hemodynamic effects of primary closure of omphalocele/ gastroschisis in human newborns. Anesthesiology 1988; 69(1):84-88.
14. Puffinbarger NK, Taylor DV, Tuggle DW et al. End-tidal carbon dioxide for monitoring primary closure of gastroschisis. J Pediatr Surg 1996; 31(2):280-282.
15. Chin TW, Wei CF. Prediction of outcome in omphalocele and gastroschisis by intraoperative measurement of intravesical pressure. J Formosan Med Assoc 1994; 93-691-693.
16. Yaster M, Scherer TL, Stone MM et al. Prediction of successful primary closure of congenital abdominal wall defects using intraoperative measurements. J Pediatr Surg 1989; 24(12):1217-1220.
17. Oldham KT, Coran AG, Drongowski RA et al. The development of necrotizing enterocolitis following repair of gastroschisis: A surprisingly high incidence. J Pediatr Surg 1988; 23(10):945-949.
18. Blane CE, Wesley JR, DiPietro MA et al. Gastrointestinal complications of gastroschisis. Am J Roentgenol 1985; 144(3):589-591.
19. Kidd J, Jackson RJ, Smith S et al. Evolution of staged versus primary closure of gastroschisis. Ann Surg 2003; 237:759-765.

20. Lynch FP, Ochi T, Scully JM et al. Cardiovascular effects of increased intra-abdominal pressure in newborn piglets. J Pediatr Surg 1974; 9(5):621-626.
21. Beck R, Halberthal M, Zonis Z et al. Abdominal compartment syndrome in children. Pediatr Crit Care Med 2001; 2:51-56.
22. Cooney DR. Defects of the abdominal wall. In: O'Neill Jr JA, Rowe MI, Grosfeld JL et al, eds. Pediatric Surgery. 5th ed. St. Louis: Mosby, 1998:1060.
23. Epelman M, Soudack M, Engel A et al. Abdominal compartment syndrome in children: CT findings. Pediatr Radiology 2002; 32(5):319-322.
24. Glick RD, Hicks MJ, Nuchtern JG et al. Renal tumors in children less than six months of age. J Pediatr Surg 2004; 39(4):522-525.
25. Hobson KG, Young KM, Ciraulo A et al. Release of abdominal compartment syndrome improves survival in patients with burn injury. J Trauma 2002; 53:1129-1134.
26. Markley MA, Mantor PC, Letton R et al. Pediatric vacuum packing wound closure for damage-control laparotomy. J Pediatr Surg 2002; 37:512-514.
27. Wetzel RC, Burns RC. Multiple trauma in children: Critical care overview. Crit Care Med 2002; 30(11):S468-S477.
28. Sharpe RP, Pryor JP, Gandhi RR et al. Abdominal compartment syndrome in the pediatric blunt trauma patient treated with paracentesis: Report of two cases. J Trauma 2002; 53:380-382.
29. Latenser BA, Kowal-Vern A, Kimball D et al. A pilot study comparing percutaneous decompression with decompressive laparotomy for acute abdominal compartment syndrome in thermal injury. J Burn Care Rehabil 2002; 23:190-195.

CHAPTER 18

Prevention of Abdominal Compartment Syndrome

John C. Mayberry*

B ecause the Abdominal Compartment Syndrome (ACS) is associated with considerable morbidity and mortality, the development of strategies to detect and moderate intra-abdominal hypertension (IAH) before ACS is fully developed is warranted. Modern surgeons have considered and studied the probability that the warning signs of impending ACS can be identified and that a single surgical decision or intervention at a crucial time will be preventative. The contributing factors that lead to the development of ACS are now, after decades of IAH research, well-described. This chapter seeks to codify the contributing factors, warning signs, and prevention strategies that have been described across a wide variety of disciplines including trauma surgery, emergency surgery, vascular surgery, burn surgery as well as surgical and medical critical care. Prevention strategies discussed will include the surveillance of intra-abdominal pressure, the limitation of unnecessary fluid resuscitation, the use of prophylactic temporary abdominal closure, and pharmacological neuromuscular blockade.

Contributing Factors and Warning Signs

Although ACS occurs in a wide variety of clinical scenarios, several common contributing factors have been identified (Table 1). In cases of primary ACS where the intra-abdominal process causing the IAH is in the early stages of recognition, several of these factors including intraperitoneal or retroperitoneal hemorrhage, would not be contributing factors but rather causes of ACS. In cases where the ACS develops many hours after an injury or an inflammatory insult each of these factors listed will contribute to further elevations in IAH. Shock, no matter what the origin, is an important denominator because many clinicians believe that the shock state and the subsequent reperfusion of ischemic tissue causes endothelial permeability.[1-3] In addition, the dilution of solute and protein in the intravascular space by fluid resuscitation will promote the movement of water into the interstitium.[4,5] Edema of the intra-abdominal viscera and the abdominal wall will crowd organs within a compartment that has an expandable but finite volume. Resuscitation ascites has also been described.[6,7] Presumably, the more intense the shock state and the more aggressive the ensuing fluid resuscitation are, the more likely ACS will appear. This is the predominate cause of ACS in burn patients and patients who develop ACS from "supra-normal" fluid therapy.[8,9]

The two main contributors to the development of ACS that can be readily determined are the volume of fluid resuscitation and the magnitude of intrathoracic pressure. McNelis et al performed univariate and multivariate analysis of variables associated with the development of the ACS in the surgical ICU in nontrauma patients.[10] Although the Acute Physiology and Chronic Health Evaluation (APACHE) II and III, the Simplified Acute Physiology Score (SAPS),

*John C. Mayberry— Department of Surgery/L223A, Oregon Health and Science University, 3181 SW Sam Jackson Park Rd., Portland, Oregon, 97239 U.S.A. Email: mayberrj@ohsu.edu

Abdominal Compartment Syndrome, edited by Rao R. Ivatury, Michael L. Cheatham, Manu L. N. G. Malbrain and Michael Sugrue. ©2006 Landes Bioscience.

Table 1. Contributing factors and warning signs of ACS

Shock	Low intravascular oncotic pressure
Massive fluid resuscitation	Elevated intrathoracic pressure
Multiple abdominal/pelvic injuries	Coagulopathy
Peritoneal inflammation	Hypothermia
Retroperitoneal inflammation or hematoma	Hepatic cirrhosis
Abdominal wall tension	

and the multiple organ dysfunction syndrome (MODS) score were significantly higher in the cohort of patients who developed ACS, these validated measures of critical illness were not independently predictive of ACS when compared to 24 hour net fluid balance and the peak airway pressure (PAP) variables in multivariate analysis. Patients who developed ACS had an average net positive fluid balance of 16 ± 10 liters compared to 7 ± 3.5 liters and a PAP of 58 ± 12 mm Hg compared to 32 ± 7 mm Hg in matched controls. The predictive equation developed from their analysis was $P = 1/(1 + e^{-z})$. P equals the probability of ACS formation and z equals $-18.7 + 0.17(PAP) + 0.0009(\text{fluid balance at 24 hours})$. According to this equation, a patient with a PAP of 45 mm Hg and a 12 liter net fluid balance will have a 41% chance of developing ACS.

In the series of secondary ACS collected by the trauma group at Memphis, patients with no intra-abdominal injuries who developed ACS received an average of 19 ± 5 liters of crystalloid and 29 ± 10 units of packed red blood cells.[1] Ivatury et al noted that in patients with penetrating abdominal trauma those that developed postoperative IAH received 19 ± 12 units of blood compared to 10 ± 6 units of blood in those who did not.[11] In one of the original large series of post-traumatic ACS the mean blood transfusion requirement was 29 ± 8 units and the mean PAP was 44 ± 3 cm H_2O.[12] In a series of patients with acute ascites formation after trauma resuscitation, the majority of whom did not have intra-abdominal injury, the mean PAP was 39 ± 5.8 cm H_2O and the mean crystalloid and blood product infusions prior to decompression were 16 ± 10 liters and 5.2 ± 4.8 liters, respectively.[7] In comparison, the mean volumes of crystalloid and blood product infusion among 100 contemporary, randomly selected patients undergoing trauma laparotomy were 5.1 ± 5.5 liters and 1.1 ± 2.5 liters, respectively.

The severity of visceral injury as established by the number of injuries or the presence of a high-grade injury is also a significant risk factor. The incidence of ACS in all patients with moderate to severe abdominal injury or pelvic fracture (abbreviated injury scale (AIS) ≥ 3) is only 3%.[13] But patients who develop ACS tend to have more individual intraperitoneal organs injured and higher abdominal trauma index (ATI) and abdominal pelvic trauma score (APTS) than those who do not.[14] The APTS is the product of the number of individual abdominal and pelvic injuries and the abbreviated injury scale (AIS) of the most serious injury. In contast to the ATI, it includes pelvic fracture.[15] The APTS was developed as a simple tool that the surgeon can calculate at the operating table prior to the conclusion of a trauma laparotomy.[16] In a retrospective review of our trauma laparotomy database (930 patients surviving more than 24 hours) the odds ratio of developing post-laparotomy ACS when the fascia was closed was 4.4 (95% confidence interval 1.5-12.7) if the APTS was ≥ 15. The odds ratio of developing post-laparotomy ACS (fascia closed) was 1.5 (95% confidence interval 1.2-1.8) for each 10 unit increase in ATI.

Miscellaneous other factors and pathology associated with IAH and the ACS include peritoneal inflammation (e.g., peritonitis or abscess), retroperitoneal inflammation (e.g., necrotizing pancreatitis), bowel obstruction, elevated intrathoracic pressure (e.g., mechanical ventilation and positive end-expiratory pressure (PEEP)) and hepatic cirrhosis.[7,17,18] Other nontrauma scenarios include ruptured abdominal aortic aneurysm (AAA), liver transplantation, burns, septic shock of any etiology, and major abdominal/gynecologic surgery or catastrophe.[8,17,19-22]

Any patient with 1 or more contributing factors/warning signs in a high risk category is at significant risk for preventable ACS.

Prevention by Surveillance

The most effective IAH surveillance strategy is the serial measurement of urinary bladder pressure (UBP).[23,24] Although the abdominal examination (inspection and palpation) of a critically ill or injured patient will lead the clinician to conclude that IAH is unlikely, only a very soft abdomen will rule IAH out. The subjective determination of IAH by palpation, i.e., concluding whether the abdomen is moderately or highly tense and then translating the clinical finding to a usable measurement, is not only unreliable for a single measurement, but is undependable for serial examination.[25] UBP measurement with a few caveats is now a well-established and reliable method of determining serial intra-abdominal pressures.[26] In a survey of American trauma surgeons in 1998, 85% had experience with ACS one or more times in the previous year.[27] Surgeons who had more experience with ACS and more frequently measured UBP were more likely to diagnose ACS. Each critical care unit should therefore seek to standardize the method of UBP measurement that is preferred in that unit so that both serial and random measurements reported are internally consistent and clinically valid. If UBP measurement is presented properly, critical care nurses and surgeons in training should begin to regard UBP measurements on a par with the measurement of the traditional vital signs and vascular pressure monitoring. Placing an UBP measurement option on critical care unit preprinted admission orders will help both physicians and nurses to remember its importance.

But although IAH should be considered a possibility in a noticeable percentage of critically ill or injured patients, frank ACS is still rare. The incidence of IAH and ACS in a cohort of all patients admitted to a trauma intensive care unit (ICU) with an aggressive UBP measurement surveillance policy was only 2% and 1%, respectively.[26] In other series, 4% of all trauma ICU admissions and 14% of all ICU admissions after trauma laparotomy developed ACS.[1,12] Trauma units that perform routine admission and serial UBP measurements on all admissions are therefore casting a very wide net. Nurses in our trauma unit have expressed concern that frequent UBP measurements may increase the risk of urinary infection. A UBP surveillance protocol that efficiently screens for IAH while minimizing unnecessary urinary catheter contamination risk would be ideal.

Several authors have proposed efficient strategies for the surveillance of IAH in the ICU. Maxwell et al recommended that bladder pressures be checked and acted on appropriately when resuscitation volumes approach 10 liters of crystalloid or 10 units of blood.[1] Offner et al recommended routine postoperative UBP monitoring in any trauma patient undergoing damage-control laparotomy.[28] Hong et al recommended UBP measurement in all patients with signs of organ dysfunction following trauma, even if they have an open abdomen.[26] In burn patients, serial UBP after infusion of more than 0.25 liters/kg of resuscitation fluid during the acute resuscitation phase and for PAPs > 40 cm H_2O.[8] Certainly any patient with a moderately tense or distended abdomen should have a screening UBP measurement regardless of clinical status. High-risk patients include any critically ill patient with transient or ongoing shock requiring resuscitation who manifest any of the contributing factors/warning signs listed in Table 1. Our practice is to measure UBP every 4 hours in high-risk patients while moderate and low-risk patients can probably tolerate a monitoring interval of every 6-12 hours.

Prevention by Limiting Resuscitation

Since massive fluid resuscitation is associated with both primary and secondary ACS, several authors have suggested that if the fluid resuscitation could be limited, the patient's morbidity could be reduced.[7,9,29,30] While historically investigators have ascribed the development of the ACS to the need for aggressive resuscitation of hypovolemia, more recently authors have suggested that many cases of ACS are complications of overaggressive fluid resuscitation. Ivatury has succinctly described the conundrum surgeons face with patients in shock who need fluid resuscitation – the splanchnic bed seems to be adversely affected by both inadequate and over-zealous resuscitation.[31] What has become clear, however, is that "supranormal" resuscitation, i.e., infusing crystalloid, blood, and pressor agents to achieve elevated oxygen delivery indices is associated with an increased incidence of IAH, ACS, multiple organ failure (MOF), and death.[9] The Houston and Denver groups have thus suggested further study into both the composition of fluids used for resuscitation and the endpoints of resuscitation.[30]

Prevention by Prophylactic Temporary Abdominal Closure

Since the historical treatment for established ACS is surgical decompression and the place-ment of a temporary abdominal closure (TAC), several authors have proposed the prophylactic use of TAC to prevent ACS.[11,14,20,28,32] Although there are no randomized trials of prophylac-tic TAC versus fascial closure for the prevention of ACS, all retrospective and prospective stud-ies agree that there is a subset of patients who benefit from prophylactic TAC. In a retrospective review of our experience with absorbable mesh (Dexon®, Davis and Geck, Danbury, CT) TAC, we compared two groups of injured patients that were statistically similar in demograph-ics and injury severity – those who had received TAC at their initial trauma laparotomy and those who received TAC at a subsequent laparotomy.[14] The second group without primary TAC had a significantly higher incidence of ACS (35% versus 0%), necrotizing fasciitis (39% versus 0%), abscess/peritonitis (35% versus 4%) and enterocutaneous fistula (23% versus 11%). These findings were confirmed by Ivatury et al in a series of trauma patients with severe pen-etrating abdominal injuries.[11] Patients who had absorbable mesh (Vicryl®, Ethicon, Somerville, NJ) placed as prophylaxis for IAH had a much lower incidence of IAH (22% versus 52%) and mortality (11% versus 36%) than similar patients who had primary fascial suture. Those pa-tients who developed IAH (the majority of whom had primary facial closure) had 43.5% mor-tality compared to 8.5% mortality in those patients who did not develop IAH. Offner et al divided a series of patients requiring damage control laparotomy into three groups – those who had TAC (Bogota bag), those with skin closure only, and those with primary fascial closure.[28] Primary fascial closure was associated with an 80% incidence of ACS and a 90% risk of acute respiratory distress syndrome (ARDS) and/or MOF while skin closure resulted in 24% ACS and 36% MOF. The Bogota bag TAC cohort had 18% ACS and 47% ARDS/MOF.

Oelschlager et al and Rasmussen et al reported similar results in series of patients with ruptured AAA.[20,33] In the case of ruptured AAA, the persistent retroperitoneal hematoma height-ens the competition for a limited space in the abdominal cavity (Fig. 1). Rasmussen et al noted that the compromised abdominal space may not be obvious at the conclusion of the AAA repair.[33] By dividing those patients with ruptured AAA into cohorts of those who received primary TAC with silastic or polytetraflouroethylene mesh (early mesh group) and those who required an abdominal decompression with TAC later (late mesh group), the authors found significant differences in postoperative MOF even though the two groups had similar preop-erative physiology (Fig. 2). No patient in the early mesh group developed ACS. Warning signs of postoperative IAH included preoperative hemoglobin < 8 grams, prolonged hypotension (> 18 minutes), need for cardiopulmonary resuscitation (CPR), intra-operative acidosis (base deficit > 14 mEq), hypothermia (< 32 °C), and requirement for massive fluid resuscitation (> 4 liters/hour).

There are thus two points in the clinical course of a critically ill or injured patient requiring laparotomy where the surgeon can act decisively to prevent ACS: one is at the conclusion of a laparotomy for multiple or severe injuries and the second is during the postoperative resuscita-tion phase of a patient with a primary fascial closure. Clearly the data supports the prophylactic placement of TAC in selected patients as preferable to primary fascial suture and selective repeat laparotomy decompression. In a survey conducted in 1998, we polled expert trauma surgeons on their willingness to close the fascia at the conclusion of trauma laparotomy.[27] The vast majority of surgeons expressed unwillingness to primarily close the fascia in the settings of pulmonary or hemodynamic deterioration with closure, massive bowel edema, subjectively

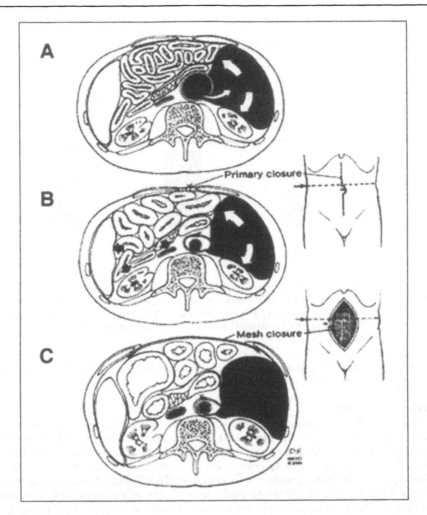

Figure 1. A) Ruptured abdominal aortic aneurysm with associated retroperitoneal hematoma and reduction of abdominal domain. B) Standard primary abdominal closure of compromised abdominal space after ruptured aneurysm repair with compression of abdominal structures. C) Closure of abdomen with mesh expands abdominal domain and reduces pressure on abdominal structures. Reproduced with permission from Rasmussen TE, Hallett JW Jr, Noel AA et al. Early abdominal closure with mesh reduces multiple organ failure after ruptured abdominal aortic aneurysm repair: guidelines from a 10-year case-control study. J Vasc Surg 2002; 35:247.

tight closure, planned reoperation, and packing (Fig. 3). The more prophylactic indications for primary TAC including coagulopathy, acidosis, multiple injuries, hypothermia, massive trans-fusion, and fecal contamination, however, did not at that time influence the majority's decision whether or not to close the fascia. Based on these results we divided the indications for prophy-lactic TAC into two categories: mandatory and discretionary (Table 2). There was substantial agreement that patients with the mandatory indications should have primary TAC while the discretionary indications were controversial. This survey will need to be repeated to ascertain whether trauma surgeons have subsequently trended toward more prophylactic use of TAC to prevent ACS than the 1998 survey. Our current practice is to place TAC following trauma laparotomy prophylactically in intubated patients with signs of shock who are requiring

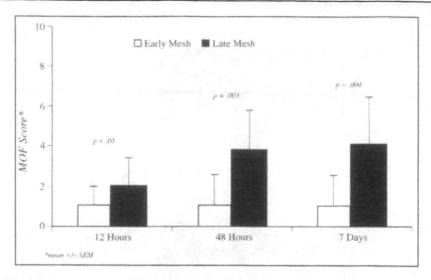

Figure 2. Multiple organ failure (MOF) scores at 12 hours, 48 hours, and 7 days after ruptured abdominal aortic aneurysm repair compared by early versus late temporary abdominal closure. Reproduced with permission from Rasmussen TE, Hallett JW Jr, Noel AA et al. Early abdominal closure with mesh reduces multiple organ failure after ruptured abdominal aortic aneurysm repair: guidelines from a 10-year case-control study. J Vasc Surg 2002; 35:250.

massive fluid resuscitation, those with multiple abdominal/pelvic injuries (APTS >15), and those requiring damage-control surgery.

Medical Prevention

Aside from changing resuscitation practices to limit edema formation, the only medical therapy that has been proposed to prevent IAH from becoming ACS is pharmacologic neuro-muscular blockade (NMB). Two case reports have described the potential beneficial effects of total body muscle paralysis although one report cautioned that surgical decompression may still be required for definitive treatment.[34,35] Certainly NMB will diminish both intra-abdominal and intrathoracic pressures, but the harmful effects of IAH on gastrointestinal anastomoses may persist.[36,37] NMB may be an effective treatment of mild to moderate IAH, but in severe cases (grade III or IV) NMB should at this time be reserved as a temporary measure for patients awaiting surgical or percutaneous decompression.

Summary

The cornerstone of prevention of the ACS is the prompt recognition and treatment of IAH. The amount of attention that the ACS has received in trauma and critical care professional meetings over the last decade is hopefully now insuring that surgeons, medical intensivists, and critical care nurses will recognize IAH when it develops. Certainly the severely injured patient with intraperitoneal trauma whether managed operatively or nonoperatively will stimulate the clinician to look for IAH. Contrastingly, clinicians who under-appreciate the fact that the ACS will develop in patients who initially lack visceral pathology will not be looking for it. Although UBP monitoring of every critically ill patient is unnecessary, a low threshold and the inclusion of UBP monitoring option on preprinted ICU admission orders will prevent IAH from progressing undetected. Limiting fluid resuscitation to avoid excessive edema in patients recovering from shock and medical management of IAH with NMB are controversial preven-tative measures. Prophylactic TAC is an effective means of preventing ACS in selected high-risk patients with multiple abdominal and pelvic injuries who require laparotomy.

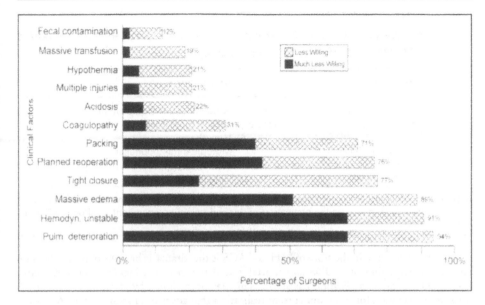

Figure 3. Percentage of surgeons choosing "much less willing" or "less willing" to complete a primary fascial closure after celiotomy for trauma for each of 12 clinical factors. Reproduced with permission from Mayberry JC, Goldman RK, Mullins RJ et al. Surveyed opinion of American trauma surgeons on the prevention of the abdominal compartment syndrome. J Trauma 1999; 47:510.

Commentary

Rao R. Ivatury

Mayberry is one of the earliest authors to stress the importance of the benefits of prophylaxis against IAH and ACS. As emphasized in this chapter, one must identify the patients at risk for the development of these complications and institute prophylaxis as well as clinical monitoring of IAP. While the list is constantly expanding, the high-risk scenarios include : all trauma patients with abdominal or nonabdominal injuries of a moderate to high injury severity score, massive resuscitation volumes, "damage-control" operations, peritonitis or intra-abdominal sepsis especially with delayed diagnosis, retroperitoneal inflammation (e.g. necrotizing pancreatitis), bowel obstruction with bowel dilatation, elevated intrathoracic pressure (e.g., mechanical ventilation and positive end-expiratory pressure (PEEP)), hepatic cirrhosis, ruptured abdominal aortic aneurysm (AAA), liver transplantation, burns, and major abdominal or gynecologic surgery and so on.

Having identified such patient categories, the data support the prophylactic placement of a temporary abdominal closure and avoidance of a primary fascial suture. As recent experience has shown, such measures may not always prevent the development of IAH. Selective repeat laparotomy and decompression may still be necessary with prompt recognition of IAH by abdominal pressure monitoring. The critical level at which this may be indicated is being refined by our expanding knowledge of IAH and ACS. Although IAP monitoring of every critically ill patient is unnecessary, a low threshold for monitoring and abdominal decompression, if necessary , and judiciously limiting fluid resuscitation are reasonable recommendations to prevent IAH. It is often a fine line to tread : too little resuscitation causes tissue dysoxia and MOF; too aggressive resuscitation results in compartment syndromes and MODS.

Table 2. Clinical factors divided into mandatory or discretionary use of the temporary abdominal closure

Mandatory	Discretionary
Pulmonary deterioration on closure	Fecal contamination/peritonitis
Hemodynamic instability with closure	Massive transfusion
Massive bowel edema	Hypothermia
Subjectively tight closure	Multiple intra-abdominal injuries
Planned re-operation	Acidosis
Intra-abdominal packing	Coagulopathy

Reproduced with permission from Mayberry JC, Goldman RK, Mullins RJ et al. Surveyed opinion of American trauma surgeons on the prevention of the abdominal compartment syndrome. J Trauma 1999; 47:510.

What is intriguing in the story of IAH and ACS is the dramatic benefits of prophylaxis on subsequent reduction of MODS and mortality and the increased incidence of unfavorable outcome once the complication has set in. This has been shown by several investigators. A possible explanation for this is our current realization that adverse consequences of IAH occur at lower and lower levels of pressure, especially in the setting of sequential insults of ischemia and reperfusion injury.

Mayberry and colleagues have also given us an understanding of the attitudes of trauma surgeons towards the concept of IAP. As they pointed out, there is still a reluctance on the part of a majority of surgeons to resort to "open abdomen" in high risk patients. In a recent article, Malbrain[1] argued that it is unwise not to think about IAP in ICU patients. The weight of evidence summarized in this chapter and Malbrain's review should convince the most skeptical clinician to pay attention to the phenomenon of IAH and to attempt its prevention.

Commentary References

1. Malbrain ML. Is it wise not to think about intra-abdominal hypertension in the ICU? Curr Opin Crit Care 2004; 10(2):132-45.

References

1. Maxwell RA, Fabian TC, Croce MA et al. Secondary abdominal compartment syndrome: An underappreciated manifestation of severe hemorrhagic shock. J Trauma 1999; 47:995-9.
2. Kopelman T, Harris C, Miller R et al. Abdominal compartment syndrome in patients with isolated extraperitoneal injuries. J Trauma 2000; 49:744-7, discussion 747-9.
3. Biffl WL, Moore EE, Burch JM et al. Secondary abdominal compartment syndrome is a highly lethal event. Am J Surg 2001; 182:645-8.
4. Zarins CK, Rice CL, Peters RM et al. Lymph and pulmonary response to isobaric reduction in plasma oncotic pressure in baboons. Circulation Research 1978; 43:925-30.
5. Lucas CE. The water of life: A century of confusion. J Am Coll Surg 2001; 192:86-93.
6. Eldor J, Olsha O, Farkas A. Over-transfusion ascites. Resuscitation 1991; 21:289-91.
7. Mayberry JC, Welker KJ, Goldman RK et al. Mechanism of acute ascites formation after trauma resuscitation. Arch Surg 2003; 138:773-6.
8. Ivy ME, Atweh NA, Palmer J et al. Intra-abdominal hypertension and abdominal compartment syndrome in burn patients. J Trauma 2000; 49:387-91.
9. Balogh Z, McKinley BA, Cocanour CS et al. Supranormal trauma resuscitation causes more cases of abdominal compartment syndrome. Arch Surg 2003; 138:637-42, discussion 642-3.
10. McNelis J, Marini CP, Jurkiewicz A et al. Predictive factors associated with the development of abdominal compartment syndrome in the surgical intensive care unit. Arch Surg 2002; 137:133-6.
11. Ivatury RR, Porter JM, Simon RJ et al. Intra-abdominal hypertension after life-threatening penetrating abdominal trauma: Prophylaxis, incidence, and clinical relevance to gastric mucosal pH and abdominal compartment syndrome. J Trauma 1998; 44:1016-21, discussion 1021-3.

12. Meldrum DR, Moore FA, Moore EE et al. Prospective characterization and selective management of the abdominal compartment syndrome. Am J Surg 1997; 174:667-72, discussion 672-3.
13. Ertel W, Oberholzer A, Platz A et al. Das abdominale kompartmentsyndrom (AKS) nach schwerem bauch- und/oder beckentrauma. Langenbecks Arch Surg 1998; (Suppl)II:1189-90.
14. Mayberry JC, Mullins RJ, Crass RA et al. Prevention of abdominal compartment syndrome by absorbable mesh prosthesis closure. Arch Surg 1997; 132:957-61, discussion 961-2.
15. Prince RA, Hoffman CJ, Scanlan RM et al. The distinct and secondary harmful effect of pelvic and extremity injury on the outcome of laporotomy for trauma. J Surg Res 2005; 124(1):3-8.
16. Mayberry JC, Mayberry MR, Brand DM et al. Retrospective validation of the abdominal pelvic trauma score: A simple, bedside predictor of outcome. Seattle, WA: Association for Academic Surgery Annual Meeting, 1998.
17. Morken J, West MA. Abdominal compartment syndrome in the intensive care unit. Curr Opin Crit Care 2001; 7:268-74.
18. Gecelter G, Fahoum B, Gardezi S et al. Abdominal compartment syndrome in severe acute pancreatitis: An indication for a decompressing laparotomy? Dig Surg 2002; 19:402-4, discussion 404-5.
19. Katz R, Meretyk S, Gimmon Z. Abdominal compartment syndrome due to delayed identification of a ureteral perforation following abdomino-perineal resection for rectal carcinoma. Int J Urol 1997; 4:615-7.
20. Oelschlager BK, Boyle Jr EM, Johansen K et al. Delayed abdominal closure in the management of ruptured abdominal aortic aneurysms. Am J Surg 1997; 173:411-5.
21. von Gruenigen VE, Coleman RL, King MR et al. Abdominal compartment syndrome in gynecologic surgery. Obstet Gynecol 1999; 94:830-2.
22. Sullivan KM, Battey PM, Miller JS et al. Abdominal compartment syndrome after mesenteric revascularization. J Vasc Surg 2001; 34:559-61.
23. Fusco MA, Martin RS, Chang MC. Estimation of intra-abdominal pressure by bladder pressure measurement: Validity and methodology. J Trauma 2001; 50:297-302.
24. Lee SL, Anderson JT, Kraut EJ et al. A simplified approach to the diagnosis of elevated intra-abdominal pressure. J Trauma 2002; 52:1169-72.
25. Sugrue M, Bauman A, Jones F et al. Clinical examination is an inaccurate predictor of intraabdominal pressure. World J Surg 2002; 26:1428-31.
26. Hong JJ, Cohn SM, Perez JM et al. Prospective study of the incidence and outcome of intra-abdominal hypertension and the abdominal compartment syndrome. Br J Surg 2002; 89:591-6.
27. Mayberry JC, Goldman RK, Mullins RJ et al. Surveyed opinion of American trauma surgeons on the prevention of the abdominal compartment syndrome. J Trauma 1999; 47:509-13, discussion 513-4.
28. Offner PJ, de Souza AL, Moore EE et al. Avoidance of abdominal compartment syndrome in damage-control laparotomy after trauma. Arch Surg 2001; 136:676-81.
29. Balogh Z, McKinley BA, Cocanour CS et al. Secondary abdominal compartment syndrome is an elusive early complication of traumatic shock resuscitation. Am J Surg 2002; 184:538-43, discussion 543-4.
30. Balogh Z, McKinley BA, Cox Jr CS et al. Abdominal compartment syndrome: The cause or effect of postinjury multiple organ failure. Shock 2003; 20:483-92.
31. Ivatury RR. Supranormal trauma resuscitation and abdominal compartment syndrome. Arch Surg 2004; 139:225-6, author reply 226-7.
32. Ciresi DL, Cali RF, Senagore AJ. Abdominal closure using nonabsorbable mesh after massive resuscitation prevents abdominal compartment syndrome and gastrointestinal fistula. Am Surg 1999; 65:720-4, discussion 724-5.
33. Rasmussen TE, Hallett Jr JW, Noel AA et al. Early abdominal closure with mesh reduces multiple organ failure after ruptured abdominal aortic aneurysm repair: Guidelines from a 10-year case-control study. J Vasc Surg 2002; 35:246-53.
34. Macalino JU, Goldman RK, Mayberry JC. Medical management of abdominal compartment syndrome: Case report and a caution. Asian J Surg 2002; 25:244-6.
35. De Waele JJ, Benoit D, Hoste E et al. A role for muscle relaxation in patients with abdominal compartment syndrome? Intensive Care Med 2003; 29:332.
36. Behrman SW, Bertken KA, Stefanacci HA et al. Breakdown of intestinal repair after laparotomy for trauma: Incidence, risk factors, and strategies for prevention. J Trauma 1998; 45:227-31, discussion 231-3.
37. Polat C, Arikan Y, Vatansev C et al. The effects of increased intraabdominal pressure on colonic anastomoses. Surg Endosc 2002; 16:1314-9.

CHAPTER 19

Medical Management of Abdominal Compartment Syndrome

Michael J. A. Parr* and Claudia I. Olvera

Introduction

The medical management of intra-abdominal hypertension (IAH) has been described as of limited efficacy making expedient surgical decompression the treatment of choice for abdominal compartment syndrome (ACS).[1A] There is little doubt that prompt surgical decompression will decrease IAH and can lead to improvement in the consequences of abdominal compartment syndrome. Currently however, the lack of satisfactory randomised controlled trials prevents strong evidence based recommendations on optimal management of ACS. There are good reasons to explore the potential for medical and less aggressive surgical management in selected patients. Important considerations in this regard include:

There are significant risks associated with surgical decompression

- despite surgical decompression survival rates after abdominal compartment syndrome range from 38-71%.[1A]
- temporary abdominal closure is associated with increased risk of intra-abdominal sepsis and graft infection and often necessitates continued management in a critical care area with all associated risks
- there are groups of patients for whom surgical decompression may not be necessary or appropriate

There are currently few data on the medical management of ACS. A Medline literature search combining the key words "abdominal compartment syndrome" and "medical management" only yields 3 articles[1A-3] as of May 2004, two in English and one in French.[1A] This is clearly an area where more investigation and data are required. Medical nonoperative management of abdominal compartment syndrome has not been studied when trying to avoid the complications of surgery especially in the absence of haemodynamic instability. For these reasons much of what is discussed in this chapter is based on anecdotal and limited experience and restricted by this lack of clinical data.

There are patients for whom aggressive surgery is not appropriate or who have declined this option but for whom continued aggressive support is appropriate. As with all medical interventions there is a need to assess and balance the risks and benefits of any particular strategy. Most of these patients will already be in a critical care facility and on ventilatory support, having had recent surgery, major injury or intra-abdominal pathology.

Medical management of patients with ACS may cover several aspects of care and can be considered under the headings of:

*Corresponding Author: Michael J. A. Parr— Intensive Care Units, Liverpool and Campbelltown Hospitals, University of New South Wales, Sydney, Australia.
Email: Michael.Parr@swsahs.nsw.gov.au

Abdominal Compartment Syndrome, edited by Rao R. Ivatury, Michael L. Cheatham, Manu L. N. G. Malbrain and Michael Sugrue. ©2006 Landes Bioscience.

- specific procedures to reduce IAP and the consequences of ACS
- general support (intensive care) of the critically ill patient
- optimisation after surgical decompression to perhaps counteract some of the specific adverse effects associated with decompression

There are three types of ACS depending on aetiology. Primary ACS exists in association with injury or disease in the abdominal/pelvic region that often requires surgery or interventional radiology, or develops following abdominal surgery (such as damage control surgery, peritonitis, bleeding pelvic fractures or massive retroperitoneal hematoma). Secondary ACS refers to conditions that do not require early surgical or radiological intervention (such as sepsis and generalized capillary leak syndome, severe acute pancreatitis, major burns and other conditions requiring massive fluid resuscitation). Tertiary ACS occurs following prophylactic or therapeutic surgical or medical treatment of primary or secondary ACS (e.g., persistence of ACS after decompressive laparotomy or the development of a new ACS episode following the definitive closure of the abdominal wall after the previous utilization of temporary abdominal wall closure).

Specific Procedures to Reduce IAP and the Consequences of ACS

There are numerous causes of intra-abdominal hypertension and abdominal compartment syndrome. Several specific therapies may rationally be directed towards these specific causes and include:

- neuromuscular blockade
- medical management to reduce gastrointestinal ileus and promote gastrointestinal decompression
- prokinetics (erythromycin, metoclopramide, cisapride)
- gastric tube
- colonic tube
- endoscopic decompression of the large bowel
- neostigmine bolus/infusion
- rectal enemas in the cases of fecal impaction or intractable constipation
- percutaneous tube decompression of ascites and blood
- externally applied continuous negative abdominal pressure
- octreotide and melatonin in secondary abdominal compartment syndrome
- diuretics, dialysis and or ultrafiltration to remove excess oedema and for renal replacement therapy
- targeted abdominal perfusion pressure (APP)

Neuromuscular Blockade

Neuromuscular blockade (NMB) is necessary for those patients who are being mechanically ventilated where intrinsic abdominal tone may be adding to the IAH. In this situation NMB allows confirmation of the true IAP. While a fall in IAP may be seen following NMB administration, it is unlikely to be sustained or of enough significance to reverse the adverse effects of true ACS. A case has been reported where NMB administration reversed the cardiovascular and pulmonary effects of IAH. Surgical decompression of ACS was, therefore, postponed, but the patient required reoperation for intra-abdominal sepsis several days later and subsequently died. The authors make the point that although medical management of ACS with NMB may lower IAH and reverse its negative cardiopulmonary effects, surgical decompression may still be required for definitive treatment.[3]

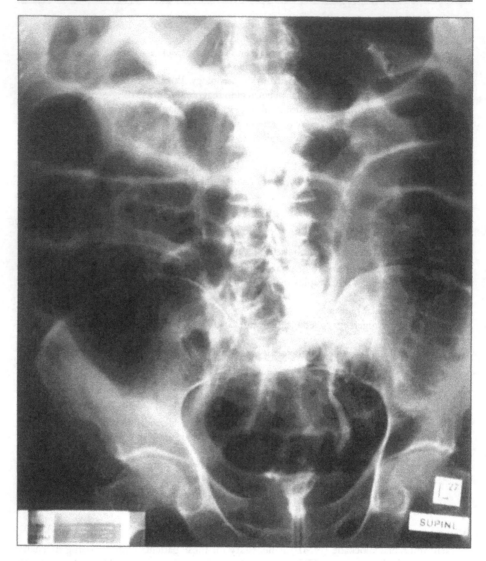

Figure 1. Massive colonic dilatation exacerbating IAH and ACS following multitrauma due to a fall.

Reducing Gastrointestinal Ileus and Promoting Gastrointestinal Decompression

Ileus is common in most critically ill patients and in particular in those who have had abdominal surgery, peritonitis, major trauma, massive fluid resuscitation, electrolyte abnormality and the administration of narcotic and sedative drugs. These factors characterise the patients commonly at risk from ACS. Measures to counter and prevent the adverse effects of ileus include gastric and rectal decompression with tubes, or in more advanced cases of colonic distention an endoscpic decompression may be required. Correcting electrolyte abnormalities, particularly potassium and magnesium is important. Prokinetic agents may be useful and erythromycin (200 mg IV 8 hourly) has a direct contractile effect via the motilin receptor. It has been shown to be an effective prokinetic in the critically ill and is the drug of choice. Metoclopromide (10 mg IV 8 hourly) as an alternative may promote gastric emptying by dopamine antagonism. Neostigmine (2 mg diluted in up to 50 mL IV given slowly by infusion) is not a true prokinetic but may be an effective therapy for pseudo-obstruction (Ogilvie's syndrome, Fig. 1), which can cause massive bowel dilatation and worsen or generate IAH and ACS.[4] Because the bowel may be significantly compromised by these processes there is small but significant risk of bowel perforation associated with the use of neostigmine and other decompressive procedures.

Constipation may also be so severe that it adds to IAH and ACS and aggressive use of aperients and enemas are required. While cases of massive faecal impaction and constipation are rarely reported they are seen in clinical practice and may be associated with rectal and colonic necrosis.[5,6]

Percutaneous Tube Decompression of Ascites and Blood

Drainage of tense ascites by insertion of a small tube may result in a reduction in IAH. Cirrhotic patients with tense ascites may develop a circulatory dysfunction syndrome after massive paracentesis, manifested by an increase in plasma rennin activity, a decrease in systemic vascular resistance and peripheral arterial vasodilatation. If IAP is maintained at its original level, during the process of paracentesis these haemodynamic changes may be avoided despite large volume paracentesis.[7]

Haemoperitoneum large enough to cause IAH and ACS may be seen in patients for whom a major surgical decompression may be a poor option. For example, from the authors (MP) recent experience, an oncology patient with a highly malignant lymphoma while being staged for aggressive chemotherapy had a liver biopsy. This was followed by a slow haemorrhage with pathological coagulopathy that failed to respond to conventional therapy but which eventually responded to treatment with recombinant activated factor VII (rFVIIa, Novoseven). He received a massive transfusion and IAH with ACS and oliguric then anuric renal failure ensued. A large laparotomy and leaving the abdomen open in this situation would have almost certainly resulted in his being too unwell for the chemotherapy he urgently required or death from the septic consequences following chemotherapy. He was managed with a small tube drainage of the intraperitoneal haemorrhage, his ACS and renal function improved and aggressive chemotherapy directed at the lymphoma was commenced. Other cases of percutaneous decompression have been reported as case reports and pilot studies and are described as a safe and effective modality for intra-abdominal hypertension and abdominal compartment syndrome in burn patients with less than 80% of total body surface area, and without inhalation injury.[7,9]

Externally Applied Continuous Negative Abdominal Pressure

Externally applied continuous negative pressure has been applied in the clinical setting and is the subject of Chapter 20.[10]

Octreotide and Melatonin in Secondary Abdominal Compartment Syndrome

Octreotide is a long-acting somatostatin analog widely used in the treatment of metastatic neuroendocrine tumors, acute pancreatitis, and gastrointestinal and pancreatic fistulas. The impact of somatostatin and octreotide on intestinal microcirculation has not been studied in relation to IAP-induced oxidative multiorgan damage. Reperfusion of the ischaemic tissue may release reactive oxygen metabolites that can mediate the microvascular abnormalites that precede organ damage induced by ischaemia and reperfusion These mediators also trigger and activate leukocytes which generate oxygen free radicals that cause further tissue injury. Preventing this sequence with agents such as octreotide may have a protective effect against reperfusion injury.[11] Octreotide has been studied primarily in animals and has shown ability to control neutrophil infiltration and improve the reperfusion-induced oxidative damage after decompression of intra-abdominal hypertension.[12]

Following the same principles for the use of octreotide, experimental studies have focused on melatonin, a secretory product of the pineal gland known to have free radical scavenging

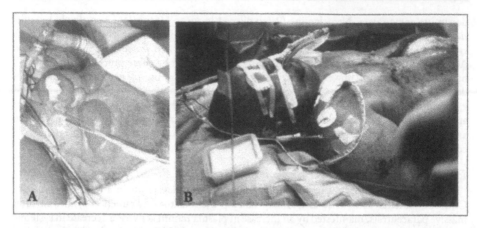

Figure 2A,B. Massive oedema is a frequent complication of aggressive fluid resuscitation, may be increased with some forms of goal directed therapy and increase the risk of all forms of ACS.

and antioxidative properties. It was recently shown that melatonin can reduce lipid peroxidation in cell membranes, a process that promotes cell death as the functional integrity of these structures is damaged. Melatonin also has anti-inflammatory effects and inhibits the activation of neutrophils by free radicals. In rats melatonin reduces reperfusion-induced oxidative organ damage.[13]

Diuretics, Dialysis and/or Ultrafiltration to Remove Excess Edema
Because of the nature of the illness and injury associated with ACS these patients retain large volumes of sodium and water, which exacerbates tissue oedema, IAP and ACS. In the early stages diuretic therapy is often not a viable option because the patients are intravascularly depleted secondary to large capillary leakage due to systemic inflammatory response syndrome (SIRS) and despite their grossly oedematous states. Administration of large volumes of fluid resuscitation is usually unavoidable in the early stages (Figs. 2A, 2B). As the acute SIRS resolves the use of diuretics may be appropriate to reduce oedema and perhaps IAP. For many patients as IAH progresses oliguria and then anuria occur as renal blood flow is reduced. The further administration of fluid in this situation will clearly add to tissue oedema and worsen the IAP, and it becomes mandatory to initiate renal replacement therapy with fluid removal if a secondary or tertiary ACS is to be avoided despite the abdomen being left open.

Targeted Abdominal Perfusion Pressure (APP)
In a similar manner to targeting cerebral perfusion pressure it may be appropriate to target abdominal perfusion pressure (APP), where APP = MAP - IAP, to a level that reduces the risk of worsened splanchic perfusion and subsequent organ dysfunction. Clear guidance on this subject is currently lacking as are levels for APP that should be targeted. It is likely that individual variation and requirements will be different and patients with preexisting hypertension and splanchnic arterial disease (e.g., renal artery stenosis and mesenteric arterial disease) are likely to be high risk groups. There may also be a valid argument for suggesting a role for gastric tonometry as a monitor of the splanchnic circulation.[14] An outcome benefit is yet to be demonstrated by this strategy and when used as part of a oxygen delivery goal directed therapy there is also the potential to increase adverse effects.[15]

General Support (Intensive Care) of the Critically Ill Patient

Management of a patient with an open abdomen mandates high quality intensive care to manage the pulmonary, cardiovascular, splanchnic, urinary and central nervous system effects. Given the clinical scenarios of most patients with raised IAP and ACS most will already be in a critical care facility. If the patient is not intubated it is likely to be appropriate to intubate and ventilate. The administration of appropriate levels of sedation and neuromuscular blockade will then allow accurate assessment of the IAP and optimise ventilation and the patient status for any advanced intervention. This will allow appropriate selection of patients for surgical intervention or further decompression of an already open abdomen. Once abdominal decompression is achieved, it is the usual practice to leave the abdominal fascia and skin open with foreign material at skin level to prevent evisceration. These have the problems of secondary infection, fluid shifts from the exposed bowel and peritoneum, and injury to the bowel.

The nature of acute respiratory failure in these patients may be multifactorial and high airway pressures may be required to achieve satisfactory oxygenation and carbon dioxide elimination. As many of the patients are at high risk for acute lung injury and acute respiratory distress syndrome a protective strategy of ventilation should also be considered.[16] Excessive airway pressure while potentially increasing the risk of volume and barotraumas may also increase IAP.

The haemodynamic consequences of IAH must be addressed. As mentioned above NMB may improve haemodynamic status but the true role of muscle relaxant and compliance of the abdominal wall in patients with ACS has not been studied.[3,17] Hypotension due to hypovolaemia requires rapid correction. Fluid resuscitation may consist of natural or artificial colloids or crystalloids and blood products. As the volume of distribution for crystalloids is larger than for colloids, crystalloid resuscitation needs more fluid to achieve the same end-points and results in more edema. There is however little evidence to support one type of fluid over another. The safety of colloids and albumin in particular has been questioned in patient groups at risk of IAH and ACS.[18,19]

A recent large prospective randomized trial of albumin versus saline for fluid resuscitation (the SAFE study) while demonstrating no excess mortality associated with albumin use found an excess mortality in the albumin group for patients with multitrauma and a head injury. The study also suggests that the current teaching of a 1:3 ratio of volume equivalence is incorrect and that a ratio of 1:1.3 for saline versus albumin.[20]

If following correction of hypovolaemia there is still evidence of hypotension then inotropic/vasopressor administration is appropriate. Human and animal studies suggest potential adverse effects of epinephrine on the splanchnic circulation. Dobutamine, dopamine and noradrenaline have been advocated to improve splanchnic perfusion but there is no good human evidence in IAH and ACS to recommend any particular agent. There is no evidence to support the use of low-dose dopamine for renal protection in a large study that included patients at risk of IAH and ACS.[21]

The aggressive management of acidosis, coagulopathy and hypothermia is required and this is particularly the case for patients who are undergoing damage control surgery. This is recently covered elsewhere.[22]

In the past arguments have been made for trying to achieve supranormal resuscitation levels, aiming for oxygen delivery index higher than 600 mL/min/m². In a retrospective analysis however, Balogh found that a supranormal resuscitation strategy was associated with more fluid infusion (Ringers lactate), decreased intestinal perfusion, intra-abdominal hypertension, abdominal compartment syndrome incidence, multiple organ failure and mortality.[15] In common with the management of septic patients the current goal of resuscitation should instead be to achieve adequate levels of oxygen delivery while avoiding flow dependant tissue hypoxia.[23] Much of what has been recently described as the basis for evidence based intensive care management of patients with sepsis will apply to patient who is at risk of IAH and ACS and this document acts as a useful resource.[23]

Optimisation after Surgical Decompression to Counteract Adverse Effects Associated with Decompression and Prevent Recurrence

Reperfusion syndrome refers to the damage done by restoration of blood flow to ischaemic tissues and is distinct from the original ischaemic insult itself. Reperfusion syndrome may occur at the time of abdominal decompression in some patients with ACS. During decompression, abdominal organs reperfusion may produce arterial hypotension and asystole.[24] Pretreatment or treatment with mannitol and bicarbonate infusions has been recommended. It is likely that this can be avoided if decompression is performed at lower levels of IAP.[25]

Management of ACS by opening the abdomen and temporary abdominal closure does not prevent the development of recurrent ACS and the mortality is high when ACS occurs in this scenario.[26] Optimal medical management may reduce this potential. From what has been presented above some key strategies to prevent ACS or its recurrence would include:

- prevent ileus
- reduce excessive fluid resuscitation
- avoid resuscitative strategies that may increase the incidence of ACS (some forms of goal directed therapy)
- therapy to reduce the inflammatory response

Conclusion

Although medical management of abdominal compartment syndrome may lower intra-abdominal hypertension, reduce systemic effects, and prevent a secondary ACS, surgical decompression currently represents definitive treatment. As there is morbidity and mortality associated with the decompression procedures, temporary abdominal closure and abdominal wall reconstruction, assessment of nonsurgical treatment options is warranted. Some progress has been made in describing the medical and surgical management of ACS but further studies are needed to fully understand the clinical implications and appropriate management.

Commentary

Michael Sugrue

The non-operative medical management of abdominal compartment syndrome (ACS) remains in its infancy. Surgeons have led the crusade relating to ACS and this is reflected in the international literature. Between 1994 and 2003, 28 review articles relating to perioperative renal failure identified only one article that makes reference to intra-abdominal hypertension (IAH).[1] The medical management of the ACS is of crucial importance in optimising abdominal perfusion, and yet avoiding the adverse effects of fluid overload. The abdominal compartment syndrome occurs in medical patients albeit less frequently than in surgical patients. Recognition is vital for early treatment modalities such as aggressive gastrointestinal decompression and facilitation of release of rectal gaseous and faecal material. The role of diuretics and inhibitors of the angiotensin renal vascular system remain to be verified and further exciting research will help in this area. Continuously applied negative abdominal pressure may have a role in a certain sub-group of patients. Until randomised control trials identify key issues in relation to fluid resuscitation and surgery it is important that at least the concept of ACS is recognised in our medical ICUs and that common sense prevail ensuring avoidance of abdominal distension through known basic treatment modalities.

References

1. Reddy VG. Prevention of postoperative acute renal failure. J Postgrad Med 2002; 48:64-70.
1A. Decker G. Abdominal compartment syndrome. J Chir 2001; 138:270-6.
2. Walker J. Criddle LM. Pathophysiology and management of abdominal compartment syndrome. Am J Crit Care 2003; 12(4):367-71, quiz 372-3.
3. Macalino JU, Goldman RK, Mayberry JC. Medical management of abdominal compartment syndrome: Case report and a caution. Asian J Surg 2002; 25:244-6.
4. van der Spoel JI, Oudemans-van Straaten HM, Stoutenbeek CP et al. Neostigmine resolves critical illness-related colonic ileus in intensive care patients with multiple organ failure-a prospective, double-blind, placebo-controlled trial. Intensive Care Med 2001; 27:822-7.
5. Gorecki PJ, Kessler E, Schein M. Abdominal compartment syndrome from intractable constipation. J Am Coll Surg 2000; 190:371.
6. Lohlun J, Margolis M, Gorecki P et al. Fecal impaction causing megarectum-producing colorectal catastrophes. A report of two cases. Digestive Surgery 2000; 17:196-8.
7. Cabrera J, Falcon L, Gorriz E et al. Abdominal decompression plays a major role in early postparacentesis haemodynamic changes in cirrhotic patients with tense ascites. Gut 2001; 48:384-9.
8. Corcos AC, Sherman HF. Percutaneous treatment of secondary abdominal compartment syndrome. J Trauma 2001; 51:1062-4.
9. Latenser BA, Kowal-Vern A, Kimball D et al. A pilot study comparing percutaneous decompression with decompressive laparotomy for acute abdominal compartment syndrome in thermal injury. J Burn Care Rehabil 2002; 23:190-5.
10. Bloomfield G, Saggi B, Blocher C et al. Physiologic effects of externally applied continuous negative abdominal pressure for intra-abdominal hypertension. Journal of Trauma-Injury Infection & Crit Care 1999; 46:1009-14.
11. Kaçmaz A, Polat A, User Y et al. Octreotide: A new approach to the management of acute abdominal hypertension. Peptides 2003; 24:1381-6.
12. Kaçmaz A, Polat A, User Y et al. Octreotide improves reperfusion-induced oxidative injury in acute abdominal hypertension in rats. J Gastrointest Surg 2004; 8:113-9.
13. Sener G, Kaçmaz A, User Y et al. Melatonin ameliorates oxidative organ damage induced by acute intra-abdominal compartment syndrome in rats. J Pineal Res 2003; 35:163-8.
14. Ivatury RR, Porter JM, Simon RJ et al. Intra-abdominal hypertension after life-threatening penetrating abdominal trauma: Prophylaxis, incidence, and clinical relevance to gastric mucosal pH and abdominal compartment syndrome. J Trauma 1998; 44:1016-21.
15. Balogh Z, McKinley BA, Cocanour CS et al. Supranormal trauma resuscitation causes more cases of abdominal compartment syndrome. Arch Surg 2003; 138:637-42.
16. The Acute Respiratory Distress Syndrome Network. Ventilation with lower tidal volumes as compared with traditional tidal volumes for acute lung injury and the acute respiratory distress syndrome. N Engl J Med 2000; 342:1301–1308.
17. De Waele JJ, Benoit D, Hoste E et al. A role for muscle relaxation in patients with abdominal compartment syndrome? Intensive Care Medicine 2003; 29:332.
18. Cochrane injuries group albumin reviewers human albumin administration in critically ill patients: Systematic review of randomised controlled trials. Br Med J 1998; 317:235-40.
19. Schierhout G, Roberts I. Fluid resuscitation with colloid or crystalloid solutions in critically ill patients: A systematic review of randomized trials. Br Med J 1998; 316:961–964.
20. The SAFE study investigators. A comparison of albumin and saline for fluid resuscitation in the intensive care unit. N Engl J Med 2004; 350:2247-56.
21. Bellomo R, Chapman M, Finfer S et al. Australian and New Zealand intensive care society clinical trials group. Low dose dopamine in patients with early renal dysfunction: A placebo controlled randomised trail. Lancet 2000; 356:2139-2143.
22. Parr MJA, Alabdi T. Damage control surgery and intensive care. Injury 2004; 35:713-722.
23. Dellinger RP, Carlet JM, Masur H. Surviving sepsis campaign guidelines for management of severe sepsis and septic shock. Intensive Care Med 2004; 30:536–555.
24. Wysocki A. Abdominal compartment syndrome: Current view. Przeglad Lekarski 2001; 58:463-5.
25. Morris Jr JA, Eddy VA, Blinman TA et al. The staged celiotomy for trauma. Issues in unpacking and reconstruction. Ann Surg 1993; 217:576-86.
26. Gracias VH, Braslow B, Johnson J et al. Abdominal compartment syndrome in the open abdomen. Arch Surg 2002; 137:1298-300.

CHAPTER 20

Continuous Negative Abdominal Pressure

Franco Valenza* and Luciano Gattinoni

Abstract

In this chapter we will focus on the possibility of artificially decreasing intra-abdominal pressure by applying a continuous negative pressure around the abdomen.

We will start from the rationale of this potential noninvasive tool to treat intra-abdominal hypertension, to subsequently describe our initial experience with the use of continuous negative extra abdominal pressure (NEXAP).

The results of a trial conducted using different levels of NEXAP on 30 patients admitted to our intensive care unit will be presented and discussed together with the insights of animal studies we and others conducted to investigate the cardio-respiratory effects of NEXAP.

The bulk of data will be put together so to give to the reader a general view of our understanding of the possibility of using NEXAP to treat the abdominal compartment syndrome.

Rationale

There is an increasing appreciation of the importance of intra-abdominal hypertension (IAH) in critically ill patients. In fact, as many as 50% of patients in the ICU present with an intra-abdominal pressure (IAP) higher than 12 mm Hg, 8% of which are characterized by the abdominal compartment syndrome.[2] The effects of IAH on organ function are well known[3-5] and increasingly recognized by physicians who treat critically ill patients.[6-8]

At higher levels of pressure surgical decompression is an accepted treatment,[3,6,9] while somewhere in between 12 and 25 mm Hg the detrimental consequences of IAH are present but surgical decompression is not indicated. Patients with these values of IAP represent the great majority of those labeled as having IAH, however there is no definite treatment modality for these patients, except for supportive therapy for failing organs.

Recently non surgical decompression of the abdomen has been proposed to treat intracranial hypertension associated with abdominal compartment syndrome: a decrease of IAP with the use of a continuous application of negative pressure around the abdomen of experimental animals has been described.[10,11]

The rationale and the initial knowledge to treat non invasively IAH in critically ill patients was posed. However, the appealing possibility of using negative extra abdominal pressure (NEXAP) in the ICU setting was not yet tested.

Fascinated by this hypothesis, we set out to test if NEXAP could decrease IAP in critically ill patients and if there were major side effects in this population. The data obtained left us with open questions on the cardio-respiratory effects of NEXAP, that prompted us to a subsequent animal study.

In this chapter, after having briefly presented the protocols of the human and animal studies, we will take each of the following questions into consideration:

*Corresponding Author: Franco Valenza —Universita degli Studi di Milano, Istituto di Anestesia e Rianimazione, Ospedale Maggiore di Milano, Milano, Italy.
Email: Franco.valenza@unimi.it

Abdominal Compartment Syndrome, edited by Rao R. Ivatury, Michael L. Cheatham, Manu L. N. G. Malbrain and Michael Sugrue. ©2006 Landes Bioscience.

Table 1. Patient demographics

Number of patients	30
Age (years)	57 ± 17
Sex (M/F)	18/12
Body Mass Index (kg/m^2)	26.1 ± 4
SAPS II	41.8 ± 17
Ramsay score	4.6 ± 1.8
Muscle relaxation	6
Vasoactive drugs	7
Outcome (D/S)	4/26

1. Does NEXAP decrease intra-abdominal pressure?
2. Are there problems with NEXAP application in the critically ill?
3. Does NEXAP alter general hemodynamics, and to what extent?
4. Does NEXAP alter respiratory mechanics?

We will discuss these questions taking into consideration the bulk of data coming from ours and others' investigations.

Human Study

We recruited 30 consecutive patients the characteristics of which are presented below in Table 1. To generate NEXAP, we used a shell (Life Care – Nev 100, Respironics) traditionally used to apply negative pressure around the thorax. To fit the shell over the abdomen, it was rotated by 180°. The apparatus used for the study is described in Figure 1.

Patients were investigated in the supine position. Once basal measurements were taken (Basal), NEXAP was applied on the abdomen, in a random order, at a pressure equal to IAP (NEXAP0), 5 cm H_2O (NEXAP-5) or 10 cm H_2O (NEXAP-10) more negative than NEXAP0.

Measurements included IAP (bladder pressure) and cardio-respiratory parameters. These measurements were taken at each of the four steps after 30 minutes of stabilization.

Animal Study

Eight pigs were sedated and paralysed. They were randomized with respect to abdominal insufflation: four animals had their abdomen insufflated with heium (intra-abdominal hypertension - *IAH*), while basal (*basal*) measurements were taken afterwards. Four animals followed the reverse order.

Cardio-respiratory measurements were taken in both *basal* and *IAH* condition before NEXAP was applied (*pre*), immediately after the transition to NEXAP (*NEXAP 1*), 15 minutes after NEXAP was applied (*NEXAP 2*), and 15 minutes after NEXAP was relieved (*post*).

The figure shows the experimental protocol (Fig. 2).

To generate NEXAP (-20 cm H_2O), we used a shell similar to that of the human study, but smaller (Life Care – Nev 100, Respironics) (Fig. 3).

Does NEXAP Decrease Intra-Abdominal Pressure?

Saggi et al[10] and Bloomfield et al[11] found in animal models that negative pressure around the abdomen significantly decreased IAP. However, to generate negative pressure around the abdomen they used a "large poncho connected to a vacuum into which the entire animal was placed".[11] This is somewhat difficult to obtain in humans; therefore we decide to use the shell decribed above.

The effects of NEXAP on IAP in critically ill patients we found are shown in Figure 4 below:

Basal IAP ranged from 4 to 22 mm Hg. NEXAP decreased IAP from 8.7± 4.3 mm Hg to 6.0 ± 4.2 (Basal vs. NEXAP0, P<0.001). Changes were greater when more negative pressure

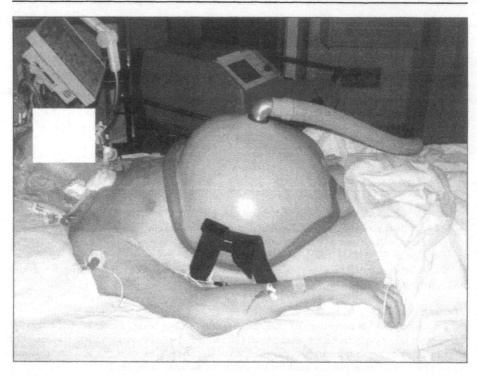

Figure 1. NEXAP applied by shell on abdomen. Reprinted with permission from Valenza F, Bottino N, Canavesi K et al. Intra-abdominal pressure may be decreased non-invasively by continuous negative extra-abdominal pressure (NEXAP). Intensive Care Med 2003; 29(11):2063-7. ©2003 Springer Science and Business Media.

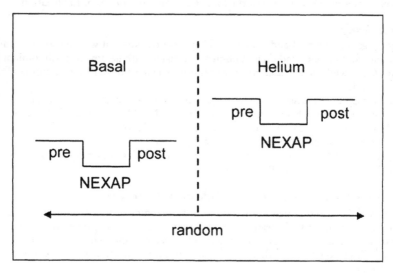

Figure 2. Animal study protocol. Helium for creating IAH.

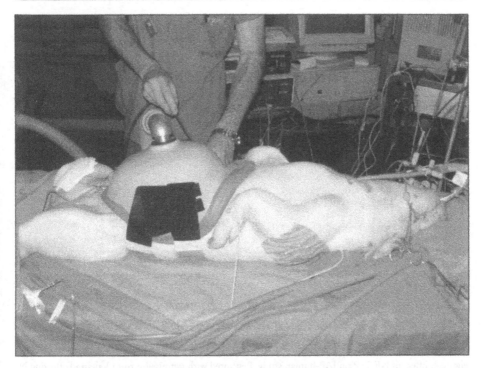

Figure 3. Smaller shell used to create NEXAP in animal model. Reprinted with permission from Valenza F, Irace M, Guglielmi M et al. Effects of continuous negative extra-abdominal pressure on cardiorespiratory function during abdominal hypertension: an experimental study. Intensive Care Med 2005; 31(1):105-11. ©2005 Springer Science and Business Media.

was applied (P<0.001). Average decrease of IAP with NEXAP-10 was 4.5 ± 2.9 mm Hg (maximum decrease 11 mm Hg).

Therefore, as clearly shown, the first of our questions was answered: yes, NEXAP may be used to decrease IAP in critically ill patients.

Are There Problems with NEXAP Application?

The application of NEXAP was well tolerated by patients and there were no major side effects. At the end of the protocol, there was mild and reversible erythema on the skin of the abdomen where the shell was applied. Awake patients suffered some degree of discomfort at higher levels of NEXAP, but only one patient was clearly uncomfortable. However, at NEXAP levels equal to baseline IAP (NEXAP0), all patients were comfortable.

These results are similar to those reported by Sugerman et al who used negative pressure to treat headhache in obese patients affected by pseudotumor cerebri. Patients were intermittently treated several hours a day without major complaints or adverse effects.[12]

Does NEXAP Alter General Hemodynamics, and to What Extent?

The third question we were interested in originated from the kind of patients we were about to treat with NEXAP (ICU patients) and the knowledge that, in animal models, cardiac index remained essentially unchanged during IAH,[11] but decreased when IAP was normal.[10]

We found that NEXAP did not cause any severe impairment of cardiovascular function: MAP did not change, so did HR, if not for a slight significant increase at highest levels of

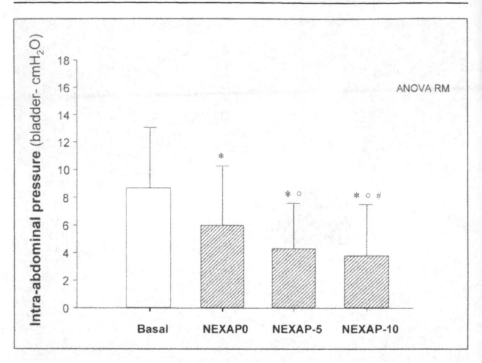

Figure 4. Effect of NEXAP on IAP-human study. Reprinted with permission from Valenza F, Bottino N, Canavesi K et al. Intra-abdominal pressure may be decreased non-invasively by continuous negative extra-abdominal pressure (NEXAP). Intensive Care Med 2003; 29(11):2063-7. ©2003 Springer Science and Business Media.

Table 2. Cardiovascular effects of NEXAP in animals

	Basal	NEXAP0	NEXAP-5	NEXAP-10
Cardiac output (L/min)	6.6 ± 1.5	6.3 ± 1.3	5.7 ± 0.9	6.4 ± 1.7
Stroke volume (mL)	75 ± 17	73 ± 15	69 ± 5	76 ± 14
Heart rate (bpm)	88 ± 12	87 ± 12	83 ± 9	85 ± 12
Mean arterial pressure (mm Hg)	85 ± 13	83 ± 12	85 ± 9	82 ± 13
Wedge pressure (mm Hg)	16.6 ± 3.5	17.0 ± 4	15.8 ± 5.0	15.0 ± 4.2
Central venous pressure (mm Hg)	12.0 ± 4.5	11.8 ± 4.4	10.7 ± 4.5	10.5 ± 5.8
Intra-abdominal pressure (mm Hg)	10.2 ± 6.8	7.4 ± 5	5.2 ± 3.3	5.0 ± 4.2

ANOVA RM. Reprinted with permission from Valenza F, Irace M, Guglielmi M et al. Effects of continuous negative extra-abdominal pressure on cardiorespiratory function during abdominal hypertension: an experimental study. Intensive Care Med 2005; 31(1):105-11. ©2005 Springer Science and Business Media.

negative pressure around the abdomen. When measured, cardiac output was not affected by NEXAP (Table 2).

However, even if patients were stable during NEXAP, central venous pressure decreased and HR slightly increased when CVP changes were greater (NEXAP-10), as from a compensatory response of a preload decrease (Table 3).

Table 3. Cardiovascular effects of NEXAP in patients

	Basal	NEXAP0	NEXAP-5	NEXAP-10
Central venous pressure (cm H_2O)	9.2 ± 3.4	8.4 ± 3.4*	7.5 ± 3.5*°	7.5 ± 3.8*°
Heart rate (bpm)	87 ± 16	89 ± 18	88 ± 17	91 ± 19*
Mean arterial pressure (cm H_2O)	88 ± 13	86 ± 14	86 ± 14	87 ± 15

ANOVA RM, * $p < 0.05$ vs Basal; ° $p < 0.05$ vs. NEXAP0. Reprinted with permission from Valenza F, Bottino N, Canavesi K et al. Intra-abdominal pressure may be decreased non-invasively by continuous negative extra-abdominal pressure (NEXAP). Intensive Care Med 2003; 29(11):2063-7. ©2003 Springer Science and Business Media.

These results were in line with those from the above mentioned animal studies. However, the changes of pleural pressure occurring with IAH impose us to use transmural measurements when dealing with pressure indicators of preload such as CVP (CVP tm=CVP-Pleural pressure), which we did not have in our human study. Therefore, we were not really able to discriminate a possible artefact of measurement from a likely to happen blood shift out of the thoracic cage secondary to NEXAP.

This possible effect of NEXAP is relevant to know when dealing with a critically ill patient, especially in a scenario of IAH so delicate with respect to volume status.

To answer the "blood shift" question we moved to the animal laboratory and measured CVP (as in the previous human study). However, we also used volume indicators of preload, as these are more valuable methods in this context[13] free from any interference derived from changes in pleural pressures. We measured, as indexes of preload, intrathoracic blood volumes (PiCCO system - Pulsion, Medical System AG, Munich, Germany) together with pulmonary vein diameter by means of echocardiography.

The results are shown in Table 4 below. These data prove that NEXAP causes a blood shift away from the thoracic compartment (preload effect). This effect is similar, even if contrary, to that observed during extra-thoracic negative pressure[14-17]

Interestingly, this is possibly the effect that generates the beneficial effect of abdominal decompression in the setting of experimental intracranial hypertension[10,11] or in the setting of pseudotumor cerebri.[12] Notably, the same effect may be deleterious in the absence of IAH when cerebral perfusion pressure is borderline.

Does NEXAP Alter Respiratory Mechanics?

This was not the main focus of our human study; nevertheless, respiratory compliance was computed as tidal volume/(plateau-PEEP) during mandatory volume controlled breaths.

As shown in the table below, NEXAP application was associated with slightly higher airway pressure and lower respiratory system compliance (Table 5).

When looking at these data, the following question rose: Is the effect of NEXAP on respiratory mechanics beneficial or detrimental?

To answer this question we started from the model described in Figure 5, that we applied to our experimental protocol (described above) by measuring the single variables. An example of the acute effects of NEXAP on the measured variables is shown in Figure 6.

As one can see (Table 6), the application of NEXAP caused a drop of Pga and Pes, while end-expiratory airway pressure was similar. Therefore, lung volume (i.e., Ptp=Paw-Pes) increases during NEXAP, as we directly measured in three pigs with the closed helium dilution technique (Fig. 7).

The rise of Paw may thus be explained by a similar tidal volume over-imposed on a greater lung volume generated by NEXAP. In fact, P-V curve analysis showed that cord compliance

Table 4. Animal study: preload indicators after NEXAP

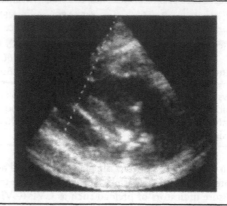

	ITBV	CVP$_{exp}$	AP diam
Basal	462.7 ± 54.2	6.16 ± 1.1	9.17 ± 0.29
NEXAP	422.7 ± 43.2*	3.66 ± 0.7*	7.93 ± 0.12*

* p < 0.05 vs. basal; ITBV: Intrathoracic blood volume; CVPexp: central venous pressure; AP dcam: pulmonary vein dcameter

(considering volume changes equals to individual tidal volumes) decreased from 22.8 ± 5.3 mL/cm H_2O to 18.0 ± 5.0 (P< 0.05) when NEXAP was applied. However, when we considered the increase in lung volume induced by NEXAP from pleural pressure drop, assuming unchanged characteristics of the lung (W/D ratio at the end of the experiment was within normal range: 5.4 ± 0.5), the decrease in cord compliance was no more apparent (23.7 ± 8.6 mL/cm H_2O, P=n.s. vs. basal).

The apparent decrease of compliance is exemplified in Figure 8 below. However, the increase in lung volume is not the only explanation. The rise of Paw may also be explained by a higher chest wall elastance, that was in fact greater during NEXAP when IAP was normal (Fig. 9).

This may be due to the stiffer diaphragm, pulled downwards by NEXAP, or may be due to the shell used to apply NEXAP. In fact, at least in a case, shell positioning on the abdomen, before NEXAP application, was associated with a small increase of pleural pressure.

Table 5. Respiratory mechanisms after NEXAP

	Basal	NEXAP0	NEXAP-5	NEXAP-10
Peak airway pressure (cm H_2O)	28.6 ± 5.9	29.6 ± 6.3	30.8 ± 6.2	30.2 ± 5.8*°
Plateau airway pressure (cm H_2O)	21.1 ± 4.9	21.5 ± 4.9	22.2 ± 4.3	22.0 ± 4.6
Mean airway pressure (cm H_2O)	8.7 ± 3.3	8.9 ± 3.3	8.7 ± 3.5	8.8 ± 3.3
Respiratory compliance (mL/cm H_2O)	50.7 ± 16	49.5 ± 15	46.2 ± 12*	47.3 ± 13*°

ANOVA RM, * p < 0.05 vs Basal; ° p < 0.05 vs NEXAP0

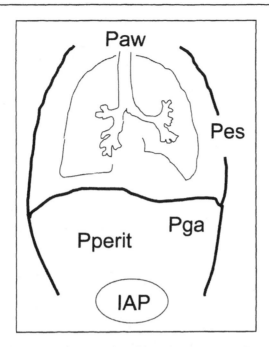

Figure 5. Pressure measurements in the animal model. Paw: airway pressure; Pes: esophageal pressure; Pga: gastric pressure; Pperit: intraperitoneal pressure; IAP: intra-abdominal pressure. Reprinted with permission from Valenza F, Irace M, Guglielmi M et al. Effects of continuous negative extra-abdominal pressure on cardiorespiratory function during abdominal hypertension: an experimental study. Intensive Care Med 2005; 31(1):105-11. ©2005 Springer Science and Business Media.

Additionally, during NEXAP application, rib cage and/or xiphoid are partly squeezed. This is clearly shown in the figure below that shows the recordings from a human subject whose thorax and abdomen was bent with a Respitrace during NEXAP application. This device allows to asses the relative movements of the chest wall (xiphoid in this case) and the abdomen. One can clearly see that rib cage volume decreases immediately after the application of NEXAP, just before the abdomen and the xiphoid itself are pulled up (Fig. 10).

Therefore, despite the net effect of NEXAP that is an increased lung volume, the shell may possibly be better designed, in order to overcome the above mentioned problems.

Perhaps a better design of the tool will also improve its utility in humans and possibly its tolerability over time, by reducing the skin alterations that, in a short time frame, were only minor.

Is NEXAP Effect Different before and after IAH?

On a first analysis of human data, it looked that there was no major different response between sub-groups of patients. In fact, the difference between basal IAP and IAP at NEXAP0 was not correlated with basal IAP as assessed by linear regression (R^2=0.08, P=0.115) or by ANOVA considering basal IAP quartiles as groups (P=0.163). Moreover, changes were not different when patients were stratified according to sedation and paralysis (P=0.964), controlled versus assisted ventilator mode (P=0.849), presence of vasoactive drugs (P=0.142) or body mass index (P=0.226, considering median BMI as cut-off).

However, the number of patients was not great and some of them were characterized by a low-normal IAP.

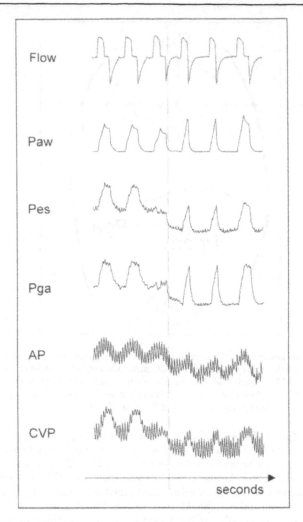

Figure 6. Aortic effects of NEXAP.

To overcome these limitations we investigated the use of NEXAP in the animal experiments before and after IAH induction by means of helium insufflation into the peritoneal space (target pressure: 25 mm Hg).

When we performed ANOVA for repeated measures taking into consideration the effects of IAH, NEXAP and the interaction of the two, we found that NEXAP significantly modified several variables during basal condition, but despite the fact that average values followed the same trend during IAH, these changes were not statistically different (Table 7).

This observation was similar to that of Saggi.[10] It was not due to gas compression, since in one animal whose abdomen was also inflated with water we obtained results similar to those with gas insufflation.

Interestingly, the effect of NEXAP during IAH became evident when more negative pressure was applied. In fact, effective negative pressure (the true distending abdominal pressure, calculated as the difference between negative pressure and IAP, see Fig. 11 below) was linearly correlated with percent changes of ITBV ($R^2 = 0.648$, $P<0.001$) and CVP ($R^2 = 0.522$, $P<0.05$), being the greater the effective NEXAP, the greater the drop.

Table 6. Acute effects of NEXAP

	Pre	NEXAP1	NEXAP2	Post
Flow (L/min)	0.314 ± 0.077	0.311 ± 0.078	0.314 ± 0.093	0.323 ± 0.102
Vt (mL/kg)	11.28 ± 1.14	10.89 ± 0.92	11.31 ± 1.17	11.94 ± 1.39[#]
RR (bpm)	19 ± 5	20 ± 5	20 ± 5	19 ± 5
Paw (cm H_2O)	16.0 ± 3.5	20.2 ± 4.4*	18.3 ± 3.2*[#]	15.3 ± 2.0[#o]
PEEP (cm H_2O)	5.7 ± 1.1	5.8 ± 0.9	5.8 ± 0.8	5.7 ± 0.9
Pes$_{exp}$ (mm Hg)	5.4 ± 1.9	4.1 ± 2.0*	4.2 ± 2.1*	5.4 ± 1.7[#o]
Pga$_{exp}$ (mm Hg)	6.1 ± 1.7	4.3 ± 2.5	4.1 ± 2.4	6.9 ± 1.8[#o]
GEDV (mL)	289 ± 38	257 ± 44*	254 ± 38*	291 ± 31[#o]
ITBV (mL)	358 ± 47	318 ± 54*	314 ± 47*	361 ± 38[#o]
EVLW (mL/kg)	10.9 ± 2.4	9.8 ± 1.4	10.2 ± 1.5	10.6 ± 1.5
CO (mL/min)	2.9 ± 0.5	2.4 ± 0.5*	2.3 ± 0.6*	2.7 ± 0.6°
HR (bpm)	119 ± 12	125 ± 28	122 ± 22	112 ± 15
SV (mL)	25.6 ± 2.8	18.1 ± 8.3*	20.6 ± 4.6*	24.2 ± 4.5
AP (mm Hg)	105 ± 15	92 ± 17*	91 ± 18*	100 ± 14[#o]
CVP$_{exp}$ (cm H_2O)	6.4 ± 1.6	4.0 ± 1.8*	4.6 ± 1.7*[#]	7.0 ± 1.6*[#o]

ANOVA for repeated measures; * $P < 0.05$ vs. pre; [#] $P < 0.05$ vs NEXAP1; ° $P < 0.05$ vs. NEXAP2. Reprinted with permission from Valenza F, Irace M, Guglielmi M et al. Effects of continuous negative extra-abdominal pressure on cardiorespiratory function during abdominal hypertension: an experimental study. Intensive Care Med 2005; 31(1):105-11. ©2005 Springer Science and Business Media.

On the contrary, the behavior of respiratory mechanics was opposite before and after IAH was induced. In fact, NEXAP improved chest wall elastance during IAH, as shown in Figure 12 below.

The decrease of chest wall elastance may be of particular value in the setting of IAH in the ICU. In fact, IAH interferes with respiratory function and even a minor improvement of respiratory function in a mechanically ventilated patient would be desirable, even if it does not necessarily translates into better oxygenation.

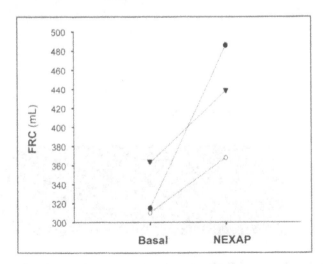

Figure 7. Functional residual capacity (FRC) measured by closed helium dilution.

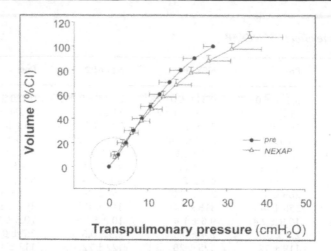

Figure 8. Change in lung compliance before and after NEXAP. Reprinted with permission from Valenza F, Irace M, Guglielmi M et al. Effects of continuous negative extra-abdominal pressure on cardiorespiratory function during abdominal hypertension: an experimental study. Intensive Care Med 2005; 31(1):105-11. ©2005 Springer Science and Business Media.

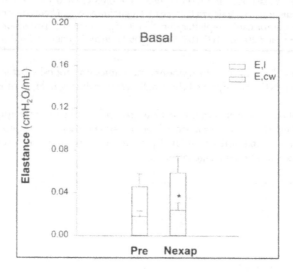

Figure 9. Chest wall elastance before and after NEXAP.

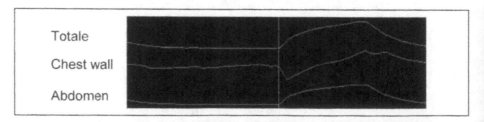

Figure 10. Relative movements of chest wall and abdomen after NEXAP (human subject).

Table 7. Effects of NEXAP in animals with IAH

	Pre	NEXAP1	NEXAP2	Post
Flow (L/min)	0.293 ± 0.095	0.293 ± 0.091	0.296 ± 0.095	0.298 ± 0.094
Vt (mL/kg)	9.85 ± 1.42	9.81 ± 1.29	10.15 ± 1.31	10.13 ± 1.21
RR (bpm)	20 ± 5	20 ± 5	20 ± 5	20 ± 5
Paw (cm H_2O)	30.2 ± 5.2 $	29.7 ± 5.1	29.0 ± 4.5	28.6 ± 4.9*
PEEP (cm H_2O)	5.2 ± 0.8	5.4 ± 0.6	5.4 ± 0.8	5.3 ± 0.8
Pes_{exp} (mm Hg)	5.4 ± 1.2	5.5 ± 1.2	5.1 ± 1.3	5.2 ± 1.4
Pga_{exp} (mm Hg)	12.6 ± 6.9 $	11.7 ± 6.4	10.9 ± 6.2	12.9 ± 5.6
GEDV (mL)	271 ± 36	264 ± 41	258 ± 27	272 ± 36
ITBV (mL)	336 ± 45	329 ± 52	319 ± 34	337 ± 45
EVLW (mL/kg)	11.3 ± 1.6	11.0 ± 1.8	10.3 ± 1.2	11.0 ± 1.7
CO (mL/min)	2.7 ± 0.7	2.5 ± 0.5	2.5 ± 0.5	2.7 ± 0.5
HR (bpm)	121 ± 28	123 ± 25	118 ± 22	126 ± 14
SV (mL)	22.3 ± 3.6	21.2 ± 4.7	21.6 ± 3.9	21.3 ± 4.1
AP (mm Hg)	105 ± 14	105 ± 15	102 ± 16	101 ± 16
CVP_{exp} (cm H_2O)	6.7 ± 2.1	6.8 ± 2.1	6.8 ± 2.1	6.8 ± 2.1

ANOVA for repeated measures; * $P < 0.05$ vs. pre; $ $P < 0.05$ vs. basal

In conclusion, to generate an effective negative pressure during IAH one should target NEXAP to a value at least equal to IAP. The negative effects on hemodynamics are less prominent during IAH, making the use of NEXAP safer in the critically ill patients. Moreover, NEXAP improved chest wall elastance during IAH.

Clinical Perspectives

As detailed above, the use of negative extra-abdominal pressure is a fascinating tool with a great potential as a noninvasive method for decompression of IAH or ACS.

In this chapter we have tried to give the reader a concise view through our attempts to learn as much as possible on NEXAP. However, there is much work still to be done, before NEXAP can be translated into clinical practice, if ever. In fact, even if understanding the underlying mechanisms and learning how to use this tool is mandatory, the next and definitive question we need to answer is the following: does NEXAP improve organ function (or even outcome) if used in patients with IAH in the "gray zone" of IAP between 12 and 25 mm Hg?

We are far from this answer now, but we're on the way to get there.

Commentary

Manu L. N. G. Malbrain

Today we don't have to wait anymore for high intra-abdominal pressure (IAP)-levels above 25-30 mm Hg at which the classic clinical manifestations of abdominal compartment syndrome (ACS) become evident before acting as we used to do 10 years ago. We probably should intervene already at the relatively low levels of 10-15 mm Hg at which subtle changes take place. These alterations remain undetectable by global indices of perfusion but yet can compromise splanchnic perfusion. Prevention is better then cure but surgeons are still reluctant to open the abdomen simply on the basis of an IAP-value without obvious clinical signs of ACS. Therefore we need other nonsurgical, less invasive interventions to lower IAP and stabilize cardiorespiratory dynamics. Whether a critical level of 10-15 mm Hg will remain the same in all patients or whether it depends upon the grading of the response (chronic versus (subacute),

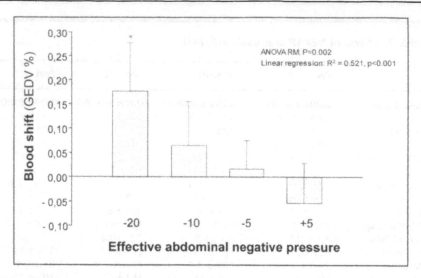

Figure 11. Blood shift during NEXAP and IAH (see text).

underlying and predisposing conditions, filling status, etiologic factors, or comorbidities remain subject for further study. Merging the results from recent retro- and prospective studies with the available literature data on the pathophysiologic implications of intra-abdominal hypertension (IAH), it is probably wise to limit the extent from IAH to ACS. Recently a lot of treatment options have been suggested, such as gastric suctioning, rectal enemas and evacuation, ascites drainage and evacuation (especially in burn patients), albumin with furosemide, ultrafiltration, gastroprokinetics (erythromycin, cisapride, metoclopramide), colonoprokinetics

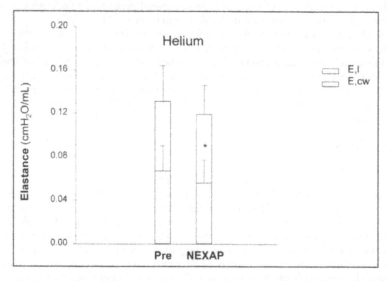

Figure 12. Improved chest wall elastance with NEXAP during IAH. Reprinted with permission from Valenza F, Irace M, Guglielmi M et al. Effects of continuous negative extra-abdominal pressure on cardiorespiratory function during abdominal hypertension: an experimental study. Intensive Care Med 2005; 31(1):105-11. ©2005 Springer Science and Business Media.

(prostygmine), and curarisation. Amongst these the use of external negative abdominal pressure has been advocated.

This chapter's focus is on this last medical noninvasive treatment option for IAH or ACS. Initially used in animal studies, a form of continuous negative extra-abdominal pressure (NEXAP) was recently described in intubated patients with ARDS, using a tank respirator covering the whole body during 2 hour periods.[1] During NEXAP tank pressures of -32.5 cm H_2O at inspiration and -15 cm H_2O at expiration were applied. This resulted in equivalent end-expiratory lung and tidal volumes at lower airway pressures and lower inspiratory transpulmonary pressures between NEXAP periods and conventional positive pressure ventilation (PPV) periods. The use of NEXAP resulted in a drop in IAP from 20 to 1 mm Hg together with better oxygenation and ventilation and stable hemodynamic parameters. The tank respirator covering the whole body was more effective to perform lung-protective ventilation compared to PPV at equivalent lung volumes.

Since the use of a whole body tank respirator is quite difficult, this chapter describes on a novel technique using a modified abdominal shell. A general description of the technique is given together with the results from recent animal and human studies. For the more experienced reader it also gives deeper insights into the cardiorespiratory implications following the use of negative extra-abdominal pressure (NEXAP).

References

1. Raymondos K, Capewell M, Knitsch W et al. Continuous external negative pressure ventilation (CENPV) versus continuous positive pressure ventilation (CPPV) in intubated ARDS patients. Intensive Care Med 2002; 28(suppl 1):S33.
2. Malbrain ML, Chiumello D, Pelosi P et al. Prevalence of intra-abdominal hypertension in critically ill patients: A multicentre epidemiological study. Intensive Care Med 2004.
3. Bailey J, Shapiro MJ. Abdominal compartment syndrome. Critical Care 2000; 4(1):23-29.
4. Barnes GE, Laine GA, Giam PY et al. Cardiovascular responses to elevation of intra-abdominal hydrostatic pressure. Am J Physiol 1985; 248:R208-R213.
5. Mutoh T, Lamm WJ, Embree LJ et al. Abdominal distension alters regional pleural pressures and chest wall mechanics in pigs in vivo. J Appl Physiol 1991; 70:2611-2618.
6. Cullen DJ, Coyle JP, Teplick R et al. Cardiovascular, pulmonary, and renal effects of massively increased intra-abdominal pressure in critically ill patients. Crit Care Med 1989; 17:118-121.
7. Pelosi P, Aspersi M, Chiumello D et al. Measuring intra abdominal pressure in the intensive care setting. Intensivmedizin und Notfallmedizin 2002; 39:509-519.
8. Malbrain ML. Abdominal pressure in the critically ill. Curr Opin Crit Care 2000; 6:17-29.
9. Balogh Z, McKinley BA, Holcomb JB et al. Both primary and secondary abdominal compartment syndrome can be predicted early and are harbingers of multiple organ failure. J Trauma 2003; 54:848-859.
10. Saggi BH, Bloomfield GL, Sugerman HJ et al. Treatment of intracranial hypertension using nonsurgical abdominal decompression. J Trauma 1999; 46(4):646-651.
11. Bloomfield GL, Saggi BH, Blocher C et al. Physiologic effects of externally applied continuous negative abdominal pressure for intra-abdominal hypertension. J Trauma 1999; 46(6):1009-1014, discussion 1014-1016.
12. Sugerman HJ, Felton III WL3, Sismanis A et al. Continuous negative abdominal pressure device for to treat pseudotumor cerebri. Int J Obes Relat Metab Disord 2001; 25(4):486-490.
13. Cheatham ML, Safcsak K, Block EF et al. Preload assessment in patients with an open abdomen. J Trauma 1999; 46:16-22.
14. Borelli M, Benini A, Denkewitz T et al. Effects of continuous negative extrathoracic pressure versus positive end-expiratory pressure in acute lung injury patients. Crit Care Med 1998; 26(6):1025-1031.
15. Adams J, Osiovich H, Goldberg R et al. Hemodynamic effects of continuous negative extrathoracic pressure and continuous positive airway pressure in piglets with normal lungs. Biol Neonate 1992; 62(2-3):69-75.
16. Pierce J, Jenkins I, Noyes J et al. The successful use of continuous negative extrathoracic pressure in a child with Glenn shunt and respiratory failure. Intensive Care Med 1995; 21(9):766-768.
17. Torelli L, Zoccali G, Casarin M et al. Comparative evaluation of the haemodynamic effects of continuous negative external pressure (CNEP) and positive end-expiratory pressure (PEEP) in mechanically ventilated trauma patients. Intensive Care Med 1995; 21(1):67-70.

CHAPTER 21

Anesthetic Considerations in Abdominal Compartment Syndrome

Ingrid R. A. M. Mertens zur Borg,* Serge J. C. Verbrugge and Karel A. Kolkman

Introduction

The primary pathogenesis of intra-abdominal hypertension (IAH) involves an increase in intra-abdominal abdominal pressure (IAP) ranging from 12 mm Hg to over 25 mm Hg in the abdominal compartment syndrome (ACS).[1,2] Increased IAP depresses hemodynamics, increases pleural and intrathoracic pressure and obstructs venous outflow, thereby increasing intracranial pressure (ICP).[3] This leads to cardiovascular and pulmonary problems, splanchnic and renal dysfunction and decreasing cerebral perfusion pressure (CPP).[4-7] This chapter discusses the cardiovascular and pulmonary effects and their relationship to anaesthetic management and monitoring. Focus is on the provision of anaesthetic care and specific anesthetic problems in the patient with IAH/ACS.

Anesthetic Management

Hemodynamic and/or humoral changes are not observed in association with laparoscopic surgery provided that normovolemia, and adequate depth of general anesthesia are continuously maintained together with high plasma level opiate administration.[8] With this anesthetic regime, only if IAP exceeds 12 mm Hg are hemodynamic changes observed.[9,10]

In patients with IAH, IAP is per definition higher than 12 mm Hg, comorbidity exists and altered pharmacokinetics make a straightforward anesthetic regime unlikely. Perioperative myocardial ischemia has to be prevented and O_2 consumption of the heart should be low. This means that the heart rate should be kept below 80 beats/min^{-1}, coronary perfusion pressure above 50 mm Hg, Hb above 6 mmol/L^{-1} and normocarbia should be maintained. Sympathetic stimulation has to be minimal (good sedation and analgesia, and β-blockade if needed). Hypothermia or hypovolemia, and all medication which induces tachycardia should be avoided.[11] If vasopressors are needed, restoring fluid volume should be considered first.

Therapeutic maneuvers in patients with elevated ICP are aimed at maintaining CPP and oxygen delivery. This includes normalization of systemic blood pressure (mean blood pressure >80 mm Hg) and arterial oxygenation (SaO_2>95), sedation and paralysis. If necessary, mannitol or hypertonic saline are administered and possibly a loop diuretic to decrease ICP.

*Corresponding Author: Ingrid R. A. M. Mertens zur Borg—Department of Anesthesiology (Room HN 1279), Erasmus Medical Center, Dr. Molenwaterplein 40, 3015 GD Rotterdam, The Netherlands. Email: i.mertenszurborg@erasmusmc.nl

Abdominal Compartment Syndrome, edited by Rao R. Ivatury, Michael L. Cheatham, Manu L. N. G. Malbrain and Michael Sugrue. ©2006 Landes Bioscience.

Altered Pharmacokinetics and Pharmacodynamics

Dosage of medication depends on the pharmacodynamics of the drug and the pharmacokinetics of the patient. In an ICU patient with IAH and organ dysfunction, absorption, distribution, metabolism and especially excretion of injected drugs and their metabolites can be disturbed. Plasma protein concentration, degree of ionization, volume distribution and hepatic, biliary or renal clearance of the drug are all altered. Biotransformation necessary to inactivate drugs is dependent on enzyme activity which also is altered in the ICU patient. The anesthetist should be familiar with the pharmacodynamics of the drugs being used and should try to create a pharmacokinetic picture of the patient and the interactions with drugs the patient is using during the preoperative evaluation; this in order to estimate what kind of reaction the drug will induce and for how long.

Balanced anesthesia comprises a combination of anesthesia, analgesia and muscle relaxation. Its specific use and problems in a patient with ACS are discussed below.

Premedication

Most IAH/ACS patients are already sedated and mechanicaly ventilated before they are transported to theatre. If the patient is not yet sedated or sedation is not adequate it is recommended to do so, this to reduce stress. Stress causes release of vasoactive hormones which will trouble induction of anesthesia. To sedate a patient who is already hemodynamically unstable and is going to be transported, during which time monitoring and hemodynamic management is not optimal, is not simple. The drug already chosen to sedate the patient in the ICU, can be given as a bolus before preparing transport, amount and timing depending on the pharmacodynamics of the drugs. It is best to select a drug with a limited duration of action, so that modifications can be made if required.

Intravenous Agents, Total Intravenous Anesthesia (TIVA)

Intravenous agents are chosen to induce anesthesia, but can also be used to maintain anesthesia during the procedure (TIVA). Most anesthetics have a direct myocardial depressant and vasodilatory effect, and with the loss of sympathetic tone there will be a reduction of mean arterial pressure (MAP).

Among intravenous (IV) agents, etomidate has the least[12] and thiopenthal, propofol and midazolam the greatest direct cardiovascular depressant activity.[13,14] Ketamine has stimulatory effects when the autonomic nervous system is intact.[15] Therefore ketamine and etomidate are the preferred induction agents,[14,16-18] if hypoperfusion is expected. Ketamine is, however, known to induce elevated ICP which is not preferable if ICP is already expected to be elevated as in IAH/ACS. But at low titrated induction doses, all modern IV anesthetics are unlikely to produce gross hypotension. Titration of the lowest dose of hypnotics with adequate sedation can be achieved safely with BIS monitoring. If there is concern that inadequate anesthesia will result from reducing the dose of anaesthetics, patients may be pretreated with small doses of opioids.

The hypnotics used mostly nowadays for TIVA, are propofol and midazolam. After prolonged use midazolam is eliminated faster than propofol, this because of the high volume of distribution and reduced clearance of propofol especially in patients with organ failure (Table 1).

Etomidate is known to diminish steroid syntheses after prolonged use, inducing an Addison's crisis which can last long after the use of etomidate. This is the reason it is not used anymore for TIVA or prolonged sedation in the ICU.[19] Patients with IAH most likely are in the situation of generating radicals, especially if the compromised organs during decompressive surgery are reperfused. Propofol may be a good choice of anaesthetic when an ischemia-reperfusion injury is anticipated, due to its antioxidant properties.[20,21]

Table 1. Comparative characteristics of (normal) induction doses of Propofol, Midazolam, Diazepam, Etomidate and Ketamine

	Propofol	Etomidate	Thiopenthal	Ketamine
Elimination Half-time (*h*)	0.5-1.5	2-5	11.6	1-2
Volume of Distribution (L/kg^{-1})	3.5-4.5	2.2-4.5	2.5	2.5-3.5
Clearance (mL/kg^{-1}/min^{-1})	30-60	10-20	3.4	16-18
Blood Pressure	decreased	no change	decreased	increased
Heart Rate	decreased	no change	increased	increased
Cardiac output	no change	no change	no change	increased

	Midazolam	Lorazepam	Diazepam	
Elimination Half-time (*h*)	1-4	10-20	21-37	
Volume of Distribution (L/kg^{-1})	1-1.5	0.8-1.3	1-1.5	
Clearance (mL/kg^{-1}/min^{-1})	6-8	0.7-1	0.2-0.5	
Blood Pressure	decreased	decreased	decreased	
Heart Rate	increase	no change	no change	
Cardiac output	no change	no change	no change	

Elimination Half-time (*h*); the time necessary for the plasma concentration of drug to decline 50%, this is directly proportional to its volume of distribution (Vd) and inversely proportional to its clearance. Renal or hepatic disfunction that alters Vd and/or clearance will alter the elimination half-time.

Volume of distribution (Vd); the dose of drug administered intravenously divided by the resulting plasma concentration of drug before elimination starts. Vd is influenced by: (1) lipid solubility; (2) binding to plasma proteins; (3) molecular size and (4) ionization.

Clearance is the volume of plasma cleared of drug by renal excretion and/or metabolism in the liver or other organs. Clearance is one of the most important pharmacokinetic variables to be considered when defining a constant drug infusion regime. Renal clearance can be impaired grossly in ACS patients.

Table includes data from references 81, 82, 83.

Inhalation Agents

Among the inhalation agents, isoflurane has less of an inhibitory effect on the baroreflex mechanism than halothane or enflurane[22] and isoflurane depresses the myocardium less. Isoflurane produces a decrease in mean arterial pressure because of vasodilatory action comparable with that of halothane and enflurane. Therefore, isoflurane offers a distinct advance over the other two.[23] Desflurane and sevoflurane are not significantly better than isoflurane in this regard. However, because of their low solubility in blood they will be rapidly eliminated from the body if severe hemodynamic depression is encountered.[24] Sevoflurane also shows some anti-oxidant properties like propofol.[25] Overall volatile anesthetics can impair immunologic defense mechanisms.[26] Some volatile anesthetics have myocardial protective properties in ischemia reperfusion studies.[27,28]

Nitrous oxide (N_2O) has a direct depressant action on the myocardium, but in vivo it increases sympathetic activity (except for the stressed patient), resulting in a negligible overall change in systemic blood pressure.[29] In addition, N_2O is commonly used with a high percentage of the agents mentioned. This incurs a risk of hypoxemia in patients with reduced CI or pulmonary compromise. Also the use of N_2O has been questioned[30] due to the possibility that it may cause bowel distention. N_2O usage in ACS patients is therefore not advisable.

Analgesia

Opioid agents have little direct cardiovascular or baroreflex depressant effect. This is why in hemodynamically unstable patients high doses of opioids with low doses of hypnotics are chosen. However, these agents can cause hypotension by inhibiting central sympathetic activity, especially in patients whose apparent hemodynamic stability is maintained by a hyperactive sympathetic tone.[31] Ketamine also has analgesic effects but has in contrast, stimulatory effects on central sympathetic activity. Morphine can induce bradycardia and histamine release and associated hypotension. Fentanyl and sufentanyl do not evoke release of histamine.[31,32] Compared with large doses of morphine or fentanyl, sufentanyl results in more rapid induction of anesthesia, earlier emergence from anesthesia, and earlier extubation of the trachea.[33] The new short acting remifentanyl is independent of organ function for its elimination, plasma esterases convert the active drug as inactive metabolites, and it is hemodynamic stable. This is why high doses can be given and hormonal stress response can be maximally depressed, without long acting postoperative side effects like morphine induced ventilatory depression. When remifentanyl is discontinued there will be no more analgesia and other analgesics should be given in adequate doses.

Muscle Relaxation

Adequate muscle relaxation is recommended to minimize abdominal pressure. Succinylcholine, a depolarizing neuromuscular blocking agent, has a rapid onset and short duration of action. It is very useful for rapid sequence induction, but has a few side effects which can be disturbing for the IAH/ACS patient. These are: (1) cardiac dysrhythmias, (2) hyperkalemia, (3) myoglobinuria, (4) increased intracranial pressure (ICP).

There is a variety of nondepolarizing neuromuscular blocking agents, all of them with slower onset and longer duration of action than succinylcholine. Most of them are more or less dependent on hepatic metabolism or biliary- or renal excretion (Table 2), functions which can all be impaired in the ACS patient. Atracurium, the newer cisatracurium and mivacurium are not dependent on these organs but depend on hydrolysis in the plasma. Neuromuscular blocking agents may exert cardiovascular effects through drug-induced release of histamine or other vasoactive substances. Atracurium, mivacurium and succinylcholine are known for this, but especially the older agents like d-tubocurarine and metocurine. Pancuronium and the old drug galamine induce tachycardia, because of selective cardiac vagal blockade and activation of the sympathetic nervous system. Nondepolarizing neuromuscular blocking agents like vecuronium, rocuronium and cisatracurium have an intermediate duration of action and do not interfere with hemodynamics.

IAP Elevation Is Present, the Patient Is Not Yet Intubated/Ventilated

If a patient is presented for operation with elevated IAP, the things of vital interest to the anesthesiologist are; state of hydration, electrolyte disturbance and intubation qualification.

Prehydration, adjustment of electrolytes, decompression by proper naso-gastric tubes (NGT) and adequate pain relief are the objectives of preoperative treatment.

Every acute patient and especially the patient with expected elevated IAP, which has to be operated needs to be intubated. This to protect the lungs from aspiration. The procedure necessary for intubation is the Rapid Sequence Induction (RSI). Besides the "full stomach" of the patient with elevated IAP, every patient in stress has increased gastric acid secretion.[34] Threshold values of a pH <2.5 or gastric volume >25 mL are widely accepted as placing the patient at risk of sequelae should aspiration occur. The combination of cimetidine (a selective and competitive H-2 receptor antagonist, which increases gastric fluid pH and prevents the increase in gastric fluid secretion) and metoclopramide (a dopamine antagonist, that stimulates motility of the upper gastrointestinal tract and increases lower esophageal sphincter tone by 10 to 20 cm H_2O) preoperatively given to the patient, reduce the risk for aspiration.

Table 2. Comparative pharmacology of nondepolarizing muscle relaxants, in intubation dose

	Galamine	Pancuronium	Vecuronium	Atracurium	Mivacurium
Onset to maximum twitch suppression (min)	3-5	3-5	3-5	3-5	2-3
Duration to return control twitch height (min)	60-90	60-90	20-35	20-35	12-20
Renal excretion (%)	> 95	80	15-25	NS	NS
Bliliary excretion (%)	0	5-10	40-75	NS	NS
Hepatic degradation (%)	NS	10-40	20-30	Modest?	NS
Hydrolyses in plasma	0	0	0	yes	yes
Histamine release	0	0	0	Slight	Slight
Cardiac muscarine receptor blockade	moderate	moderate	0	0	0

Table includes data from references 81 and 82.

Before induction of anesthesia the patient is preoxygenated for at least 3 minutes. RSI is performed to gain control of the airway in the least amount of time after the ablation of protective airway reflexes with the induction of anesthesia. Administration of an intravenous anesthetic induction agent is immediately followed by a rapidly acting neuromuscular blocking drug. Direct laryngoscopy and intubation are performed as soon as muscle relaxation is confirmed. Cricoid pressure (Sellick's maneuver) is applied by an assistant from the beginning of induction until confirmation of endotracheal tube placement. After neuromuscular blockade the diaphragm tends to move upwards (this will be pronounced by elevated IAP) and gravity can help to reduce small lung volume. For this reason intubation is performed in anti-Trendelenburg position.

Mechanical Ventilation

The increase in IAP in IAH/ACS will inevitably lead to decreased pulmonary-thoracic compliance, decreased functional residual capacity (FRC), atelectasis, enlargement of the functional right-to-left shunt, hypoxemia with anaerobic metabolism and metabolic acidosis, and pulmonary edema. The application of high inspiratory pressures and volumes with over distension of open alveoli for a long time is associated with an increased risk for barotraumas.[35-39] On the other hand, low levels of positive end-expiratory pressure (PEEP) may contribute to ventilation-induced lung injury by allowing alveoli to collapse and reopen during each respiratory cycle.[36,37,39-41]

Both experimental and clinical data have demonstrated that ventilation settings that prevent lung injury in both healthy and diseased lungs should prevent alveolar over distension and recruit all alveoli and prevent their collapse at end-expiration.[37] The lung should be opened and the lung should be kept open with the least possible pressure swings to ensure the required gas exchange. Hemodynamic side effects are thus minimized.[42,43] The open lung is characterized by an optimal gas exchange.[42] The intrapulmonary shunt is ideally less than 10%, which corresponds to a PaO_2 of more than 450 mm Hg on pure oxygen.[44,45]

A rational treatment concept is the following:[42,43]

1. One must overcome a critical opening pressure during inspiration.
2. This opening pressure must be maintained for a sufficiently long period of time.
3. During expiration, no critical time that would allow collapse of lung units should pass.

The goal of the initial increase in inspiratory pressure is to recruit collapsed alveoli and to determine the critical lung opening pressure. Then, the minimum pressures that prevent the lung from collapse are determined. Finally, after an active reopening maneuver sufficient pressure is implemented to keep the lung open.

A clinical study by Amato et al showed that a ventilation strategy aimed at opening atelectatic lungs and keeping them open at all times, in combination with a treatment strategy of permissive hypercapnia and a restriction on the size of tidal volume and limited peak inspiratory pressures, resulted in a higher rate of weaning from mechanical ventilation, lower rate of barotrauma, and improved 28 day survival in ARDS patients compared to conventional ventilation.[46] The authors stratified the patients according to PEEP levels and concluded that PEEP levels higher than 12 cm H_2O and especially higher than 16 cm H_2O significantly improved survival of these acute respiratory distress syndrome (ARDS) patients.[47]

Alveolar recruitment should usually be possible during the first 48 hours on mechanical ventilation (which may be more difficult if the disease exists for a longer period of time).[48] Even if not all of the lung tissue may be fully recruited for gas exchange, as in consolidating pneumonia, this ventilatory strategy may prevent further damage to the injured portion of the lung.

Weaning and Extubation

Weaning and extubation can be considered if the cause for the elevated IAP is solved and IAP is normalized. Fluid and electrolyte balance should be optimal. In most cases the patient will have been ventilated for several days and first a period of weaning is warranted. Again during the period of weaning it is important that atelectasis is avoided with adequate PEEP levels.

Fluid Management

Hypovolemia

In IAH/ACS there may be occult hypoperfusion that may not be detected by traditional hemodynamic monitoring such as blood pressure, heart rate, and urine output.[49,50] Unrecognized or untreated hypovolemia preoperatively may lead to gross hypotension and hypoperfusion after induction of anesthesia. Brain and kidney have an autoregulation system but the intestinal mucosa lacks autoregulatory control in splanchnic ischemia and is particularly vulnerable to hypoperfusion in combination with IAH. Ischemia and resulting acidosis in the intestinal wall then permits the passage of luminal micro-organisms into the circulation and release of inflammatory mediators, causing sepsis and multiorgan failure.[50] Hypovolemic patients are sensitive to the cardiovascular depressant action of IV and inhalation anesthetics. These not only have direct effects, but they inhibit hemodynamic compensatory mechanisms such as central catecholamine output and baroreflex (neuroregulatory) mechanisms, which maintain systemic pressure in hypovolemia. Two important principles in the use of anaesthetic agents are accurate estimation of the degree of hypovolemia and reduction of the dose accordingly. Intra-vascular volume must be restored before their use. The use of any of the anesthetic drugs in reduced doses is probably more important than the particular agent chosen.[51,52] Usually, reducing the dose of the anaesthetic drug does not result in inadequate anesthesia because the dose needed for induction is decreased in the hypovolemic patient secondary to a variety of effects, such as decreased volume of distribution, preferential distribution of the cardiac output to the brain and the heart, brain hypoxia, dilutional hypoproteinemia, acidosis, and increased sensitivity of the brain to anesthetics.[53]

Fluid Management during IAH/ACS

Sufficient fluid load is required to achieve physiologic endpoints in ICU patients.[54-56] Patients with IAH are even more in need of adequate volume, because of compression within the abdomen, flow towards but especially from the organs has declined and perfusion of the abdominal organs is at risk. Inadequate delivery of oxygen and increased consumption of oxygen preceeds the appearance of multiple organ failure. Fluids and IV cathecholamines are used to achieve supra normal oxygen delivery, but recent meta-analyses does not show clear reduction in mortality with this goal directed therapy and in many cases the goal is not achieved.[57] On the other hand Balogh showed that supranormal trauma resuscitation causes even more cases of ACS.[58]

Infusion Fluids

The ongoing discussion about best type of fluid for volume resuscitation has resulted in several meta-analyses with mortality as an end-point.[59-63] Considering all the critics on these studies, we can conclude that still no consensus about best fluid resuscitation is reached.

In perioperative volume substitution, fluid replacement with neither crystalloids nor colloids has proven to be superior as far as the volume relation of 1:3-4 (colloid to crystalloid) is used.

Balogh who showed an increase in ACS cases with goal directed fluid loading used only crystalloids.[58] Isotonic crystalloid solutions are rapidly dispersed into the extracellular volume of the body. Only one quarter of an isotonic electrolyte solution remains in the intravascular space, so large volumes have to be infused to substitute for intravascular fluid losses. Since fluid shifts from the interstitium into the vascular space occur as a physiological response to a reduced intravascular volume state, infusion of isotonic volume substitutes replace both intravascular and interstitial fluid losses. Increased extracellular fluid may manifest itself as significant tissue edema. This could result in compromised end-organ perfusion both from compression of small vessels and capillaries leading to reduced flow and by increasing diffusion distance within tissues. This will result in reduced oxygen delivery.

Colloidal volume substitutes contain high molecular weight molecules, which maintain or even increase the oncotic pressure of plasma after infusion. Compared with crystalloids, colloids remain in the vascular space for longer, and are thus more effective in restoring hemodynamics, even after infusion of lower volumes. Colloid might leak out into the extravascular space during capillary leakage syndrome. This would result in storage of water in the interstitium, fluid that is difficult to mobilize.[64] In contrast, new hydroxethyl starch (HES) preparations may have capillary sealing properties.[65,66] Depending on their molecular weight profile, structural organ lesions and tissue edema were reduced during inflammation and in reperfusion injury.[67,68]

In recent years hypertonic (600-1800) mosmL-1) crystalloid and colloid solutions have been introduced for certain clinical conditions. Hypertonic saline solutions (HSS) and hypertonic hyperoncotic solutions (HHS) like Hyperhess, expand intravascular volume by shifting endothelial and intracellular water into the intravascular space. Because less volume of HHS is needed, less of the side effects of massive infusion are seen (less edema, increase in nutritional capillary blood flow, reduction of leukocyte adhesion, decrease in bacterial translocation from the gut and reduction of post-ischemia capillary permeability and reperfusion injury).[69,70] During sepsis, hypovolemia by peripheral vasodilatation may be aggravated by increased capillary permeability and subsequent fluid shifts to the extravascular space. HES improves splanchnic microcirculation in septic patients, and HHS might give a reduction in gut hypoperfusion and an increase in oxygen delivery.[69,71] To describe a best fluid protocol for patients with elevated IAP, who have a variety of codiseases like sepsis, capillary leakage syndrome, and ARDS, is not possible. HSS or HHS may treat hypovolemia and disturbed microcirculatory blood flow efficiently. HHS has a low potential for complications;[72] it restores hemodynamic stability faster and without volume overload[73] and it reduces otherwise therapy-resistant intracranial hypertension and improves cerebral perfusion without negatively affecting blood pressure or causing a rebound ICP increase.[74,75] Thus the new HES 140, HSS and HHS seem to be useful fluids in IAH/ACS patients.

Laparotomy

Surgical decompression should be considered in the presence of IAH/ACS. The extent of cardiovascular changes associated with IAH/ACS depend on the interaction of several factors. The amount of IAH, preoperative cardio-respiratory status, the intravascular volume,[9,76-78] and comorbidity all are relevant.

Patients undergoing decompression laparotomy have shown intractable asystole and hypotension during this procedure, therefore vigilance must be maintained during the release of the increased IAP. During decompression surgery the anesthetist might be faced with a deterioration of the patient because of release of inflammatory mediators due to recirculation. In the period following decompression, there is an increase in cardiac index, oxygen delivery, urine output and renal function, and static compliance with a decrease in pulmonary artery occlusion pressure (PAOP) and systemic vascular resistance (SVR).[79] In this phase inotropic use can be reduced and ventilator settings readjusted appropriately.

Conclusion

Manifestations of IAH/ACS include cardiovascular, pulmonary, renal, splanchnic, and neurologic impairment. Patients at risk for IAH/ACS warrant close monitoring and aggressive hemodynamic and pulmonary management.

Haemodynamic and/or humoral changes should be minimized; normovolemia (perhaps hypervolemia), adequate depth of general anesthesia, and high plasma level opiate administration should be continuously maintained. Protective lung strategies may prevent the transfer of inflammatory mediators and the transfer of bacteria and bacterial endotoxins from the bloodstream into the lung and vice versa. There is a vast number of ventilatory strategies available, but all these techniques should comply with one rational concept which prevents further damage due to artificial ventilation itself. It should produce minimal pressure swings during the ventilation cycle and keep the lung open during the whole ventilatory cycle. Open up the whole lung and keep it open with the least possible influence on the cardio-circulatory system.

Anesthesia induces direct myocardial depression and a loss of sympathetic tone with a vasodilatory effect, with concomitant decline in blood pressure, and therefore should be titrated and monitored.

Although many precautions and medical adjustments in anaesthetic management are desirable and possible, especially in ventilation and fluid management; the main issue is to alleviate the IAP by surgical decompression of the abdomen.[80]

Commentary

Michael L. Cheatham

The patient with intra-abdominal hypertension (IAH) or abdominal compartment syndrome (ACS) may manifest significant abnormalities in cardiac, pulmonary, renal, hepatic, gastrointestinal, and cerebral perfusion and function that complicate their minute-to-minute resuscitation as well as long-term morbidity and mortality. Appropriate management is now widely recognized to include open abdominal decompression (OAD) in many patients who develop ACS as a means of reducing intra-abdominal pressure (IAP) and mitigating these life-threatening physiologic derangements. Such a procedure, however, removes the patient from the relative safety of the intensive care unit (ICU) in most hospitals requiring transport to an operating room. The expertise and clinical judgment of the anesthetist thus becomes essential in the team approach necessary to improve survival of the patient with IAH/ACS.

In this chapter, the authors highlight the pertinent issues related to the pre, intra-, and postoperative anesthetic management of these critically ill patients. The relative risks and benefits of various anesthetic regimens and agents with respect to the pathophysiology of IAH/ACS are presented. Preoperative assessment and preparation are mandatory to minimizing the risk of both patient transport as well as subsequent intraoperative complications. The importance of ensuring a safe and patent airway, adequate central venous access, and appropriate hemodynamic monitoring cannot be overemphasized. As discussed, the majority of these patients require elevated levels of positive end-expiratory pressure (PEEP) and increased inspired oxygen fractions (FiO_2) as a result of their decreased pulmonary compliance and increased intrapulmonary shunt. This tenuous pulmonary status can hamper safe transport to the operating room and may require the use of a battery-powered transport ventilator to ensure adequate oxygenation and ventilation during transfer. As a result of the morbidity and mortality associated with intra-hospital transport of the critically ill, performance of OAD and other lifesaving operative procedures at the patient's bedside in the ICU should always be considered. Active communication between the intensivist and anesthetist regarding the potential risks and benefits of transport as well as the patient's response to intraoperative resuscitation and management is essential.

Following transport to the operating room, a number of critical issues may be encountered. The severity of the patient's pulmonary dysfunction may require ventilatory pressures that exceed the capabilities of the typical anesthesia machine. The resulting hypoventilation may result in hypercarbia and hypoxemia in a patient who is ill-prepared to tolerate such changes. As a result, consideration should be given to using the patient's ICU mechanical ventilator in the operating room, adopting a total intravenous anesthetic (TIVA) regimen to maintain anesthesia during the procedure while simultaneously ensuring adequate alveolar ventilation and oxygenation. As discussed by the authors, adequate resuscitation and maintenance of intravascular volume is mandatory. Anesthesia and analgesia may both cause hypotension and systemic malperfusion in the under-resuscitated patient. Further, at the time of decompression, release of IAP results in an acute return of acidotic, hyperkalemic blood containing a variety of pro-inflammatory mediators from both the peripheral and mesenteric vasculature. In the hypovolemic patient, this can result in a sudden drop in blood pressure and occasionally cardiac arrest if intravascular volume is not maintained and acidosis is not corrected. As the diaphragm is now free to move caudally, pulmonary barotrauma may occur if PEEP is not rapidly decreased by approximately 50% at the time of decompression. Following decompression, continued volume resuscitation to maintain systemic perfusion and account for both hemorrhagic and insensible losses will be necessary. Patient transport back to the ICU is generally less perilous for the patient as their ACS has been treated and IAP reduced restoring systemic perfusion and improving pulmonary function.

Emergent anesthetic management of the patient with ACS can be both a physical and mental challenge for the anesthetist who must not only administer a general anesthetic, but also continue the active resuscitation of a profoundly ill patient with numerous pathophysiologic changes. With advances in our understanding and management of IAH and ACS, however, this challenge can also be extremely gratifying as a critically ill patient is snatched from the "jaws of death" presented by ACS.

References

1. Ivatury RR, Diebel L et al. Intra-abdominal hypertension and the abdominal compartment syndrome. Surg Clin North Am 1997; 77(4):783-800.
2. Pelosi P. Measuring intra-abdominal pressure in the intensive care setting. Intensivmed 2002; 39:509-519.
3. Bloomfield GL, Ridings PC et al. A proposed relationship between increased intra-abdominal, intrathoracic, and intracranial pressure. Crit Care Med 1997; 25(3):496-503.
4. Bloomfield GL, Ridings PC et al. Effects of increased intra-abdominal pressure upon intracranial and cerebral perfusion pressure before and after volume expansion. J Trauma 1996; 40(6):936-41; discussion 941-3.
5. Cullen DJ, Coyle JP et al. Cardiovascular, pulmonary, and renal effects of massively increased intra-abdominal pressure in critically ill patients. Crit Care Med 1989; 17(2):118-21.
6. Diebel LN, Wilson RF et al. Effect of increased intra-abdominal pressure on hepatic arterial, portal venous, and hepatic microcirculatory blood flow. J Trauma 1992; 33(2):279-82; discussion 282-3.

7. Sugrue M, Jones F et al. Intra-abdominal hypertension is an independent cause of postoperative renal impairment. Arch Surg 1999; 134(10):1082-5.
8. Lentschener C, Axler O, Fernandez H et al. Haemodynamic changes and vasopressin release are not consistently associated with carbon dioxide pneumoperitoneum in humans. Acta Anaesthesiol Scand 2001; 45(5):527-35.
9. Ishizaki Y, Bandai Y et al. Safe intraabdominal pressure of carbon dioxide pneumoperitoneum during laparoscopic surgery. Surgery 1993; 114(3):549-54.
10. Mertens zur Borg IRAM, Verbrugge S et al. Effect of intraabdominal pressure elevation and positioning on hemodynamic responses during carbon dioxide pneumoperitoneum for laparoscopic for laparoscopic donor nephrectomy. Surg Endosc 2004; 18(6):919-23.
11. Poldermans D, Boersma E et al. The effect of bisoprolol on perioperative mortality and myocardial infarction in high-risk patients undergoing vascular surgery. Dutch Echocardiographic Cardiac Risk Evaluation Applying Stress Echocardiography Study Group. N Engl J Med 1999; 341(24):1789-94.
12. Wauquier A, Ashton D et al. Anti-hypoxic effects of etomidate, thiopental and methohexital. Arch Int Pharmacodyn Ther 1981; 249(2):330-4.
13. Adams P, Gelman S et al. Midazolam pharmacodynamics and pharmacokinetics during acute hypovolemia. Anesthesiology 1985; 63(2):140-6.
14. Gauss A, Heinrich H, Wilder-Smith OH. Echocardiographic assessment of the haemodynamic effects of propofol: a comparison with etomidate and thiopentone. Anaesthesia 1991; 46(2):99-105.
15. Lippmann M, Appel PL et al. Sequential cardiorespiratory patterns of anesthetic induction with ketamine in critically ill patients. Crit Care Med 1983; 11(9):730-4.
16. Doenicke A, Lorenz W et al. Histamine release after intravenous application of short-acting hypnotics. A comparison of etomidate, Althesin (CT1341) and propanidid. Br J Anaesth 1973; 45(11):1097-104.
17. Ebert TJ, Muzi M et al. Sympathetic responses to induction of anesthesia in humans with propofol or etomidate. Anesthesiology 1992; 76(5):725-33.
18. Priano LL, Bernards C, Marrone B. Effect of anesthetic induction agents on cardiovascular neuroregulation in dogs. Anesth Analg 1989; 68(3):344-9.
19. Crozier TA, Schlaeger M et al. TIVA with etomidate-fentanyl versus midazolam-fentanyl. The perioperative stress of coronary surgery overcomes the inhibition of cortisol synthesis caused by etomidate-fentanyl anesthesia. Anaesthesist 1994; 43(9):605-13.
20. De La Cruz JP, Zanca A et al. The effect of propofol on oxidative stress in platelets from surgical patients. Anesth Analg 1999; 89(4):1050-5.
21. Aldemir O, Celebi H et al. The effects of propofol or halothane on free radical production after tourniquet induced ischaemia-reperfusion injury during knee arthroplasty. Acta Anaesthesiol Scand, 2001; 45(10):1221-5.
22. Takeshima R, Dohi S. Comparison of arterial baroreflex function in humans anesthetized with enflurane or isoflurane. Anesth Analg 1989; 69(3):284-90.
23. Theye RA. Effects of anesthetics on whole-body oxygen uptake. Clin Anesth 1975; 11(1):49-60.
24. Gupta A, Stierer T et al. Comparison of recovery profile after ambulatory anesthesia with propofol, isoflurane, sevoflurane and desflurane: a systematic review. Anesth Analg 2004; 98(3):632-41.
25. Allaouchiche B, Debon R et al. Oxidative stress status during exposure to propofol, sevoflurane and desflurane. Anesth Analg 2001; 93(4):981-5.
26. Kotani N, Hashimoto H et al. Exposure to cigarette smoke impairs alveolar macrophage functions during halothane and isoflurane anesthesia in rats. Anesthesiology 1999; 91(6):1823-33.
27. Schlack W, Preckel B et al. Effects of halothane, enflurane, isoflurane, sevoflurane and desflurane on myocardial reperfusion injury in the isolated rat heart. Br J Anaesth 1998; 81(6):913-9.
28. Preckel B, Schlack W et al. Effects of enflurane, isoflurane, sevoflurane and desflurane on reperfusion injury after regional myocardial ischaemia in the rabbit heart in vivo. Br J Anaesth 1998; 81(6):905-12.
29. Ebert TJ, Kampine JP. Nitrous oxide augments sympathetic outflow: direct evidence from human peroneal nerve recordings. Anesth Analg 1989; 69(4):444-9.
30. Taylor E, Feinstein R et al. Anesthesia for laparoscopic cholecystectomy. Is nitrous oxide contraindicated? Anesthesiology 1992; 76(4):541-3.
31. Flacke JW, Bloor BC et al. Comparison of morphine, meperidine, fentanyl, and sufentanil in balanced anesthesia: a double-blind study. Anesth Analg 1985; 64(9):897-910.
32. Rosow CE, Moss J et al. Histamine release during morphine and fentanyl anesthesia. Anesthesiology 1982; 56(2):93-6.
33. Sanford TJ, Jr., Smith N et al. A comparison of morphine, fentanyl, and sufentanil anesthesia for cardiac surgery: induction, emergence, and extubation. Anesth Analg 1986; 65(3):259-66.

34. Bresnick WH, Rask-Madsen C et al. The effect of acute emotional stress on gastric acid secretion in normal subjects and duodenal ulcer patients. J Clin Gastroenterol 1993; 17(2):117-22.
35. Verbrugge SJ, Lachmann B. Mechanisms of ventilation-induced lung injury: physiological rationale to prevent it. Monaldi Arch Chest Dis 1999; 54(1):22-37.
36. Webb HH, Tierney DF. Experimental pulmonary edema due to intermittent positive pressure ventilation with high inflation pressures. Protection by positive end-expiratory pressure. Am Rev Respir Dis 1974; 110(5):556-65.
37. Dreyfuss D, Saumon G. Ventilator-induced lung injury: lessons from experimental studies. Am J Respir Crit Care Med 1998; 157(1):294-323.
38. Dreyfuss D, Basset G et al. Intermittent positive-pressure hyperventilation with high inflation pressures produces pulmonary microvascular injury in rats. Am Rev Respir Dis 1985; 132(4):880-4.
39. Dreyfuss D, Soler P et al. High inflation pressure pulmonary edema. Respective effects of high airway pressure, high tidal volume, and positive end-expiratory pressure. Am Rev Respir Dis 1988; 137(5):1159-64.
40. Tremblay LN, Slutsky AS et al. Role of pressure and volume in ventilator-induced lung injury. Appl Cardiopulm Pathophys 1997; 6:179-190.
41. Ashbaugh DG, Petty TL et al. Continuous positive-pressure breathing (CPPB) in adult respiratory distress syndrome. J Thorac Cardiovasc Surg 1969; 57(1):31-41.
42. Lachmann B. Open up the lung and keep the lung open. Intensive Care Med 1992; 18(6):319-21.
43. Lachmann B, Jonson B et al. Modes of artificial ventilation in severe respiratory distress syndrome. Lung function and morphology in rabbits after wash-out of alveolar surfactant. Crit Care Med 1982; 10(11):724-32.
44. Sjostrand UH, Lichtwarck-Aschoff M et al. Different ventilatory approaches to keep the lung open. Intensive Care Med 1995; 21(4):310-8.
45. Kesecioglu J, Mechanical ventilation in ARDS. Adv Exp Med Biol 1996; 388:533-8.
46. Amato MB, Barbas CS et al. Effect of a protective-ventilation strategy on mortality in the acute respiratory distress syndrome. N Engl J Med 1998; 338(6):347-54.
47. Barbas CSV, Magaldi RB et al. High PEEP levels improved survival in ARDS patients. AM J Resp Crit Care Med 2002; 165:A 218.
48. Grasso S, Mascia L et al. Effects of recruiting maneuvers in patients with acute respiratory distress syndrome ventilated with protective ventilatory strategy. Anesthesiology 2002; 96(4):795-802.
49. Ivatury RR, Porter JM et al. Intra-abdominal hypertension after life-threatening penetrating abdominal trauma: prophylaxis, incidence, and clinical relevance to gastric mucosal pH and abdominal compartment syndrome. J Trauma 1998; 44(6):1016-21; discussion 1021-3.
50. Blow O, Magliore L et al. The golden hour and the silver day: detection and correction of occult hypoperfusion within 24 hours improves outcome from major trauma. J Trauma 1999; 47(5):964-9.
51. Weiskopf RB, Bogetz MS et al. Cardiovascular and metabolic sequelae of inducing anesthesia with ketamine or thiopental in hypovolemic swine. Anesthesiology 1984; 60(3):214-9.
52. Brown DL, Anesthetic agents in trauma surgery: are there differences? Int Anesthesiol Clin 1987; 25(1):75-90.
53. Weiskopf RB, Bogetz MS. Haemorrhage decreases the anaesthetic requirement for ketamine and thiopentone in the pig. Br J Anaesth 1985; 57(10):1022-5.
54. Lobo SM, Salgado PF et al. Effects of maximizing oxygen delivery on morbidity and mortality in high-risk surgical patients. Crit Care Med 2000; 28(10):3396-404.
55. Shoemaker WC, Apple PL et al. Prospective trial of supranormal values of survivors as therapeutic goals in high-risk surgical patients. Chest 1988; 94(6):1176-86.
56. Boyd O, Grounds RM, Bennett ED. A randomized clinical trial of the effect of deliberate perioperative increase of oxygen delivery on mortality in high-risk surgical patients. Jama 1993; 270(22):2699-707.
57. Mackenzie SJ. Should perioperative management target oxygen delivery? Br J Anaesth 2003; 91(5):615-8.
58. Balogh Z, McKinley BA et al. Supranormal trauma resuscitation causes more cases of abdominal compartment syndrome. Arch Surg 2003; 138(6):637-42; discussion 642-3.
59. Bunn F, Roberts I et al. Hypertonic versus isotonic crystalloid for fluid resuscitation in critically ill patients. Cochrane Database Syst Rev 2000(4):CD002045.
60. Bunn F, Alderson P, Hawkins V. Colloid solutions for fluid resuscitation. Cochrane Database Syst Rev 2001; (2):CD001319.
61. Alderson P, Schierhout G et al. Colloids versus crystalloids for fluid resuscitation in critically ill patients. Cochrane Database Syst Rev 2000; (2):CD000567.
62. Choi PT, Yip G et al. Crystalloids vs. colloids in fluid resuscitation: a systematic review. Crit Care Med, 1999; 27(1):200-10.

63. Schierhout G, Roberts I. Fluid resuscitation with colloid or crystalloid solutions in critically ill patients: a systematic review of randomised trials. BMJ 1998; 316(7136):961-4.
64. Roberts JS, Bratton SL. Colloid volume expanders. Problems, pitfalls and possibilities. Drugs 1998; 55(5):621-30.
65. Oz MC, FitsPatrick MF et al. Attenuation of microvascular permeability dysfunction in postischemic striated muscle by hydroxyethyl starch. Microvasc Res 1995; 50(1):71-9.
66. Wisselink W, Patetsios P et al. Medium molecular weight pentastarch reduces reperfusion injury by decreasing capillary leak in an animal model of spinal cord ischemia. J Vasc Surg 1998; 27(1):109-16.
67. Traber LD, Brazeal BA et al. Pentafraction reduces the lung lymph response after endotoxin administration in the ovine model. Circ Shock 1992; 36(2):93-103.
68. Webb AR, Moss RF et al. A narrow range, medium molecular weight pentastarch reduces structural organ damage in a hyperdynamic porcine model of sepsis. Intensive Care Med 1992; 18(6):348-55.
69. Reed LL, Manglano R et al. The effect of hypertonic saline resuscitation on bacterial translocation after hemorrhagic shock in rats. Surgery 1991; 110(4):685-8; discussion 688-90.
70. Nolte D, Bayer M et al. Attenuation of postischemic microvascular disturbances in striated muscle by hyperosmolar saline dextran. Am J Physiol 1992; 263(5 Pt 2):H1411-6.
71. Hannemann L, Korell R et al. [Hypertonic solutions in the intensive care unit]. Zentralbl Chir 1993; 118(5):245-9.
72. Schimetta W, Schochl H et al. Safety of hypertonic hyperoncotic solutions—a survey from Austria. Wien Klin Wochenschr 2002; 114(3):89-95.
73. Christ F, Niklas M et al. Hyperosmotic-hyperoncotic solutions during abdominal aortic aneurysm (AAA) resection. Acta Anaesthesiol Scand 1997; 41(1 Pt 1):62-70.
74. Hartl R, Ghajar J et al. Treatment of refractory intracranial hypertension in severe traumatic brain injury with repetitive hypertonic/hyperoncotic infusions. Zentralbl Chir 1997; 122(3):181-5.
75. Hartl R, Ghajar J et al. Hypertonic/hyperoncotic saline reliably reduces ICP in severely head-injured patients with intracranial hypertension. Acta Neurochir Suppl (Wien), 1997; 70:126-9.
76. Safran D, Sgambati S, Orlando R. Laparoscopy in high-risk cardiac patients. Surg Gynecol Obstet 1993; 176(6):548-54.
77. Safran DB, Orlando R. Physiologic effects of pneumoperitoneum. Am J Surgery 1994; 167(2):281-6.
78. Ho HS, Saunders CJ et al. Effector of hemodynamics during laparoscopy: CO2 absorption or intra-abdominal pressure? J Surg Research 1995; 59(4):497-503.
79. Meldrum DR, Moore FA et al. Prospective characterization and selective management of the abdominal compartment syndrome. Am J Surg 1997; 174(6):667-72; discussion 672-3.
80. Morken J, West MA. Abdominal compartment syndrome in the intensive care unit. Curr Opin Crit Care 2001; 7(4):268-74.
81. Stoelting RK. Pharmacology and physiology in anesthetic practice. New York: Lippincott, Williams & Wilkins, 2005:Tables 5.1, 5.2, 6.1, 6.2, 8.5, 8.8.
82. McCaramon RL, Hilgenburg JC, Stoelting RK. Hemodynamic effects of diazepam-nitrous oxide in patients with coronary artery disease. Anesth Analg 1980; 59:438-41.
83. Tweed WA, Minuck MS, Cymin D. Circulatory response to ketamine anesthesia. Anesthesiology 1972; 37:613-9.

CHAPTER 22

Surgical Management
of Abdominal Compartment Syndrome

Zsolt Balogh, Frederick A. Moore, Claudia E. Goettler,
Michael F. Rotondo, C. William Schwab and Mark J. Kaplan

Part A: The Surgical Management of Abdominal Compartment Syndrome

Zsolt Balogh* and Frederick A. Moore

Introduction

With the evolution of "damage control" laparotomy and "goal directed" ICU resuscitation as standards of care for trauma patients arriving with life threatening hemorrhage, abdominal compartment syndrome (ACS) has emerged to be a virtual epidemic in busy trauma centers worldwide. Many alternative management strategies have been described to minimize risk of ACS and to manage the consequent open abdomens. To date, virtually all of this information is based on retrospective data analysis; no prospective comparative data exists. The purpose of this chapter is to briefly describe alternative methods of temporary abdominal closure (TAC) and review how we incorporate these techniques into the surgical management of ACS, acknowledging that our approach to this vexing problem continues to evolve.

Temporary Abdominal Closure (TAC)

Surgical care of ACS includes prevention, decompression and treatment of the open abdomen. All three aspects of care involve TAC. TAC techniques should be accessed by the nine criteria: (1) easiness (simple, straightforward, available everywhere), (2) cost (inexpensive), (3) time (quick application and removal), (4) drainage (controls body peritoneal fluid and blood), (5) barrier function (protects from evisceration and contamination), (6) facilitate closure (keeps or even brings fascial edges closer to each other), (7) tissue friendly (does not destroy the skin and/or fascia with multiple applications), (8) prevents ACS and (9) prevents fistula formation.

TAC methods we employ are listed in Table A1. None of these are clearly superior to the others and they are not mutually exclusive. In fact, most of our patients are managed with more than one of these techniques and thus it is important that the bedside ICU physicians and nurses be familiar with all of them.

*Corresponding Author: Zsolt Balogh—Department of Traumatology, University of Szeged, Szeged, Hungary. Email: zsoltbalogh@yahoo.com

Abdominal Compartment Syndrome, edited by Rao R. Ivatury, Michael L. Cheatham, Manu L. N. G. Malbrain and Michael Sugrue. ©2006 Landes Bioscience.

Table A1.The comparison of temporary abdominal closure methods

	Towel Clip	Bogotá Bag	Mesh	VAC
Easiness	+ + + +	+ +	+ +	+
Cost	+ + + +	+ + +	-	- -
Time	+ + + +	+ +	+	+
Drainage	+ / -	+ / -	+/ -	+ + + +
Barrier	+ + + +	+ +	+	+ + +
Facilitates closure	+ +	- -	- -	+ + +
Tissue friendly	- - -	- - - -	- -	+ + + +
Prevents ACS	- - -	+ + +	+ + +	+
Prevents fistulas	+ + + +	- - - -	- - -	+ +

ACS: abdominal compartment syndrome; VAC: vacuum-assisted closure

Towel Clip Closure

This is the easiest, cheapest technique and permits quick abdominal reexploration. The towel clips are placed through the skin edges at 2 to 3 cm intervals to approximate the midline wound. Problems include damage to the skin and poor drainage of accumulating intra-abdominal fluid and/or blood. Additionally, because the fascia is almost totally reapproximated there is no room for the increases in abdominal contents and thus patients are at increased risk for ACS. Temporary towel clip closure during damage control laparotomy is a valuable adjunct in patients who present in profound hemorrhage shock.[1] Once the initial hemorrhage control is achieved, the abdomen is packed and the towel clips are applied to generate tamponade effect in the abdominal cavity. This slows ongoing bleeding giving the anesthesiologist time to catch up with the resuscitation. The operating team can also assemble the resources necessary for a second exploration at which time more definitive hemorrhage control can be achieved. This minimizes the need for trips back to the operating room (OR) during early ICU resuscitation for uncontrolled bleeding. For patients being triaged from the OR to ICU, towel clip closure should only be utilized in cases where the wound edges are easily approximated and vigorous ICU resuscitation is not anticipated.

Bogotá Bag

The Bogotá bag (named after the Columbian city) is a large saline infusion bag cut and folded open. This plastic bag is sewn to the skin with strong nylon or polypropylene suture material. Drains and self adhesive foil can be used to achieve better control of the peritoneal fluid and barrier function. The Bogotá bag is efficient in minimizing the occurrence of ACS in those patients who require vigorous ICU resuscitation.[2]

Mesh Closure

Several different types of mesh closures have been described.[3,4] The major role for mesh closure is to provide interposition material between the separated fascial edges to prevent bowel evisceration and to increase abdominal volume to prevent ACS. It should only be employed when early fascial closure is not feasible. Use of mesh over a large defect will result in a ventral hernia which will require complex delayed reconstruction.[5,6] Interposition of omentum between the bowel and the mesh is optimal to minimize the risk of fistula formation. Absorbable mesh is preferred because if a fistula occurs the mesh will ultimately be absorbed and this simplifies subsequent abdominal wall reconstruction. If inadequate omentum is available, a Goretex interposition patch is a reasonable choice to minimize the risk of fistula formation. It, however, will need to be removed in 10 to 14 days. Fluid will begin to accumulate under the patch and it can become infected. At this point the Goretex can be gently peeled off the defect. The underlying bowel will be fixated and covered with granulation tissue. Once granulation tissue has incorporated the mesh and is adequately matured after Goretex patch removal, a split thickness skin graft (STSG) is applied. Unless the defect is small, delayed ventral hernia repairs will be necessary.

Vacuum Assisted Wound Closure (VAWC)

While alternative vacuum pack techniques are described,[7] we utilize a modification of the technique developed by Meredith and colleagues.[8,9] This uses a commercially available VAWC device (Kinetics Corporation Inc., San Antonio, TX, U.S.A.) Diagrammatic depiction of this technique can be found in the manuscript by Garner et al.[10] In brief, a nonadherent plastic barrier (Steri-Drape, 3M Healthcare, St. Paul, MN, U.S.A.) is perforated multiple times with a scalpel. It is then placed over the bowel and extends laterally under the anterior abdominal wall. This is followed by a polyurethane sponge cut to the appropriate size to fit the wound. The sponge is then secured in the wound by closing the skin over it (as much as possible) with a running monofilament nylon suture. Bites are taken close to the skin edge of the wound and are spaced 4 to 5 cm apart. The skin surrounding the wound is coated with benzoin and an occlusive dressing is then applied to the entire abdomen, creating a seal over the wound. The airtight dressing is then placed at -175 mm Hg using an intermittent vacuum system (VAC Therapy, Kinetic Concepts, San Antonio, TX, U.S.A.) Once the sponge is connected to vacuum suction, tension is taken off of the suture that was used to retain the sponge. Generally, this procedure is performed in the OR, but may also be performed at the bedside in the ICU if necessary.

Our Approach

Prevention

ACS has consistently been reported to have a high morbidity and mortality. Recent studies have shown that despite early recognition and decompression outcome remains unacceptable. Prevention, therefore, is the best strategy.[11,12] We believe that patients at high risk to develop ACS can be accurately identified within the first 3 to 6 hours after hospital admission. Hemorrhage control is of paramount importance. Indiscriminant crystalloid infusion should be minimized. We have developed a massive transfusion protocol to insure ready access to blood products and emphasize the early administration of fresh frozen plasma. In damage control surgery, packing is a key method to tamponade hemorrhage, but it also obstructs venous and lymphatic outflow from the gut which exacerbates gut edema with ongoing resuscitation. We, therefore, discourage bulky packing and advocate early pack removal (usually within 24 hours).[13] At initial "damage control" laparotomy, a generous Bogotá bag is placed with anticipation that abdominal contents will increase due to resuscitation induced edema and ongoing bleeding. Patients are triaged to the ICU where resuscitation is completed concurrent with rewarming and correction of coagulopathy. Patients must be closely monitored to avoid overzealous resuscitation or resuscitation of unrecognized potentially correctable (i.e., by interventional radiology) sources of bleeding (e.g., pelvic fractures or a packed liver). Continuous monitoring of urinary bladder pressure and gastric regional CO_2 levels by tomometry is desirable for early detection of intra-abdominal hypertension (IAH).

Decompression Laparotomy

To date surgical decompression is the accepted therapeutic intervention for full blown ACS (see definitions). When the organ dysfunctions (cardiac, pulmonary, renal) are present and due to IAH, abdominal decompression is a lifesaving intervention. Presumptive decompression for IAH without organ dysfunctions has been advocated by some authorities. At this point in time, given the hazards of managing the open abdomen, we do not recommend this. Surgical decompression usually entails a full midline incision, evacuation of the peritoneal fluid and the application of a TAC. The procedure can be done on the ICU, which is especially advisable in critically ill patients on maximal ventilatory and renal support.[14] On-site decompression should be avoided when the cause of the ACS potentially can not be managed outside the operative room environment (i.e., uncontrollable bleeding). If the IAH was the cause of the organ dysfunctions (except for terminal cases) marked improvement is observed in oxygenation, cardiac output, airway pressures, visceral perfusion and urine output. Among these improvements in cardiac output and urine output are associated with improved outcome.[11]

Management of the Open Abdomen

This is an organized strategy with planned reexplorations, dressing changes and progressive fascial approximation. If this is not feasible, then planned ventral hernia formation and late reconstruction will be needed. Patients undergoing "damage control" or decompressive laparotomy have a Bogotá bag closure. At their second laparotomy the fascial is closed if there is no excessive tension. If this is not feasible the VAWC devise is applied. The dressing, sponge, and barrier are changed at 2 to 3 day intervals. At each dressing change, the abdomen is explored and washed out as much as possible. The fascia is then closed inferiorly and superiorly as much as possible using interrupted sutures, and the sponge component is down sized to match the defect size of the fascia. The dressing changes are repeated until fascia is completely closed. Once fascia is closed, the subcutaneous tissue is allowed to heal by secondary intention. Patients are removed from mechanical ventilation, extubated, and discharged from the ICU when they meet standard criteria. Extubated patients are returned to the OR and undergo general anesthesia for dressing changes and fascial approximation.

We have published two reports of our use of the VAWC device. The first, by Garner et al included 14 selected general surgery and trauma patients with open abdomens.[10] In this initial experience, early definitive fascial closure was achieved in 13 (92%) patients with associated morbidity of two superficial wound infections. The second series described 104 consecutive reported trauma patients who met specific high risk criteria and were resuscitated by a standardized process.[15] Seventy four required emergency laparotomies of which 55 were initially closed with a Bogotá bag. At the second laparotomy, the midline fascia could be primarily closed in 19 (35%), the remaining 36 (65%) required application of the VAWC device. There were six early deaths. Of the remaining 29 discharged patients, we achieved early fascial closure in 25 (86%) at a mean of 7 ± 1 days (range 3 to 18 days). Four patients failed VAWC, two developed fistulas. There were no intra-abdominal infections.

Other groups have described the use of vacuum-assisted closure. Barker et al have published a series of articles describing their vacuum pack technique.[7] Similar to our technique, they place a perforated polyethylene sheet over the bowel that extends laterally under the anterior abdominal wall. However, instead of a sponge, they place a moistened, folded, sterile surgical towel over the polyethylene sheet. Two 10-French flat silicone drains are placed on top of the towel followed by an occlusive dressing that seals the wound. The drains are then connected via a Y-adaptor to continuous negative wall suction. Their reported success of obtaining fascial closure is less then ours. They reported on 112 patients, of which 88 (79%) survived. Of these survivors, 62 (70%) achieved primary fascial closure, 25 (28%) underwent mesh repairs, 1 was closed with skin only, and 2 were closed by secondary intervention. They had five fistulas and five intra-abdominal abscesses. Meredith and colleagues from Wake Forest, using the same technique as we do, have two recent reports which document a success rate similar to ours. In their combined series they report 116 survivors in whom the VAWC device was used of which 97 (84%) achieved primary fascial closure at a mean of 9.5 days.[8,9]

Late Reconstruction

Despite our success with the VAWC device, we still have a subset of patients who end up with fascial defects and large disabling hernias. The presence of large ventral hernias significantly interferes with professional and social life resulting poor quality of life. Once, however, the abdominal wall is reconstructed, these patients get back to their normal life and regain their preACS quality of life.[16] As with any difficult problem, multiple techniques of abdominal wall reconstruction of large ventral hernias have been described. In a noncontaminated surgical field, we use nonabsorbable mesh if we can interpose abdominal wall and/or omentum between the mesh and the underlying bowel. If this is not feasible, we then perform the component separation technique as recently reported by Jernigan et al.[6] If this is not feasible due to previous loss of the abdominal wall, we enlist the assistance of our plastic surgery colleague to mobilize pedicle flaps. An elegant solution is the full thickness innervated latissimus dorsi flap, these requires microsurgical skills and 5 to 6 hours operating room time. Ninkovic et al had excellent results with this technique; patients regained enough contractile power in the full thickness flap to support their abdominal wall.[17]

Summary

With prospective awareness, better resuscitation and advanced hemorrhage control techniques the incidence and hopefully the mortality of ACS can be decreased.[12] This approach, however, will lead to increased number of open abdomens. To decrease the open abdomen related morbidity and mortality, we need to continue to focus on primary fascial closure. After the initial decompression or preventive open abdomen treatment TAC that does not involve fascial sutures is recommended. We do not advocate VAWC as first time TAC. It is expensive and one third of the patients can have their fascia primarily closed at the second laparotomy. Additionally, there are reported cases in the literature when the utilization of VAWC immediately after ACS decompression resulted in recurrent ACS.[18] If primary closure can not be achieved at the second look procedure VAWC is our method of choice for TAC. With regular 72 hours changes of the VAWC primary fascial closure can be achieved up to 88% of the severe shock/trauma patients with open abdomen. Mesh interposition is recommended only if primary fascial closure can not be performed. There is no high quality data concerning the ideal interposition material. Based on case series and our local expert opinion Goretex seems to be the best choice. The classic method of planned ventral hernia formation still has a role in the toughest cases. Open granulation may be expedited with the VAWC, which is followed by STSG is the standard method. At 6-12 month delayed abdominal wall reconstruction recommended to a regain the original quality of life.

References

1. Moore EE. Staged laparotomy for the hypothermia, acidosis and coagulopathy syndrome. Am J Surg 1996; 172:405-410.
2. Offner PJ, de Souza AL, Moore EE et al. Avoidance of abdominal compartment syndrome in damage-control laparotomy after trauma. Arch Surg 2001; 136:676-681.
3. Schachtrupp A, Fackeldey V, Klinge U et al. Temporary closure of the abdominal wall (laparostomy). Hernia 2002; 6:155-62.
4. Nagy KK, Fildes JJ, Mahr C et al. Experience with three prosthetic materials in temporary abdominal wall closure. Am Surg 1996; 62:331-5.
5. Cohen M, Morales Jr R, Fildes J et al. Staged reconstruction after gunshot wounds to the abdomen. Plast Reconstr Surg 2001; 108:83-92.
6. Jernigan TW, Fabian TC, Croce MA et al. Staged management of giant abdominal wall defects: Acute and long-term results. Ann Surg 2003; 238:349-55.
7. Barker DE, Kaufman HJ, Smith LA et al. Vacuum pack technique of temporary abdominal closure: A 7-year experience with 112 patients. J Trauma 2000; 48:201-6.
8. Miller PR, Thompson JT, Faler BJ et al. Late fascial closure in lieu of ventral hernia: The next step in open abdomen management. J Trauma 2002; 53:843-9.
9. Miller PR, Meredith JW, Johnson JC et al. Prospective evaluation of vacuum-assisted fascial closure after open abdomen: Planned ventral hernia rate is substantially reduced. Ann Surg 2004; 239:608-14.
10. Garner GB, Ware DN, Cocanour CS et al. Vacuum-assisted wound closure provides early fascial reapproximation in trauma patients with open abdomens. Am J Surg 2001; 182:630-638.
11. Balogh Z, McKinley BA, Holcomb JB et al. Both primary and secondary abdominal compartment syndrome (ACS) can be predicted early and are harbingers of multiple organ failure. J Trauma 2003; (6)54:848-861.

12. Balogh Z, McKinley BA, Cox Jr CS et al. Abdominal Compartment Syndrome: The Cause or Effect of Postinjury Multiple Organ Failure. Shock 2003; 20:483-492.
13. Meldrum DR, Moore FA, Moore EE et al. Cardiopulmonary hazards of perihepatic packing for major liver injuries. Am J Surg 1995; 170:537-542.
14. Balogh Z, McKinley BA, Cocanour CS et al. Secondary Abdominal Compartment Syndrome: An Elusive Complication of Traumatic Shock Resuscitation. Am J Surg 2002; 184:538-544.
15. Suliburk JW, Ware DN, Balogh Z et al. Vacuum-assisted wound closure achieves early fascial closure of open abdomens after severe trauma. J Trauma 2003; 55:1155-60.
16. Cheatham ML, Safcsak K, Llerena LE et al. Long-term physical, mental, and functional consequences of abdominal decompression. J Trauma 2004; 56:237-41.
17. Ninkovic M, Kronberger P, Harpf C et al. Free innervated latissimus dorsi muscle flap for reconstruction of full-thickness abdominal wall defects. Plast Reconstr Surg 1998; 101:971-8.
18. Gracias VH, Braslow B, Johnson J et al. Abdominal compartment syndrome in the open abdomen. Arch Surg 2002; 137:1298-300.

Part B: Surgical Management of the Open Abdomen after Damage Control or Abdominal Compartment Syndrome

Claudia E. Goettler,* Michael F. Rotondo and C. William Schwab

Introduction

Indications for leaving an abdomen open include "Damage Control", abdominal compartment syndrome and prevention of abdominal compartment syndrome, and planned repeat operation or operations. Regardless of the reason for maintaining an open abdomen, multiple further decisions must be made. These include the method of temporary abdominal content containment, the timing of reoperation, when to attempt abdominal wall closure, and what method of closure is selected. Finally, the management of certain circumstances such as feeding and drainage tubes or ostomy placement and care require special considerations and planning.

Indications for Open Abdomen

A growing body of literature clearly shows that increased intra-abdominal pressures result in deleterious physiologic effects. These include the defining criteria of abdominal compartment syndrome, including ventilatory difficulties, oliguria, and hypotension. Even more poorly understood is the role ACS plays in the prolongation of the systemic inflammatory response with resultant multi-system organ failure and/or reperfusion injury.

As the consequences of abdominal compartment syndrome have been appreciated, it has become common to manage high risk patients with an open abdomen to prevent this condition. Determination of which patients are at risk remains an inexact science, however; consideration for open abdomen management should be made in all emergency laparotomy and cases associated with large resuscitation requirements, given in a short time period. Greater than 10 liters of crystalloid resuscitation and/or more than six units of blood given acutely have been suggested by some. In addition the presence of bowel and/or retroperitoneal edema protruding above the fascia has also been suggested as an operative sign requiring open abdominal management.

Any patient in whom ongoing large volume aggressive resuscitation is likely should not be closed. This group of patients includes the "Damage Control" population who undergo abbreviation of their surgical procedure after hemostasis and containment of intestinal contamination.[1,2] These patients are hypothermic, coagulopathic and acidotic and hence will require

*Corresponding Author: Claudia E. Goettler—Department of Surgery, Brody School of Medicine, East Carolina University, 600 Moye Blvd., Greenville, North Carolina, 27858 U.S.A. Email: c.goettle@pcmh.com

both rapid termination of their procedure (temporary abdominal containment) as well as ongoing, potentially massive resuscitation. They are best managed with an open abdomen and some form of transient synthetic abdominal wall closure.

Other reasons for maintaining an open abdomen is severe peritonitis requiring serial abdominal washouts, ischemic viscera requiring second-look laparotomy, removal of packs used for hemostasis or serial debridments required to manage pancreatic necrosis. Repeated opening and closing of the fascia in these cases results in fascial damage and loss, and future difficulty in definitive closure. The use of a temporary abdominal containment dressings affords an excellent option that accommodates any volume of extra abdominal viscera and leaves all layers of the abdominal wall untouched.

A subset of patients without intra-abdominal pathology will develop abdominal compartment syndrome after massive resuscitation for their disease process. This has been seen after extensive orthopedic injuries, large body surface burn injury, severe pancreatis and occasionally in medical patients. The common scenario is some massive inflammatory event requiring large crytalloid resuscitation given over a short time (12-24 hours).

Temporary Abdominal Containment

Once the decision is made to manage the patient with an open abdomen, a temporary dressing must be selected and used to keep all abdominal viscera contained to prevent further contamination of the peritoneal cavity and optimally seal the abdomen from fluid leakage. Methods of temporary containment (Damage Control part 1) vary widely but most have several important common factors. The optimal containment method is rapid and inexpensive. The dressings must have enough surface area to cover any size of extra abdominal visceral protuberence without causing tension on the abdominal wall or increasing intra-abdominal pressure. Optimally the dressing will prevent leakage of fluids but allow egress and collection of intraperitoneal fluid. This allows accurate measurements of intake and output, protects the patient's skin from maceration due to dampness, and facilitates nursing care. The material should be nonreactive to avoid adhesion formation and slippery enough to allow changes in bowel size and position as edema increases and subsequently resolves, as well as normal peristalsis. The dressing method should also allow rapid reopening for second look laparotomy or development of abdominal hypertension.

Skin closure, by towel clips or whipstitch is a rapid technique that maintains abdominal domain and avoids injury to the fascia. The use of towel clips does not create a watertight seal and the clips interfere with radiographic studies. Skin closure with a large running continuous nonabsorbable suture is more watertight and radiolucent. It does not involve use of synthetic sheets of foreign material and on occasion can be maintained as a permanent closure with planned ventral hernia repair in the future. Both methods provide little increase in abdominal volume and should be considered as temporary containment techniques used only transiently to move a patient quickly to another therapeutic modality, such as angiographic embolization.

Interposition methods of closure allow coverage of the largest and most protuberant of intra-abdominal contents. These have been shown to decrease multiple organ failure, abdominal compartment syndrome, abscess, necrotizing fasciitis and fistula, and improve outcome in a diverse group of patients.[3,4,5] These interposition closures may be done with various materials and several will be discussed. Initial placement of these temporary bridging synthetic material requires fixing them to the fascia or skin. Thus they are fixed and most are not layered or elastic enough to allow expansion as visceral edema increases. Therefore, despite a large increase in abdominal volume, recurrent abdominal compartment syndrome can occur.

The "Bogota bag" is the least expensive method[6] and is still the most commonly used technique in a survey of American trauma surgeons.[7] This method consists of an opened sterilized intravenous solution bag, usually a three liter irrigation bag, which is sewn into the abdominal defect, either to skin or fascia. This device is the most inexpensive bridging material and has the advantage of transparency, which allows the abdominal contents to be inspected without opening it. It does not provide a watertight seal, and requires time to sew it in place in the operating room. It does allow rapid repeat laparotomy, as it can be simply opened down the middle and the reclosed with a running large suture.

Mesh of all types has also been described. The disadvantages of this method, however, are numerous. None of the mesh varieties are water tight, as even Gortex (Polytetrafluoroethylene, Gore & Assoc., Flagstaff, AZ) mesh leaks around the edges. They all require added operative time to affix to the abdominal wall. If placed tightly enough to maintain abdominal domain, recurrent abdominal compartment syndrome is likely, thereby canceling this advantage entirely, as the mesh will need to be opened or reapplied. Vicryl (polyglactic acid, Ethicon, Somerville, NJ) or Dexon (polyglycolic acid, Davis & Geck, Danbury, CT) mesh have very little tensile strength and tends to tear both during suturing and as the abdominal contents expand and place pressure on it. Activity as minimal as nurses turning the patient in the bed can result in evisceration. Polypropylene mesh (Marlex, Bard, Billerica MA ; Prolene, Ethicon, Somervill, NJ; Surgipro, US Surgical, Norwalk, CT) is very strong, stiff and abrasive. It adheres to the underlying bowel, incites an inflammatory response and is associated with increased fistula rates, as high as 12-50%.[8] Gortex mesh is nonadherent and quite strong, very expensive and has a high infection rate.

Our method of choice for rapid temporary abdominal containment is the vacuum adhesive dressing ("Vac-Pac"). This has been described with many minor permutations and recently there is a commercially manufactured vac-sponge system.[9,10] In general, this dressing consists of a nonadhesive, soft, clear plastic layer tucked into the abdomen under the peritoneum against the bowels. We utilize an adhesive plastic sheet backed by a sterile towel. This is covered by an interposition layer which allows fluid out of the abdomen (we use moist roll gauze) containing closed suction drains. These drains perform two functions: they collect fluid for accurate volume assessment and patient cleanliness and provide continuous suction to maintain negative pressure within the dressing and abdomen. Over this and most of the anterior abdominal wall and opening, a large adhesive drape is attached. This dressing can be applied in minutes, is quite inexpensive (about $40), and is watertight. (Figs. B1-B4) If recurrent abdominal compartment syndrome occurs, the top adhesive dressing can be slit which allows for expansion of the abdominal contents, still covered by the plasticised towel. A third large adhesive drape can be reapplied. This dressing can be made to cover any size of abdominal visceral protuberance and does not require any suturing in fascia or skin. Infrequently minor skin de-epithelializaion due to traction of the adhesive dressing during stretch of the abdominal wall occurs.

Resuscitation Period (Damage Control Part 2)

During the resuscitation period in the Intensive Care Unit, patients typically require large volumes of fluid and/or blood products. Ongoing evaluation for abdominal hypertension is necessary as it is possible to develop "recurrent" abdominal compartment syndrome despite an initially loose temporary closure even with interposition methods.[11] It is our practice to measure bladder pressures at least every four hours until the patient shows evidence of physiologic stability. Bladder pressures are measured utilizing an arterial line transducer connected to the bladder drainage catheter. Instillation of 60 mL of sterile saline with the urinary drainage tube clamped distal to the transducer connection allows measurement of intravesical pressure. This directly reflects intra-abdominal pressure in most circumstances. Pressures greater than 15 mm Hg are considered abnormal and require further evaluation for clinical evidence of abdominal compartment syndrome (elevated ventilatory peak pressures greater than 40 mm Hg, low BP, etc).[12] In patients with signs of abdominal compartment syndrome and numerical evidence of intra-abdominal hypertension, reopening of any abdominal closure will be necessary.

Decisions as to when to return to the operating room fall generally in two categories: patients who stabilize quickly and have no signs of ongoing bleeding and those patients who do not normalize physiology and are suspected of ongoing bleeding. Patients who stabilize quickly, within 4-8 hours, (clearance of acidosis and lactate, normalize temperature and hemodynamic parameters) can be returned to the operating room in 24-36 hours from their initial operation for completion all definitive procedures (Damage Control part 3). This includes unpacking,

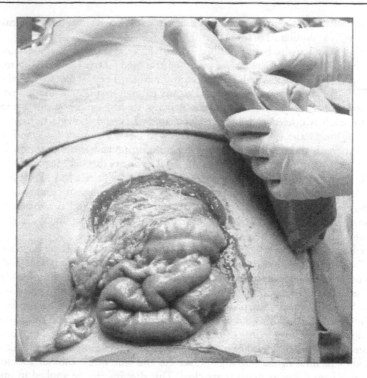

Figure B1. A surgical towel covered with adhesive plastic is placed under the fascia with the smooth side against the bowel.

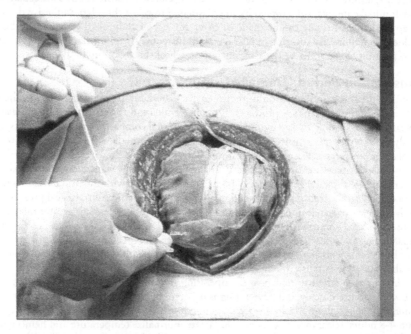

Figure B2. Drains are placed in the subcutaneous space and a roll gauze provides a separation to maintain drain function.

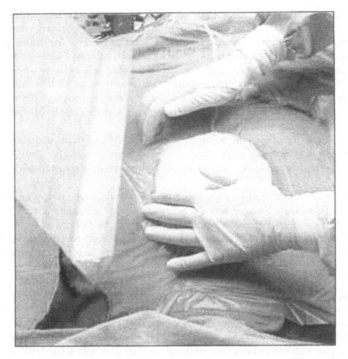

Figure B3. An adhesive dressing is applied over the entire abdomen.

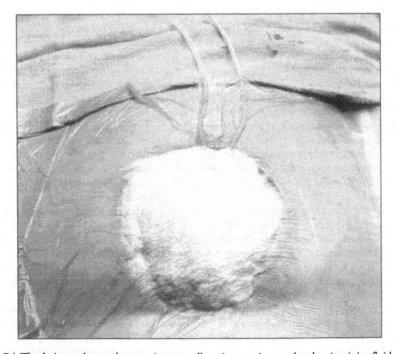

Figure B4. The drains are kept to low continuous wall suction creating a seal and maintaining fluid control.

thorough reexamination of the abdominal contents for missed injuries, definitive repair of identified injuries, and reconstruction of all gastrointestinal or genitourinary viscera. Decisions regarding definitive abdominal closure and feeding access depend on the degree of injury and on visceral and abdominal wall edema. If the abdomen can be closed without tension, this is done and appropriate feeding access is provided. If fascial closure is unlikely without tension or a third reoperation is felt necessary, another temporary dressing is applied. No consideration to fascial closure should be undertaken until the patient has resolved most of the visceral edema and the intestines no longer protrude. We use the rough gauge that all abdominal viscera must be at or below the fascia wall to attempt closure.

The second category of patients fail to have normalization of their hemodynamic parameters despite aggressive resuscitation. This occurs usually from ongoing bleeding, but can result from missed injury or a missed or progression of abdominal hypertension with a temporary dressing.

In patients with evidence of ongoing bleeding (bloody drainage, dropping hemoglobin), immediate return to the operating room is necessary for hemorrhage control. Patients with evidence of abdominal hypertension without bleeding can be opened at the bedside and a temporary dressing like a vac-pack applied. These patients should be explored for evidence of missed injury as well as iatrogenic injury. All surfaces, suture lines, and retroperitoneal and extraabdominal body regions should be inspected for site of occult blood loss.

Managing and Changing Temporary Closure

Patients with temporary closure require management of their abdominal dressing. In patients with packing for bleeding (liver or retroperitonium), or those requiring definitive visceral operative procedures (bowel anastomsis), dressing changes are done in the operating room. For some patients with intra-abdominal edema or requiring simple serial washouts, these can be done at the bedside in the Intensive Care Unit. Proper equipment, trained nursing staff and anesthesia are required.

Packs used to control bleeding can generally be removed in 24-36 hours, with replacement as necessary. Infection rates vary from 10% to 69%,[13] and probably increase with duration of pack retention.[14] Washout interval for infectious processes depends on judgement with most surgeons doing more frequent washouts (daily) early in the course and then lengthening the interval as the patient improves. For "routine" change of temporary closure over edematous contents, the dressing can be changed every other day, though there is data to support any interval. Clinically, increasing the interval seems to increase the risk of infection while too frequent manipulations seems to risk bowel injury.

A decision should be made at each dressing change regarding the potential for fascial closure. If the patient is diuresing after full resuscitation and edema and protrusion of abdominal contents is decreasing, it is reasonable to place another temporary dressing and reassess in two days. However, the longer the abdomen is left open, the greater the loss of domain. Overall, about 50-70% of these patients will be able to undergo primary fascial closure[15] during their hospitalization. 20-30% will require long-term management of an open abdomen.

It is optimal to provide a permanent closure within a week to reduce the metabolic demand on the patient and reduce the risk of intestinal injury from frequent dressings,[7] but this may not be possible. If the patient has been successfully resuscitated but is not showing signs of third space fluid mobilization, resolution of edema and return of abdominal viscera into the true abdomen, a transitional closure method should be undertaken. This decision can typically be made approximately 72-96 hours after resuscitation has been completed, and will be subsequently discussed.

A rarer subset of open abdomen patients are those maintained open for serial abdominal washouts. In these patients, placement of a prosthetic to maintain abdominal domain but prevent injury to the fascia by serial reopening and reclosing is optimal. These prosthetics are opened down the midline and reclosed at each operation, without manipulation of the fascia.

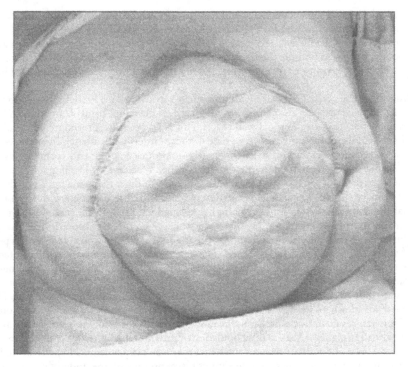

Figure B5. The large abdominal defect left after skin grafting an open abdomen.

Options for this include polypropylene mesh with an nonadherent plastic backing to prevent bowel injury,[16] zippered mesh,[17,18] and hook-and-loop fabric sewn to mesh.[19]

Transitional Closure

Several methods of transitional closure exist for patients who remain too edematous or too ill to undergo permanent abdominal closure. Each technique has relative merits.

Simplest is the placement of absorbable interposition mesh. This is typically sewn to the fascia and should not be made tight as abdominal hypertension can occur. In addition, there is little tensile strength to absorbable mesh; if it is secured tightly it will tear with resultant evisceration. Due to the laxity, absorbability, and low tensile strength of the mesh, it does little to maintain abdominal domain, and hence can be sewn to the skin edge just as well. Absorbable mesh can safely be used in an infected environment and is porous which allows egress of abdominal fluids.[7] Some believe it stimulates granulation tissue and promotes faster times to STSG application. The mesh is quite soft and has a low rate of fistula formation (5%).[8,15] Dressing changes over the mesh are undertaken until a granulation bed forms. As the mesh gradually degrades, skin graft coverage can be undertaken (usually within 2 weeks).[7] Skin closure over the mesh can rarely be accomplished. STSG over granulated mesh is simple, expeditious and a safe transient abdominal coverage method. These patients will have a very large ventral hernias requiring future complex abdominal wall closure (Fig. B5). In our experience, this has the least morbidity for these severely ill patients. It allows long periods of recovery, weight gain and proper planning for elective abdominal wall reconstruction in the future.

Polypropylene mesh (Marlex ®) is permanent and so acts to maintain abdominal domain when secured to the fascia. It is porous and hence allows extravasation of peritoneal fluids. However it has two major drawbacks, which should preclude its use. It can not be used in an infected field, and if it becomes infected will require removal. More morbid is the abrasive and adherent nature of this material making its use a prohibitive risk of intestinal fistulae.[7]

Gortex mesh is nonadherent and has a very low rate of fistula formation. It is permanent but is poorly integrated into the tissues. In addition, the pore size of Gortex allows bacteria to enter but does not permit the entry of white blood cells, resulting in high risk of mesh infection.[7]

While the use of permanent mesh with immediate soft tissue flaps for definitive abdominal closure has been described, this method has a high risk of infection, fistula formation and, if it fails, even greater loss of abdominal wall tissue. This is an extensive procedure and probably should not be done in the early recovery phase of those critically ill patients. In a few selected patients, with clean wounds and excellent metabolic proficiency, the end cosmetic result is acceptable but some long term abdominal laxity is to be expected after healing.

Various methods of serial closure have been espoused to draw the fascia back together for delayed primary closure. These include creation of a silo affixed to the fascia with a prosthetic material and daily tightening, similar to omphalocele reduction. Multiple methods have been used for this technique, including the use of binder clips![20] Various tension devices also exist including a hook-and-loop fabric system,[19] serial abdominal closure,[21] and retention type sutures[22] which can be sequentially tightened. These methods have the advantage of not requiring a second elective procedure after final delayed primary closure, though incisional hernias are relatively common.

This attempt to draw the fascia together under tension has significant risks as well. Since these methods attach to the fascia, ongoing tightening results in continuous tension on the fascia. This can result in fascial ischemia, tearing and loss with resultant even more difficult abdominal wall to manage. Also, as this population of patients remains critically ill, this method does increase intra-abdominal pressures and hence may result in abdominal hypertension. Last, the "spanning" material, whether mesh or sutures places pressure on the underlying viscera and can cause bowel erosion and fistula. The closure rate with these various methods is difficult to standardize in this very diverse population but is about 52%.[23] It must be stressed with any of these tension techniques, that tightening should only occur with decrease in abdominal contents as a method of maintaining abdominal domain. Any attempt to close more quickly by placing tension on the fascia is likely to result in any of the above complications, each of which is a major set back and at times mortal event. Hence, these sequential closure methods must be used with great caution and in our experience do not add significantly to the speed of eventual primary fascial closure.

A possible exception to this is an emerging technique utilizing vacuum assisted fascial closure. This technique utilizes vacuum created tension along the wound edges to provide a limited and constant tension without requiring suturing injury to the fascia. The vacuum also provides removal of intraperitoneal fluids, contributing to a decrease in abdominal contents. This study was randomized against absorbable mesh and skin graft and found that an additional 26% of patients could be closed in a delayed fashion (by 21 days) with no increase in fistula or mortality rate and a 9% early hernia rate.[24]

An additional method of closure is primary component separation during the initial hospitalization. This requires lateral release and/or rectus flip to bridge the abdominal gap. The procedure is lengthy and uses the natural musculo-aponeurotic layers to give strength. This should not be considered a safe method of management and is mentioned only for completeness. While this provides excellent cosmetic outcome and is a definitive procedure, the risks are extraordinary. If it fails due to infection, the development of abdominal hypertension or fistula, few options remain for any further attempts at reconstruction. While each of these methods has been individually described, there is no data comparing methods. In addition, the diversity of the populations requiring such closure hinders any useful comparisons. No prospective trials have been done.

Tubes and Stoma in the Open Abdomen

Special discussion is warranted regarding traversing the abdominal wall with tubes or stomas in patients with an open abdomen. This becomes problematic for a number of reasons. Initially, increasing edema of the abdominal viscera and the abdominal wall can result in retraction of the bowel, feeding tubes or drains. These may be lost and stomas can drop back into the abdomen or become ischemic. Later, as the edema resolves, the relative position of the viscera again changes, creating different geography, tethering points and reducing of the intestines used for stoma. For these reasons, we avoid if at all possible the placement of stomas or trans-abdominal feeding tubes in the early care of these patients. Despite the presumed risk, it may be safer to perform a primary bowel anastomosis than bring out a stoma.[2,13] Drains, if required, should be placed with considerable redundancy and be brought out as lateral as possible to avoid damage to the components of the rectus muscle. Feeding in these patients should be accomplished by a naso-duodenal feeding tube placed intra-operatively during the first return to the operating room and surgically placed gastrostomy or jejunostomy should be avoided for the above reasons. Typically, these patients require a period of TPN followed by longer periods of enteral tube feeding as they recover. Eventually patients can resume feeding by mouth. Additionally, there are case reports of percutaneous endoscopic gastrostomy in these patients after the abdominal wall edema has subsided,[25] but we have been reluctant to create openings in the bowel and have attempted to avoid trans-abdominal feeding tubes.

Later in the patient's course, sites of tubes and stomas in the abdominal wall make definitive closure, particularly if component separation is necessary, more difficult due to scarring between tissue layers and creating holes in the components. For this reason, when stoma and tubes are deemed necessary, they are brought out lateral to the rectus and well into the oblique muscles. Stomas are brought out on the flank between the costal margin and the iliac crest, at or slightly posterior to the mid-axillary line. We have formed many of these flank stomas and find that they avoid soilage in the open abdomen, are away from any dressing or transitional closure and provide significant benefit for future closure as they leave the rectus components intact.

Fistulae

The bane of open abdomen management is the enterocutaneous fistula. The rate of these varies, depending on the initial pathology and the method of abdominal containment from 1 to 15% of patients.[7] Fistulae seem to be most common in patients closed with polypropylene mesh with rates as high as 12-50%.[8] These fistulae commonly form to the mesh and have been given the name "entero-atmospheric fistula", as there is essentially no intervening tissue. Additionally, since these fistulae are within or at the edge of a granulating bed, control of contents is extremely difficult. This results in an extremely low nonsurgical closure rate (about 25%).[23]

Prevention is the key to this problem. At initial and subsequent operations, the omentum should be spared and placed in the anterior "open" position to afford intestinal coverage. Bowel anastomosis should be tucked deep in the abdomen with intervening bowel and omentum separating them from the abdominal closure. In our experience the ileo-mid transverse colon anastomoses is the most likely to fistulize as it sits anterior and high in the open abdomen and is probably injured by any transient dressing. The intestines must be handled with extreme care at interval washouts and dressing changes, since over time adhesions form and increases the chance for intestinal injury with any manipulation. Hence, it is optimal to move to a transitional closure as soon as possible. When mesh is placed, omentum should be utilized to protect the bowel. Dressings over the top of the mesh must be kept moist and are usually covered with a layer of petrolatum gauze or continuous wet to moist dressings. These dressings should never be allowed to dry out. Skin grafting should occur as soon as the wound is adequately granulated.[7]

If a fistula develops, standard therapy with bowel rest and intravenous nutrition is started. Moreover, fistula output is best controlled with a suction drain at the site to prevent further erosion of the skin and granulating bed by the caustic intestinal contents. Early skin grafting around the site is still beneficial. Some of the graft may be lost if secretions are not adequately controlled. However, if secretions are well controlled and peri-fistula healing ensues, the mature split thickness skin graft will support the placement of a stoma appliance. Enterostomal therapists can be particularly helpful in these instances. Small fistulae may close with this method, but larger fistulae, which may take on the appearance of an ostomy, will not. These will require eventual surgical resection and anastomosis. This should be done only when the patient is otherwise recovered, is nutritionally replete and is prepared for definitive abdominal closure. This typically occurs at six to twelve months after skin grafting. Operatively, the skin graft is removed, the fistula resected, bowel is anastomosed and the anastomosis is then protected from the abdominal wall closure by tucking it deep into the abdomen, and optimally under omental coverage. Should a recurrent fistula develop, closure with nonoperative therapy is commonly successful, as there is intervening native tissue that will contract and heal with time and no distal bowel obstruction.

Definitive Abdominal Wall Repair

Timing to elective definitive abdominal wall reconstruction is variable. As previously discussed, the patient must have recovered from the precipitating condition and be nutritionally replete. In addition, time should allow for intra-abdominal adhesions to soften. This typically occurs at 6-12 months after the last operations, sometimes sooner if thick omentum could be interposed between the intestines and the mesh. During this time, we require our patient to wear an abdominal binder at all times as we feel that it helps in the maintenance of abdominal domain by preventing increased protrusion of the bowels. It is also helpful by providing comfort and increased ability to move about by stabilizing the open abdominal wall. The patient should have a continuous physical therapy regime to regain mobility and strength. Weight gain should be encouraged. Last, substance abuse, smoking cessation and psychological counseling should be offered as a prerequisite for definitive reconstruction.

Determination of the timing of the definitive reconstruction depends on many things. If a transient closure has been created using a split thickness skin graft, the timing of operation depends on the separation of the mesh/skin graft from the underlying bowel. This can be determined on physical exam by the "pinch test" (Fig. B6). When the skin graft can be lifted from the bowel like the skin of the back of the hand, separation is likely to be fairly straightforward. Nutritional assessment with adequate weight gain and fat deposition is important and indicates that there is adequate subcutaneous fat to allow skin graft separation and support flap formation.

In cases of the patient with an open abdomen and a stoma, special consideration is needed. Many surgeons do not combine these procedures and will open the skin graft, create bowel continuity, and then close the skin graft. In these patients, closure of the open abdomen is then completed several months later at a second operation. We have had success with combining these procedures. We utilize a strict preoperative bowel preparation (mechanical and oral antibiotics) to decrease contamination. We close nearly all patients with native tissues by rectus component separation, with occasional extraperitoneal mesh reinforcement.

Waiting longer than one year for reconstruction is not advisable. Continued loss of abdominal domain occurs during the longer waiting periods,[8] occasionally resulting in massive skin graft hernia sacs. Clearly this makes reconstruction difficult if not impossible. Plastic surgical techniques such as tissue expanders can help in these circumstances. Again, prevention and timing are paramount and we instruct all of our patients to wear abdominal binders at all times.

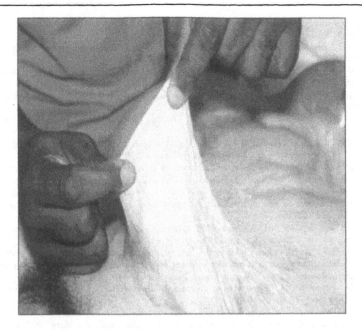

Figure B6. The "pinch test" indicates separation of the skin graft from the underlying bowel.

Summary

Managing open abdomen patients is challenging and requires careful attention to detail. All patients are best managed with a synthetic abdominal containment techniques such as a vacuum plastic dressing. Thereafter, return to the operating room for definitive surgical care is carried out as soon as the patient is resuscitated, or sooner if the patient fails to progress to physiologic stabilization. Primary closure of the abdomen is optimal but should only be considered in patients with resolved edema and no evidence of intra-abdominal hypertension. The remaining patients are closed with a transient dressing typically with absorbable mesh and skin grafting. Care must be taken to avoid fistulae. Stomas, external drainage, and feeding tubes require special considerations for placement, use and removal. Eventually, when the patient is recovered and nutritionally replete, an elective abdominal wall reconstruction can be completed.

References

1. Rotondo MF, Schwab CW, McGonigal MD et al. "Damage control": An approach for improved survival in exsanguinating penetrating abdominal injury. J Trauma 1993; 35:375-383.
2. Johnson JW, Gracias VH, Schwab CW et al. Evolution in damage control for exsanguinating penetrating abdominal injury. J Trauma 2001; 51:261-271.
3. Mayberry JC, Mullins RJ, Crass RA et al. Prevention of abdominal compartment syndrome by absorbable mesh prosthesis closure. Arch Surg 1997; 132:957-962.
4. Oelschlager BK, Boyle EM, Johansen K et al. Delayed abdominal closure in the management of ruptured abdominal aortic aneurysms. Am J Surg 1997; 173:411-415.
5. Rasmussen TE, Hallett JW, Noel AA et al. Early abdominal closure with mesh reduces multiple organ failure after ruptured abdominal aortic aneurysm repair: Guidelines from a 10-year case-control study. J Vasc Surg 2002; 35:246-253.
6. Ghimenton F, Thomson SR, Muckart DJJ et al. Abdominal content containment: Practicalities and outcome. Br J Srug 2000; 87:106-109.
7. Mayberry JC. Bedside open abdominal surgery. Critical care clinics 2000; 16:151-172.
8. Jernigan TW, Fabian TC, Croce MA et al. Staged management of giant abdominal wall defects. Ann Surg 2003; 238:349-357.
9. Markley MA, Mantor PC, Letton RW et al. Pediatric Vacuum packing wound closure for damage-control laparotomy. J Pediat Surg 2002; 37:512-514.

10. Sherck J, Seiver A, Shatney C et al. Covering the "open abdomen": A better technique. Am Surg 1998; 64:854-857.
11. Gracias VH, Braslow B, Johnson J et al. Abdominal compartment syndrome in the open abdomen. Arch Surg 2002; 137:1198-1300.
12. Burch JM, Moore EE, Morre FA et al. The abdominal compartment syndrome. Surg Clin North Am 1996; 76:833-842.
13. Morris Jr JA, Eddy VA, Blinman TA et al. The staged celiotomy for trauma. Issues in unpacking and reconstruction. Ann Surg 1993; 217:576-586.
14. Shapiro MB, Jenkins DH, Schwab CW et al. Damage control: Collective review. J Trauma 2000; 49:969-978.
15. Barker DE, Kaufman HJ, Smith LA et al. Vacuum pack technique of temporary abdominal closure: A 7-year experience with 112 patients. J Trauma 2000; 48:201-206.
16. Edwards MR, Siddipui MN. The open abdomen—a simple cost-effective technique for laparostomy management. Ann R Coll Surg Engl 2003; 85:281-282.
17. Garcia-Sabrido JL, Tallado JM, Christou NV et al. Treatment of severe intra-abdominal sepsis and/or necrotic foci by an "open-abdomen" approach. Zipper and zipper-mesh techniques. Arch Surg 1988; 123:152-156.
18. Hedderich GS, Wexler MJ, McLean APH et al. The septic abdomen: Open management with Marlex mesh with a zipper. Surgery 1986; 99:399-407.
19. Wittmann DH, Aprahamian C, Bergstein JM. Etappenlavage. Advanced diffuse peritonitis managed by planned multiple laparotomies utilizing zippers, slide fastener, and Velcro analogue for temporary abdominal closure. World J Surg 1990; 14:218-226.
20. Myers JA, Latensier BA. Nonoperative progressive "Bogota bag" closure after abdominal decompression. Am Surg 2002; 68:1029-1030.
21. Kafie FE, Tessier DJ, Williams RA et al. Serial abdominal closure technique (the "SAC" procedure): A novel method for delayed closure of the abdominal wall. Am Surg 2003; 69:102-105.
22. Koniaris LG, Hendrickson RJ, Drugas G et al. Dynamic retention. A technique for closure of the complex abdomen in critically ill patients. Arch Surg 136:1359-1362.
23. Tremblay LN, Feliciano DV, Schmidt J et al. Skin only or silo closure in the critically ill patient with an open abdomen. Am J Surg 2002; 182:670-675.
24. Miller PR, Thompson JT, Faler BJ et al. Late fascial closure in lieu of ventral hernia: The next step in open abdomen management. J Trauma 2002; 53:843-849.
25. Block EFJ, Cheatham ML, Bee TK. Percutaneous endoscopic gastrostomy in patients with an open abdomen. Am Surg 2001; 67:913-914.

Part C: Surgical Approaches to the Open Abdomen

Mark J. Kaplan*

Abstract

Surgical management of the open abdomen is a major challenge for the operating surgeon. There are numerous approaches reported to manage a patient with an open abdomen. Most reports are retrospective and uncontrolled. Most open abdomen protocols are based on surgeon preference and experience. Closure techniques should take into account the underlying status of the patient, need for multiple procedures, maintenance of fascial integrity, control abdominal fluid secretion, protection of the abdominal contents from further damage, and minimize the effects of intra-abdominal hypertension. In patients with an open abdomen, abdominal compartment syndrome is a significant risk and continues to require serial bladder pressures to prevent the development of pathologic intra-abdominal hypertension. Risks associated with an open abdominal approach include hernia formation, enterocutaneous fistulas, exacerbation of septic states and high risk of multiple organ dysfunction. Complications can be minimized with an understanding of the pathophysiology found in the open abdomen, use of materials to minimize trauma to the abdominal contents, utilization of closure techniques to

*Mark J. Kaplan—Division of Trauma and Surgical Critical Care, Albert Einstein Medical Center, Einstein Center One, Suite 205, Philadelphia, Pennsylvania, U.S.A. Email: kaplanm@einstein.edu

minimize exposure of bowel, and to control the inflammatory response found in the open abdomen.

Introduction

The open abdomen presents numerous management challenges for the operating surgeon, particularly in patients with traumatic injuries, intra-abdominal sepsis, acidosis, hypothermia, coagulopathy, bowel edema, and intra-abdominal packing.[1] Patients with open abdomens usually require multiple reexplorations and are resource intensive. The major challenge in managing the open abdomen is control of abdominal contents, intra-abdominal fluid secretion, facilitation of abdominal exploration, preservation of the fascia for abdominal wall closure, and minimizing the effects of intra-abdominal hypertension. Although there are no prospective studies documenting the effectiveness of an open abdomen retrospective studies have shown that maintaining an open abdomen is effective in the appropriate clinical setting.[2]

Currently, standard protocols have not been defined for the management of the open abdomen. Frequency of exploration, abdominal wall coverage, surgical procedures, and indications for definitive closure are based on the clinical judgment of the operating surgeon. Indications for an open abdomen include bowel edema, severe intra-abdominal infection, hypothermia, acidosis, and significant risk of developing an abdominal compartment syndrome.

The most common reason for an open abdomen is damage-control surgery.[3] An abbreviated laparotomy is performed to address life-threatening injuries, bleeding is controlled with packing, and bowel injuries are closed without anastomosis. The abdomen is closed with a temporary barrier to maintain bowel integrity and to allow for easy entry in the abdomen. Also, with the use of a temporary closure, the abdomen is decompressed and the risk of developing of abdominal compartment syndrome (ACS) is lowered. The overall result is an increase in short term survival with an associated increase in morbidity.[4]

The use of an open abdomen management strategy has increased in surgical patients.[5] A class of patients is evolving with complex problems that were not previously observed. Patients with open abdomens require multiple trips to the operating room (i.e., every 24-48 hours for as long as 2 or 3 weeks) using staged abdominal explorations. Chronic abdominal wall hernias can develop with large abdominal wall defects that have to be closed with skin grafts and require extensive surgical repair 6 months later. Patients with open abdomens have reported fistula rates between 2% and 25%, intra-abdominal abscess rates are as high as 83%, and reported abdominal wall hernia rates are around 25%.[5] A significant risk for the development of an abdominal compartment syndrome (ACS) exists in abdomens covered with a temporary closure device.[6] Patients with open abdomens run a 30% to 40% chance of developing some degree of multiple organ dysfunction syndrome (MODS).[4] These complications may develop as a result of the underlying pathophysiology, packing of the wound, bowel manipulation, abdominal pressure changes, and undergoing multiple surgical and nonsurgical procedures.

Temporary abdominal closure (TAC) is defined as a surgical technique to minimize the effects of IAH and possibly prevent the development of an ACS. TAC allows for expansion of the abdominal wall with increased abdominal pressure, allows for easy access to the abdomen for reexploration, maintains the integrity of the abdominal wall fascia for closure, controls and quantifies third space losses, and acts as an effective adjunct in the closure of the abdominal wall (Table C1). Review of the literature demonstrates many methods of TAC. However, there are no controlled studies to support the use of one method over another. Most methods are effective but need refinement. Methods used to temporarily close the open abdomen should be developed with an understanding of the pathophysiology found in this class of patients. This will allow for the development of systems that will minimize complications and increase survival.

Table C1. Abdominal wall closure may not be possible after major trauma and septic patients for many reasons

Massive intestinal edema
Risk of ACS/Treatment of ACS
Rapid conclusion of procedure in DCS
Need for multiple re-explorations of the abdomen
Fascia and abdominal wall preservation
Triad of hypothermia, coagulopathy, acidosis

Pathophysiology of the Open Abdomen

The concept of an open abdomen in surgical practice is not new. Open abdominal approaches were recognized in war surgery.[6] Since the early 80's numerous types of temporary abdominal closure techniques have been described. The earliest was gauze packing of the abdomen with formation of granulation tissue and subsequent skin grafting.[6] Numerous materials have been used for temporary closure but prosthetic materials have associated complications.[7] The most important characteristics of a temporary closure prosthesis is nonreactivity, pliability, permeability, and conformity to the shape and configuration of the abdominal contents.[8]

TAC is indicated in patients with significant bowel edema, abdominal packing, bleeding, and acute postoperative acsites formation (Table C1).[6] With the recognition of the importance of intra-abdominal hypertension (IAH is a risk factor for the development of ACS) newer techniques have evolved that control intra-abdominal contents of third space loss, lower abdominal pressure and facilitates abdominal wall closure.

The underlying pathophysiology leads to the morbidity seen in patients that have their abdomens opened and require multiple reexplorations. Consideration of systems to manage the open abdomen should therefore not only contain and protect the abdominal contents, but should not exacerbate the intraperitoneal inflammatory reaction. There may be a theoretical benefit in systems that not only minimize the potential for a compartment syndrome but removal of inflammatory substances may decrease the systemic complications noted.

The open abdomen after a traumatic or septic insult should be considered a "hostile" environment.[9] There is an extensive degree of tissue destruction, contamination, hematoma, and inflamed friable tissue. Patients that require an open abdomen have a documented higher complication rate with increased rates of SIRS/MODS, fistulas, post operative ileus, and third space fluid losses.[4] Lack of recognition of these factors can lead to an increased number of complications and death. This environment leads to a significant inflammatory response and is the underlying cause of intra-abdominal hypertension and the ACS. The control of these factors can help to lower the complications.

The peritoneal response to inflammation is a programmed response with an increased permeability of membranes, increased concentration of cytokines, and influx of inflammatory cells.[10] The end result is an intense inflammatory response with an accumulation of posttraumatic ascites and third space fluid.[11] The inflammatory response is exacerbated by shock states, ischemia reperfusion injuries, and massive fluid resuscitation.[11] The end result is a pathologic increase in abdominal pressure.

Of particular concern in the open abdomen is the increased cytokine production. The abdomen and peritoneal cavity has been described as a cytokine repository.[12] Numerous studies have shown a relationship to cytokine stimulation as a result of damage control surgery and the development of an ACS. Adams et al showed an exaggerated inflammatory response in patients that have their abdomens packed with sponges at the time of damage control surgery.[13] PMN dysfunction was shown on sponges at the time of damage control surgery that leads to an inflammatory response that could contribute to a systemic inflammatory response. Rezende-Neto and associates demonstrated a release of pro-inflammatory cytokines in animals with an abdominal compartment syndrome.[14] A significant increase in TNF-α and IL-6 was noted 30 minutes after decompression for ACS. There was also an intense neutrophil accumulation in the lung with intense pulmonary infiltration. Control of third space fluid could not only minimize the development of an ACS but theoretically reduce the systemic complications associated with an open abdomen.

Morbidity Associated with the Open Abdomen

Patients treated with an open abdomen can have significant complications. What is unclear is the cause of the complications; technique of closure or the severity of the illness? Optimization of hemodynamics is an integral part of managing a patient with an open abdomen. Control of systemic sepsis, optimal nutritional support, and meticulous care and handling of intra-abdominal organs are important factors. Intra-abdominal sepsis should be controlled with elimination of blood and gross soilage and prevention of bowel adherence to the dressings.

The most significant management problem is the development of a fistula in the open abdomen. This can occur if the bowel is exposed to air, allowed to desiccate, is abraded by dressings, or deserosalized.[15] Fistulas are also caused by "biomaterial adherence" to the bowel causing transmural changes to the bowel wall If the bowel is fixed to the abdominal wall perforation with sudden changes in abdominal pressure will occur.[7] Nonadherent dressings should be used when there is direct contact with the bowel. Omentum can be used as a protective layer. Fistula rates have been reported as high as 18.2% in patients with an open abdomen after damage control surgery.[4] Ivatury and associates were able to significantly lower fistula rates with the use of early closure of the skin and fascial defects.[16] Chavarria-Aguilar et al reported a 10.5% anastomotic breakdown in patients with an open abdomen and suggested placing the anastomosis deep in the peritoneal cavity to avoid exposure.[17]

Third space drainage can be a significant problem in patients with an open abdomen.[5] Patients can drain liters over 24 hour period that could cause cooling of the patient if the bedding was allowed to be saturated, cause additional wound breakdown, unquantifiable third space losses, and create a significant nursing problem. Vacuum systems have been developed to minimize patient soilage and quantify third space losses.

Hernia formation can be a significant problem associated with the type of closure system used to manage an open abdomen. Hernias are generally seen in patients where primary fascial closure was not possible and a skin graft applied. Hernia rates have been reported as high as 69% and vary by factors including: abdominal distention, abdominal wall tension, fistula formation, and the underlying medical status of the patient.[8] Closure of post TAC hernias are usually complex and staged for complete fascial closure.[18]

Intra-Abdominal Hypertension and the Open Abdomen

TAC is a technique developed to negate the effects of intra-abdominal pressure increases. By allowing the abdominal space to expand in proportion to the pressure exerted by intra-abdominal forces pressure can be equilibrated to allow for adequate tissue perfusion and visceral organ perfusion. The ultimate goal is to prevent the development of the abdominal compartment syndrome. Therefore, risk factors have been identified to predict the development of IAH and ACS. Factors that contribute to IAH include: free blood and clots, bowel edema and vascular congestion, excessive crystalloid resuscitation, perihepatic and retroperitoneal packing, and nonsurgical bleeding.[19,20] Controversy still exits as to the benefit of TAC and the relationship IAH to the ACS. In a survey of surgeons that were members of the AAST published by Mayberry and associates[21] reported that a majority of expert surgeons in the association would use TAC for instances of massive bowel edema, pulmonary decompression, hemodynamic instability, and packing in the abdomen. Only a minority would leave the abdomen open for massive transfusions, gross contamination, hypothermia, acidosis, multiple abdominal injuries, and coagulopathy.[21] All of these are known risk factors for the development of an ACS. In the same study only 14% of surgeons surveyed would open the abdomen on increased pressure alone without evidence of systemic signs of an ACS: oliguria, increased airway pressures, decreased cardiac perfusion, or elevated intracranial pressure.

Ivatury et al demonstrated in a retrospective study the effectiveness of TAC in preventing the development of IAH.[16] Patients defined as high risk for ACS: hemodynamic instability, multiple injuries, strong suspicion for the development of ACS, need for intra-abdominal packing were studied in two groups: group I the placement of a TAC using a prosthetic mesh applied loosely; group II had their abdomens closed at the time of surgery after bleeding was controlled and organs repaired. There was an incidence of IAH (defined as pressure greater than 25 mm Hg) in the mesh closure group of 22.2% and 52% in the fascial closure group. Survival and organ dysfunction were better in patients that did not develop IAH. There was an overall survival of 90.1% in the mesh-closed group compared to 68% in the fascial closed group. This study demonstrated the effectiveness of established guidelines to monitor and prevent the development of IAH.

In a study to evaluate the efficacy of placing a TAC to prevent ACS, Mayberry and associates compared in a retrospective study the primary placement of absorbable mesh in patients that were at high risk for an ACS at the time of the initial celiotomy or placed at a second exploration.[22] The group that had a mesh placed secondarily had an increased incidence of ACS, necrotizing fasciitis, intra-abdominal sepsis, and enterocutaneous fistulas. The conclusion from this study suggested a benefit in early placement of TAC.

Patients closed with a temporary closure are still at risk for the development of ACS. Therefore, serial bladder pressures are imperative in patients with a temporary abdominal closure. The incidence of ACS is as high as 38% with a TAC.[19] Progressive IAH and the devolvement of an ACS can be predicted by a progressive increase in serial bladder pressures. The most common cause is continued bleeding either from a surgical bleeding site or coagulopathy. Moore states that bladder pressures greater than 35 mm Hg denote ongoing bleeding and warrants immediate decompression in the OR.[20] Ivatury et al recommend immediate reexploration at the beside for increasing pressures even in the face of ongoing coagulopathy in order to evacuate clots, control bleeding, and repack the abdomen.[16] Reexploration can disrupt the "bloody vicious cycle" of acidosis, coagulopathy, and hypothermia by improving pulmonary and cardiac performance and reversing the effects of cellular shock.[20]

Gracias et al report the development of an ACS as a rapidly progressive process in the open abdomen developing between 1.5 and 12 hours and was associated with a significant mortality of 60%.[25] Patients with high crystalloid requirements and severe physiologic derangements were at risk for ACS and may be predictive of development of an ACS.[25-27]

Earlier studies used TAC in those cases where there was clinical evidence of ACS while more recent studies suggest the use of TAC to prevent IAH and therefore the sequelae of ACS. Controversy exists as to the effectiveness TAC in decompressing the abdomen and reversing the systemic effects on organ dysfunction with the onset of an ACS. A study completed by Meldrum and associates showed the detrimental effects of IAH and organ dysfunction.[26] This study demonstrated that rapid onset of the ACS can be predicted by the physiologic state of the patient with a grading system of IAH based on abdominal pressure and systemic manifestations. The paper also demonstrated the effectiveness of early decompression kin high risk patients.

Sugrue et al showed limited clinical efficacy in reversing the adverse systemic effects in patients that had an increased intra-abdominal pressure.[6] This prospective nonrandomized study used polypropylene mesh or polytetrafluroethylene (PTFE) patch for temporary closure of the abdomen. Reasons for TAC were decompression in 22 patients, inability to close the abdomen in 10 patients, the need for reexploration in 8 patients, and mutifactorial in 9 patients. Increased abdominal pressure was defined as > or = to 18 mm Hg. The mean abdominal pressure was 24.2 mm Hg reduced to 14.1 mm Hg upon decompression of the abdomen. Twenty five percent of patients in the study could not have their IAP lowered below 18 mm Hg due to inadequate drainage or severe bowel edema. Patients that were decompressed did not

Table C2. Requirements for TAC devices

Multiple abdominal re-explorations
Control of peritoneal fluid and third space loss
Minimizing increases in intra-abdominal pressure increases
Preserve fascial integrity/facilitates abdominal wall closure
Minimize dressing changes and wound exposure
Contain intra-abdominal contents
Protect intra-abdominal contents
Lower bacterial counts

have a significant change in their urinary output. Only 28% of the patients in this study had a brisk diuresis. There was a reported improvement in dynamic lung compliance but minimal improvement in gas exchange or acid base status. There was no significant improvement in renal function or oxygenation. While there was minimal benefit in decompression for IAH the study does suggest that the prevention on IAH may be the key in the reduction of complication in high-risk patients. In contrast, a retrospective study completed by Ertel et al showed that there was improvement in physiologic parameters in patients with a documented ACS with emergent decompression and defined the clinical patterns and time of onset in ACS.[28] However, with improved hemodynamics there was still a 35% mortality rate. Saggi et al reported a 59% 30 day mortality after decompression for an ACS because of multiple organ dysfunction syndrome that resulted after the systemic inflammatory process found in patients with an ACS.[29] Retrospective data suggests the prevention of the syndrome and IAH as the underlying cause could lower the complications and mortality observed.

Methods of Managing the Open Abdomen

The management of the open abdomen remains controversial in the literature. Techniques have evolved from a static approach of allowing the abdominal wall to granulate and skin graft to more dynamic systems that control peritoneal fluid and facilitate earlier closure of the abdomen. The goal of temporary closure of the abdomen is to have a tension free closure without elevating intra-abdominal pressure. This is not always possible in trauma patients or patients with intra-abdominal sepsis. To date there are no reported prospective studies comparing one form of closure to another. Methods used are chosen by the experience of the operating surgeon, materials available, retrospective studies, condition of the patient, and the clinical need for an open abdomen. Closure techniques have evolved from skin only closure for rapid completion of a damage control procedure;[6] to devices that control intra-abdominal contents; to systems that allow for treatment of the underlying physiologic state. Methods used should prevent evisceration, control third space losses, lower bacterial counts, and minimize desiccation and damage to the viscera (Table C2). However, closure methods should minimize the risk of increasing intra-abdominal pressure and the risk for developing an ACS. Material used should minimize adherence to the underling viscera and give a secure closure to minimize the chance of evisceration. Wound management systems should minimize trauma to the abdominal wall and fascia, facilitate closure of the abdomen, and quantify third space losses. With control of the physiologic causes there could be a reduction in the overall complications seen in patients with open abdomens.

TAC should also allow for rapid reexploration at the bedside. Patients at high-risk can rapidly develop high abdominal pressures secondary to bleeding, pack placement, and rapid accumulation of third space fluid. TAC devices should allow for reexploration at the bedside to allow for the evacuation of clots and to expand the size of the prosthesis if pressure should increase.[16]

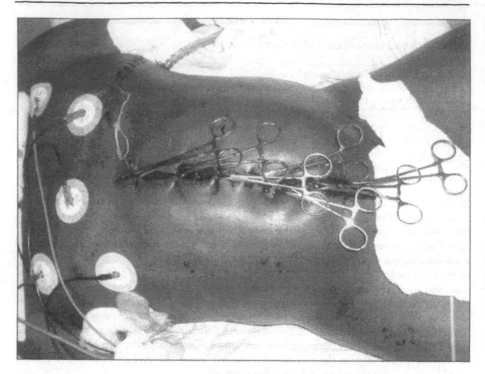

Figure C1. Towel clip closure.

Multiple types of closure have been reported. A survey of the AAST by Mayberry and associates in 1999 found the use of the Bogotá bag was 25%; absorbable mesh was 17%; polypropylene mesh was 14%; PTFE 14%; silastic mesh 7%; miscellaneous 28%; towel clip closure 1%.[21] Of note 3% of surgeons surveyed responded that they "never used open abdomen technique" (Fig. C1). Table C3 represents a summary of TAC methods.

Skin Only Closure

Skin only closure was initially described because it is a rapid method to close the skin in an unstable patient at risk.[29] A skin only closure can use towel clips (Fig. C1) or running suture to close the skin. This technique can be used for rapid closure in patients that will need a reexploration usually within 24 hours. Retrospective studies have shown efficacy with complication rates similar to other methods, however towel clips may interfere with other advanced diagnostic studies such as abdominal CT scans or MRI that many trauma patients with multiple injuries may require. Skin closure with 2-0 nylon may have an additional benefit in trauma patients requiring rapid closure to prevent interference with advanced studies. Smith et al demonstrated a high closure rate of fascial closure in patients initially closed with towel clip closure.[31] There is a 14% risk of the development of an ACS as the abdomen has a limited ability to expand with increasing abdominal pressure.[27] Skin only closure has been replaced in most centers by the placement of a prosthetic material over the abdominal wall for coverage.

Bogotá Bag

The Bogotá bag has been used in practice for over 20 years. The use of an open IV bag was popularized in Bogotá, Columbia and was initially used in the United States since 1984 (Fig. C2).[32,33] Many variations have been reported but appear to be a major advance in a prosthesis for keeping the abdomen open and is a variation of using a silo technique that allows for

Table C3. Open abdomen technique: options for temporary and permanent closure

Closure Technique	Description	Advantages	Disadvantages
Self-adhesive impermeable membranes	Abdominal dressing with gauze and coverage of the entire wound with impermeable membrane with and without placement of drains between the layers	Inexpensive Easy application	Difficult to maintain seal Potentially large volume losses
Vac Pack Closure	Bowel covered with plastic sheet and towel; jp drains attached to wall suction	Inexpensive uses material found in OR; moderate control of fluid and abdominal wall closure	Lacks ability to approximate abdominal wall; moderate fluid control; lacks safe guards to maintain constant suction and bleeding detection
Vicryl or Dexon mesh	Suturing of the mesh to the fascial edges; different options for dressing	Can be applied directly over bowel; Allows for drainage of peritoneal fluid	Rapid loss of tensile strength (in the setting of infection). Potentially large volume losses. Higher incidence of later ventral hernia development. No reopen and close option.
Polypropylene mesh	Suturing of the mesh to the fascial edges; different options for dressing	Good tensile strength; Allows for drainage of peritoneal fluid	Risk of intestinal erosion when applied directly over bowel. Potentially large volume losses. High risk of mesh infection and hernias; difficult to remove
PTFE mesh	Suturing of the mesh to the fascial edges; different options for dressing	Good tensile strength. Reopen and close option.	Potential fluid accumulation underneath the mesh. Limited tissue integration and granulation tissue formation over the mesh. Risk of mesh infection
Vacuum-assisted closure device	Sponges applied over mesh and attached to controlled low-level suction	Controlled drainage of secretions. Accelerated granulation tissue formation. Wound debridement and approximation of wound edges.	Risk of intestinal erosion when applied directly over bowel; needs to be changed in OR every 48 hours
Wittman patch	Suturing of artificial burr (i.e, Velcro) to fascia, staged abdominal closure by application of controlled tension	Good tensile strength. Allows for easy re-exploration and eventual primary fascial closure	Poor control of third space fluid; adherence of bowel to abdominal wall; potential for fistulas

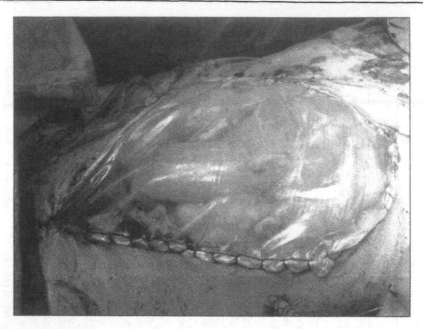

Figure C2. "Bogotá-bag" closure.

coverage as well as expansion of the abdominal wall. The Bogotá bag can be made from a pregas-sterilized 3-liter cystosocopy bag that was cut in an oval shape and sutured to the skin.[16] Other variations of the Bogotá bag include the use of sterile X-ray cassette bags, silastic drapes, and latex. Advantages of the silo closure are low cost, nonadherence, prevents evisceration, easy to apply, and is very available in the operating room.[10] Disadvantages include tearing from the skin, does not allow for easy entrance to the abdomen, and must be gas sterilized before usage.[10,28] There is also minimal control of third space loss leaking from under the bag leaving the bed wet and increasing the risk of hypothermia. Most of the effectiveness of the Bogotá bag has been through case reports and noncontrolled studies with a wide variety of patients. However, there has been a wide acceptance of the Bogotá bag a starting area for the development of new devices for the prevention and treatment of IAH and ACS.

Mesh Closure of the Open Abdomen

Mesh closure of the open abdomen has been popularized over the past years. There have been a number of variations in the type of mesh used. The advantages of mesh closure include: ease of placement, facilitates reexploration with the ability to open and reclose the abdomen, strength over Bogata bag closure, ease of reexploration at the bed side, and the ability to expand with increasing abdominal girth. Mesh closures have varied by type and material used. They appeared to be an advantage over skin alone closure because of the ability to expand with increased abdominal pressure. There have been various retrospective reports in the literature showing the benefit of mesh closure. Experience with polypropylene mesh shows an advantage of closure of the abdomen by granulation and skin grafting but the reported fistula rate when placed over bowel has been reported as high as 50-75%.[35,36] Mayberry et al showed effectiveness in placement of an absorbable mesh for the prevention of an ACS.[22] The mesh can be removed easily and facilities skin grafting because of the development of granulation tissue.[37] However, Nagy et al reported that the use of absorbable mesh resulted in significant hernia formation that was disabling with a risk of late evisceration and recommended the use of PTFE.[34] Polytetrafluroethylene (PTFE) mesh has desirable qualities in that it is nonadherent

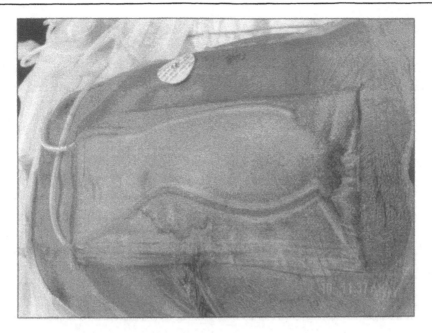

Figure C3. "Vac-Pac" dressing.

to the underlying bowel and has a low fistula rate.[5] The major draw back is the price, inability to skin graft, resistance to infection, and chronic subcutaneous infections.

Aprahamian et al in 1990 described an open abdomen technique with a Velcro-like prosthetic placed over the abdomen. This device allowed for easy entrance into the abdomen for reexploration and facilitated closure of the abdomen with serial explorations.[38] Complication rates were similar to other methods of closure but there have limited reports in the literature as to its use and effectiveness.[5]

VAC Pac Closure

Vacuum pack technique is a modification of the Bogotá concept that allows for rapid closure of the abdominal wall with application of wall suction to control abdominal secretions. Barker et al reported a 7-year retrospective experience with vacuum therapy in 112 patients (Fig. C3).[39] The technique places a fenestrated no-adherent polyethylene sheet over the viscera and is covered by moist surgical towels. Two 10-French silicone drains are placed over the towels and the wound sealed with an idofor-impregnated adhesive dressing. Wall suction is applied at 100-150 mm Hg of continuous suction. Patients are reexplored serially by changing the dressings. This method is inexpensive and effective in managing an open abdomen. There was a fascial closure rate of 55.4% with 23.3% of the patients studied had their abdomens closed with absorbable mesh and skin grafted. There was a 4.5% fistula rate which was lower than other reported techniques. Vac pack therapy appeared to be an effective technique in managing the open abdomen.

Negative Pressure Therapy (NPT)

Prior to 2001 Vac pack therapy was the author's modality of choice in managing the open abdomen with 10 year experience similar to Dr. Baker's group. Vacuum Assisted ClosureVAC therapy (KCI, Inc., San Antonio, TX) (Fig. C4) applies sub-atmospheric pressure through a reticulated polyurethane foam dressing. The negative pressure is controlled with a computer-controlled vacuum pump that applies a constant regulated pressure to the wound

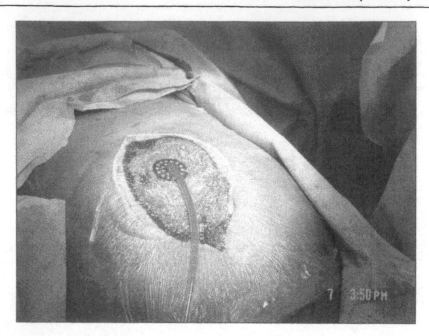

Figure C4. Vacuum assisted closure.

surface and a sensing device to prevent uncontrolled fluid drainage such as blood.[40] While initially developed at Wake Forest by Argenta and Morykwas as an adjunct to wound healing[40] the therapy has evolved to an effective device in the management of open abdomens (Table C4). There is a nonadherent polyurethane layer that is fenestrated and prevents bowel adherence to the anterior abdominal wall. Application of negative pressure to the wound allows for uniform radial pull that produces an even distribution of force over the wound. The result is an increase in blood flow, a reduction on abdominal wall tension, reduction in size of the abdominal wall defect, decreased bowel edema, and potential removal of inflammatory substances that accumulate in the abdomen during inflammatory states (Table C4).

Garner et al reported the use of negative pressure therapy in the early approximation of open wounds in patients that had a decompressive laparotomy or damage control surgery.[41] They reported a 92% fascial closure rate at an average time of 9.9 days in 14 patients. There were no fistulas or eviscerations reported. Garner placed a Bogotá bag wound for 24-48 hours and returned the patient to the OR if the abdomen could not be closed then the wound VAC was applied and used to close the abdomen. Sulburk et al expanded on Garner's experience with negative pressure therapy in trauma patients.[42] His group reported on an additional 35

Table C4. Benefits of VAC therapy in managing the open abdomen

Effective in wound dressing with control of leaking 3rd space fluid
Minimizes dressing changes
Maintains low bacterial counts
Stimulates wound healing and granulation
Improved I/O monitoring
Decreases abdominal wall tension
Facilitates closure of the abdominal wall

patients that had a primary closure rate of 86% with a mean time to closure of 7 days. Of the four patients that failed VAC therapy closure 2 developed fistulas.

Miller et al reported his group's experience using negative pressure therapy on the open abdomen. Of the 83 patients that survived their injuries for potential closure 59 patients (71%) had their fascia closed; 37 (62%) in less than 9 days and 22 (37%) greater than 9 days.[42] These results compare favorably to other modalities of closure and retrospective data shows a greater percentage of closure in patients with VAC therapy. This lowers the need for readmission to the hospital for long and complicated closure procedures.

These three combined retrospective studies show efficacy in VAC therapy and compare favorably to VAC pack therapy. There appears to be an advantage to using negative pressure therapy applied to an open abdomen wound. A dynamic closure device such as the VAC may have an advantage in that edema and third space losses can be controlled but the abdomen closed in a timelier manner. Experience of the author has been similar to other reports with additional benefits of lower ICU length of stay, higher percentage of wound closures, and lower complications when compared to other techniques.[43]

Closure of the Open Abdomen

The goal of using an open abdomen is to minimize the effects of sepsis and IAH and to minimize the secondary complications seen with an open abdomen. Early coverage of the abdominal contents will decrease complications with restoration of fascia integrity. Care must be taken to prevent desiccation and adherence of the bowel to dressings. Earlier studies using mesh closure alone had variable closure rates with a significant incidence of fistulas regardless of the material used.[20] With the introduction of dynamic closure devices such as the Velcro-closure method or VAC therapy closure rates have increased with lower complication rates.

In those cases when fascial closure is not feasible because of edema or ongoing sepsis options include: allowing the abdominal wall to granulate and place a skin graft, attempt to approximate the skin and allow a ventral hernia to develop or use a prosthetic material that can be skin grafted over granulation tissue If a skin graft is placed over granulation tissue there is usually a 6 month maturation period before attempting to close the abdomen.[18] This will allow the inflammatory process to subside and facilitate entry into the abdomen.

Late abdominal wall reconstruction can be accomplished with a number of techniques. Fabian et al have used the separation of components technique effectively in a staged procedure to close the abdomen. This procedure avoids the use of prosthetic material and therefore can be used when stomas are taken down with a lower risk of infection. There is a 33% incidence of hernia recurrence when mesh is used and 11% with the separation of components technique.[18]

Conclusions

Review of retrospective date has shown that open abdomen techniques can be effective in managing the effects of IAH. Meticulous care of the bowel and minimized trauma from systems used to cover abdominal contents can reduce the complications observed with an open abdomen. Monitoring bladder pressure should be a part of post operative management protocols with consideration of decompressing an abdomen with pressure >25-30 mm Hg, this without evidence of a clinical syndrome. While there are many reported closure techniques, a dynamic closure device as described by VAC therapy appears to have an advantage in meeting most requirements for managing an open abdomen.

Commentary

Rao R. Ivatury

The most definitive management, prophylactic or therapeutic, of IAH and ACS is surgical treatment and temporary abdominal closure (TAC) and non-suture of abdominal fascia. The preceding chapters of this book have substantiated the superiority of this "open-abdomen" approach in reducing multiple organ failure and the resulting mortality in this high-risk group of patients. Dampening the enthusiasm for this strategy is the complex management of the open abdomen in the critically ill patient. The need for extensive nursing care, the adverse effects of exposed bowel in terms of serositis and secondary hemorrhage, fluid losses from dessication, the theoretical possibility of "tertiary peritonitis", aesthetic unpalatability and the most dreaded complication of fistulization of bowel in the open abdomen, the so-called "entero-atmospheric" fistula have been just some of the deterrents to this approach. As "damage-control" operations in the trauma scenario became ever more frequent, trauma surgeons have become comfortable with the open-abdomen and are now the champions of the concept of IAH and ACS . Converts in other specialities of surgery have been scarce for many reasons, not the least of which is the need to deal with the open abdomen. They cite the convenient excuse of loss of fascial closure until a subsequent readmission for complex prosthetic and plastic hernia repairs.

The last decade has seen remarkable improvements in our ability to deal with the open abdomen in the ICU . Better prosthetic meshes have become available for hernia closure. Even more important, new concepts in vacuum-assisted closure have helped us achieve abdominal closure in a very high percentage of these patients. The complication of bowel fistulas has been reduced to less than five percent.

This chapter on surgical treatment is a distillation of this experience and is comprised of three sections. Kaplan builds a succinct summary of the problems and solutions in dealing with the open abdomen on a scientific basis of our understanding of IAH and ACS. Goettler, Rotondo and Schwab take the reader by hand step-by-step from the decision to do"damage-control" and the conduct of the stages of this approach. They provide valuable insights in the management of stomas and tubes in the open abdomen and the methods of temporary closure. Zsolt, Balogh and Moore, leaders in vacuum-assisted closure, give a succinct account, rich in pearls, of their current practice. The reader will excuse the repetition in these three sections. The editors have consciously and deliberately included all of them to underscore the surgical approach. It is fervently hoped that these sections, packed with practical pointers form pioneers in the field , will enable more clinicians to embrace the concept of the open abdomen and reap the benefits of temporary abdominal closure in the management of the critically ill or injured patient.

References

1. Asensio JA, McDuffie L, Petrone P et al. Reliable variables in the exsanguinated patient which indicate damage control and predict outcome. Am J Surg 2001; 182:743-751.
2. Shapiro MB, Jenkins DH, Schwab CW et al. Damage control: Collective review. J Trauma 2000; 49:969-978.
3. Raeburn CD, Moore EE, Biffl WL et al. The abdominal compartment syndrome is a morbid complication of postinjury damage control surgery. Am J Surgery 2001; 182:542-546.
4. Nicholas JM, Rix EP, Easley A et al. Changing patterns in the management of penetrating abdominal trauma: The more things change, the more they stay the same. J Trauma 2003; 55:1095-1110.
5. Cheatham ML, Safcsak K, Lierena LE et al. Long-term physical, mental, and functional consequences of abdominal decompression. J Trauma 2004; 56(2):237-242.
6. Mayberry JC. Bedside open abdominal surgery. Critical Care Clinics 2000; 16(1):151-172.
7. Sugrue M, Jones F, Janjua K et al. Temporary abdominal closure: A prospective evaluation of its effects on Renal and Respiratory Physiology. J Trauma1998; 45(5):914-921.
8. Losanoff JE, Richman BW, JonesJW et al. Temporary abdominal coverage and reclosure of the open abdomen: Frequently asked questions. Journal of American College of Surgeons 2002(1); 195:105-115.

9. Hirshberg A, Stein M, Adar R et al. Damage control surgery. Reoperation planned and unplanned. Surgical Clinics of North America 1997; 77(4):897-907.

10. Marshall JC, Innes M, Dellinger RP et al. Concise definitive review. Intensive care unit management of intra-abdominal infection. Critical Care Medicine 2003; 31(8):2228-2237.

11. Mayberry JC, Welker KJ, Goldman RK et al. Mechanism of acute ascites formation after trauma resuscitation. Arch Surg 2003; 138:773-776.

12. Schein M, Wittmann DH, Holzheimer R et al. Hypothesis: Compartmentalization of cytokines in intraabdominal infection. Surgery 1996; 119(6):694-700.

13. Adams JM, Hauser CJ, Livingston DH et al. The immunomodulatory effects of damage control abdominal packing on local and systemic neutrophil activity. J Trauma 2001; 50:792-800.

14. Rezende-Neto JB, Moore EE, Melo de Andrade MV et al. Systemic inflammatory response secondary to abdominal compartment syndrome. Stage for multiple organ failure. J Trauma 2002; 53:1121-1128.

15. Martin RR, Byrne M. Damage control surgery. Post operative care and compications of damage control surgery. Surgical Clinics of North America 1997; 77(4):929-942.

16. Ivatury RR, Porter JM, Simon RJ et al. Intra-abdominal hypertension after life threatening penetrating abdominal trauma. Prophylaxis, incidence, and clinical relevance to gastric mucosal pH and abdominal compartment syndrome. J Trauma 1998; 44(6):1016-1023.

17. Chavarria-Aguilar M, Cockerham WT, Barker DE et al. Management of destructive bowel injury in the open abdomen. J Trauma 2004; 56(3):560-564.

18. Fabian TC, Croce MA, Pritchard FE et al. Planned ventral hernia. Staged management for acute abdominal wall defects. Ann of Surg 1994; 219(6):643-653.

19. Ivatury RR, Diebel L, Porter JM et al. Damage control surgery: Intra-abdominal hypertension and the abdominal compartment syndrome. Surgical Clinics of North America 1997; 77(4):783-800.

20. Moore EE. Staged laparotomy for the hypothermia, acidosis, and coagulopathy syndrome. Am J Surg 1996; 172(5):405-410.

21. Mayberry JC, Goldman RK, Mullins RJ et al. Surveyed opinion of American trauma surgeons on the prevention of the abdominal compartment syndrome. J Trauma 1999; 47(3):509-514.

22. Mayberry JC, Mullins RJ, Crass RA et al. Prevention of abdominal compartment syndrome by absorbable mesh prosthesis closure. Arch Surg 1997; 132:957-962.

23. Offner PJ, Laurence de Souza A, Moore EE et al. Avoidance of abdominal compartment syndrome in damage-control laparotomy after trauma. Arch Surg 2001; 136:676-681.

24. Gracias VH, Braslow B, Johnson J et al. Abdominal compartment syndrome in the open abdomen. Arch Surg 2002; 137(11):1298-300.

25. McNeilis J, Marini CP, Jurkiewicz A et al. Predictive factors associated with the development of abdominal compartment syndrome in the surgical intensive care unit. Arch Surg 2002; 137:133-136.

26. Balogh Z, McKinley BA, Holcomb JB et al. Both primary and secondary abdominal compartment syndrome can be predicted early and are harbinger of multiple organ failure. J Trauma 2003; 54(5):848-861.

27. Meldrum DR, Moore FA, Moore E et al. Prospective characterization and selective management of the abdominal compartment syndrome. Am J Surg 1997; 174(6):667-673.

28. Ertel W, Oberholzer A, Platz A et al. Incidence and clinical pattern of the abdominal compartment syndrome after "damage-control" laparotomy in 311 patients with severe abdominal and/or pelvic trauma. Crit Care Med 2000; 28(6):1747-1753.

29. Saggi BH, Sugerman HJ, Ivatury RR et al. Abdominal compartment syndrome. J Trauma 1998; 45:595-609.

30. Tremblay LN, Feliciano DV, Schmidt J et al. Skin only or silo closure in the critically ill patient with an open abdomen. Am J Surg 2001; 182(6):670-675.

31. Smith PC, Tweddell JS, Bessey PQ et al. Alternative approaches to abdominal wound closure in severely injured patients with massive visceral edema. J Trauma 1992; 32(1):16-20.

32. Ferrada R, Birolini D. Trauma care in the new millennium. New concepts in the management of patients with penetrating abdominal wounds. Surg Clinics of North America 1999; 79(6):1331-1356.

33. Fernandez L, Norwood S, Roettger R et al. Temporary intravenous bag silo closure in severe abdominal trauma. J Trauma 1996; 40(2):258-260.

34. Nagy KK, Fildes JJ, Mahr C et al. Experience with three prosthetic materials in temporary abdominal wall closure. Am Surg 1996; 62(5):331-335.

35. Fansler RF, Taheri P, Cullinane C et al. Polypropylene mesh closure of the complicated abdominal wound. Am J Surg 1995; 170:15-18.

36. Greene MA, Mullins R, Malangoni MA et al. Laparotomy wound closure with absorbable polyglycolic acid mesh. SGO 1993; 176:213-218.

37. Buck JR, Fath JJ, Chung SK et al. Use of absorbable mesh as an aid in abdominal wall closure in the emergent setting. Am Surg 1995; 61:655-658.

38. Aprahamian C, Wittmann D, Bergstein J et al. Temporary abdominal closure (TAC) for planned relaparotomy (Etappenlavage) in trauma. J Trauma 1990; 30(6):719-723.
39. Barker DE, Kaufman HJ, Smith LA et al. Vacuum pack technique of temporary abdominal closure: A 7-year experience with 112 patients. J Trauma 2000; 48(2):201-207.
40. Argenta LC, Morykwas MJ. Vacuum-assisted closure: A new method for wound control and treatment: Clinical experience. Ann of Plastic Surg 1997; 38(6):563-576.
41. Garner B, Ware DN, Cocanour CS et al. Vacuum-assisted wound closure provides early fascial reapproximation in trauma patients with open abdomens. Am J of Surg 2001; 182(6):630-638.
42. Suliburk JW, Ware DN, Balogh Z et al. Vacuun-assisted wound closure achieves early fascial closure of open abdomens after severe trauma. J Trauma 2003; 55(6):1155-1160.
43. Miller PR, Thompson JT, Faler B et al. Late fascial closure in lieu of ventral hernias: The next step in open abdomen management. J Trauma 2003; 53(5):843-849.
44. Kaplan M. Managing the open abdomen: Acknowledging the risks, utilizing the technology. Ostomy/Wound Management 2004; 50(1A suppl):C2-8.

CHAPTER 23

Epilogue:
Options and Challenges for the Future

Michael L. Cheatham,* Rao R. Ivatury, Manu L. N. G. Malbrain and Michael Sugrue

I ntra-abdominal hypertension (IAH) and abdominal compartment syndrome (ACS) are widely believed to be relatively new disease processes that occur in direct response to exuberant crystalloid over-resuscitation.[1] The conventional wisdom dictates that IAH and ACS are largely iatrogenic in origin and would not occur were more conservative resuscitation strategies employed. If we study the past, however, we learn that elevated intra-abdominal pressure (IAP) and its detrimental impact on end-organ function was first identified almost 150 years ago in a number of pioneering studies (Fig. 1). The existence of IAH in the critically ill, therefore, clearly predates by well over a century any concept of supraphysiologic resuscitation. We must also humbly recognize that the pathophysiology of injury and reperfusion-induced increases in IAP secondary to visceral edema was largely forgotten until its "rediscovery" as IAH and ACS just over a decade ago. Perhaps in consolation, the incidence of IAH and ACS have most likely become clinically significant only within the past several decades as advancements in surgical practice and intensive care unit (ICU) management have allowed patients to survive the first 24 hours of critical illness to develop subsequent elevations in IAP. IAH and ACS, therefore, along with sepsis and multiple system organ failure, likely represent a consequence of the improved survival from shock and critical illness afforded by modern goal-directed ICU management and not necessarily a by-product thereof.

As the preceding chapters illustrate, our understanding and management of IAH and ACS as potentially life-threatening concerns in the critically ill have evolved tremendously since their rediscovery. Significant strides have been made in a relatively brief period of time in both our knowledge of the pathophysiology involved and the interventions necessary to improve patient outcome. We now recognize that the etiology of IAH and ACS is commonly multi-factorial, that there are several different forms of both IAH and ACS, and that early detection and management significantly improves survival.

To imply that clinicians now *understand* IAH and ACS, however, would be naïve. The complex relationship between IAP and organ dysfunction, the prevalence of IAH and ACS in various patient populations, the true progression of IAH to ACS following systemic injury, and the optimal management strategy for this multifaceted pathophysiologic process have yet to be fully elucidated and tested. We are clearly far from the end of the process in understanding IAH and ACS, but rather very much at the beginning. This chapter will deal with the options and challenges for the future, focusing on educating and informing clinicians, standardizing definitions and concepts, developing appropriate research programs, inventing the necessary new technologies, and finally reflecting on the need and potential for the World Society on

*Corresponding Author: Michael L. Cheatham—Department of Surgical Education, Orlando Regional Medical Center, 86 West Underwood Street, Mailpoint #100, Orlando, Florida, 32806 U.S.A. Email: michael.cheatham@orhs.org

Abdominal Compartment Syndrome, edited by Rao R. Ivatury, Michael L. Cheatham, Manu L. N. G. Malbrain and Michael Sugrue. ©2006 Landes Bioscience.

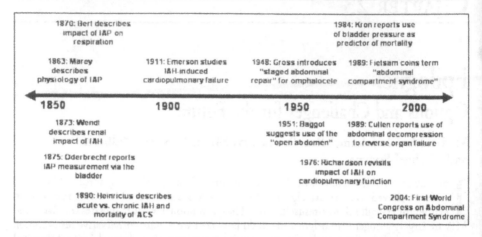

Figure 1. Intra-abdominal hypertension / abdominal compartment syndrome timeline. IAP: intra-abdominal pressure; IAH: intra-abdominal hypertension; ACS: abdominal compartment syndrome.

Abdominal Compartment Syndrome (WSACS) (Table 1). Through each of these options and challenges, we possess the potential to define and invent the future of IAH and ACS.

Education

In looking to the future, we must begin by recognizing that IAH and ACS may occur in virtually all patient populations, irrespective of age, illness, or injury. Although most commonly recognized in the traumatically-injured patient, we need to abandon the widely held belief that IAH and ACS afflict only the surgical patient and acknowledge their presence in the medical and pediatric patient populations as well.[2,3] The future of IAH and ACS must begin with educating clinicians of all disciplines as to the widespread presence, morbidity, and associated mortality of elevated IAP, IAH, and ACS within their patient populations. These efforts should focus upon three key areas: incidence, detection, and management.

A recent prospective, multi-center epidemiologic trial identified that among a mixed medical and surgical intensive care unit (ICU) patient population, IAH was present upon ICU admission in 32.1% and ACS in 4.2% of patients.[3] Further, the occurrence of IAH during the ICU stay was found to be an independent predictor of mortality (relative risk 1.85; 95% CI 1.12-3.06; p = 0.01). The clinical importance of these findings cannot be overemphasized. The incidence and associated mortality of IAH and ACS are quite similar to those associated with sepsis, a disease process that has recently received significant worldwide attention among the international medical community. Widespread educational and research efforts are currently being mounted to reduce sepsis-associated morbidity and mortality by 25% over the next five years.[4] Although sepsis-induced organ failure is a frequent cause of IAH/ACS, and IAH/ACS may well contribute to the subsequent development of sepsis, this association is infrequently recognized by many clinicians. In the absence of future educational efforts, the relative

---

Table 1. The future of intra-abdominal hypertension/abdominal compartment syndrome

- Education of physicians, nurses, and other healthcare providers
- Standardisation of definitions and concepts
- Research
- Technology Development
- World Society on Abdominal Compartment Syndrome (WSACS)

anonymity of IAH and ACS as important disease processes in the critically ill places them at risk of being forgotten yet again.

A recent survey of intensivists (those most likely to encounter IAH and ACS based upon the incidences described above) identified that 34% of medical and 32% of pediatric intensivists believed that they had "never" encountered a patient with ACS.[5] This study, among others, demonstrates that a significant lack of knowledge exists among physicians worldwide as to the presence of IAH and ACS among the critically ill. IAH and ACS are either not being recognized or are being misdiagnosed as acute respiratory distress syndrome (ARDS), mesenteric ischemia, or multiple system organ failure (MSOF) among others. If the significant morbidity and mortality of these disease processes is to be changed in the near future, the education of physicians, nurses, respiratory therapists, and others must be given considerable emphasis, much as is being done in the Surviving Sepsis Campaign.

In the novel, "The House of God" by Samuel Shem, MD, the tenth law governing the care of all patients is "If you don't take a temperature, you can't find a fever".[6] As restated by Dr. Malbrain in Chapter 3, "If you don't measure IAP you cannot make a diagnosis of IAH or ACS". In the above cited survey of intensivists, 24% of respondents were "unaware" that IAP could be measured clinically and utilized to guide therapy.[5] Clinical examination is notoriously poor in identifying the presence of IAH.[7] The safety, simplicity, cost-effectiveness, and importance of IAP monitoring in the patient at risk for IAH and ACS cannot be over-emphasized.[8,9] IAP monitoring does not require specialized equipment, but rather can be performed using materials that are readily available in any hospital. Educational efforts must encourage the application of liberal IAP monitoring to detect the presence of IAH and serial calculations of abdominal perfusion pressure (APP) to assess the adequacy of end-organ perfusion and resuscitation.[10] Institution of routine IAP monitoring in patients at risk for IAH would likely have an immediate and dramatic impact upon the detection of IAH and ACS with significant reductions in IAP-associated morbidity and mortality. Such monitoring should be implemented as a standard practice in all intensive care units.

Whereas educating clinicians on the incidence and detection of IAH and ACS will likely be relatively easy, educating them as to the appropriate management will be more difficult. IAH and ACS, like sepsis, may present in a variety of forms depending upon the patient population, the inciting cause of illness, and the resuscitative strategy required. IAH and ACS in a surgical / trauma patient must be recognized as frequently requiring a different management strategy from that of a medical or pediatric patient. The efficacy of nonsurgical treatment options in reducing elevated IAP, the relative risks and benefits of open abdominal decompression versus percutaneous paracentesis, and the merits of prophylactic decompression on patient survival and end-organ dysfunction must all be considered and carefully weighed. In the intensivist survey, 20% of medical, 33% of pediatric, and 4% of surgical intensivists stated that they would "never" consider decompressive laparotomy for the treatment of ACS.[5] This emphatic stance further emphasizes the common lack of knowledge on the part of many clinicians as to the appropriate management options for the patient with IAH and/or ACS. No therapeutic option should ever be excluded from potential consideration. The current variety of management options for the treatment of IAH and ACS will only increase in diversity in the coming years. Education on the management of IAH and ACS will no doubt require an ongoing discussion and debate among clinicians based upon the current best evidence at that point in time. Just as our understanding of IAH and ACS are evolving, so must our education efforts.

Consensus Definitions

IAH and ACS are now recognized as being dynamic, rather than static, processes characterised by a constantly changing continuum of physiological events. IAH and ACS do not appear suddenly, but rather develop over time in response to cellular ischemia and reperfusion injury. As our understanding of this continuum has evolved, so have our definitions of IAH and ACS. The "critical IAP" that was considered to mandate intervention in years past has been steadily revised downward as the detrimental effect of IAH on end-organ perfusion and patient survival has been documented. The current standard of care would now suggest the need for intervention at an IAP of 20-25 mm Hg rather than the 35-40 mm Hg that was commonly accepted a decade ago. Further, we now appreciate that absolute intra-abdominal pressures possess limited diagnostic sensitivity and are of less clinical importance than the calculated perfusion pressure across the compartment.[10,11]

As a result of our changing understanding of IAH and ACS physiology and the numerous discoveries that have been made with regard to improving patient outcome, a variety of definitions for IAH and ACS have been utilized over the years in both scientific trials and clinical reviews. This diversity has made comparison of one trial with another difficult and has, at times, led to confusion over what constitutes IAH and ACS, how IAP should be measured, when intervention is necessary, and which management strategies are associated with the lowest morbidity and mortality.

To facilitate communication regarding IAH and ACS and to allow meaningful comparison of clinical trials, it is imperative that a common terminology and data set be adopted for future discussions and research. The International ACS Consensus Definitions Conference, sponsored by the World Society on Abdominal Compartment Syndrome (WSACS), has proposed a series of definitions for IAH and ACS which will hopefully serve as an initial foundation upon which future investigations and trials will be developed. These consensus definitions, based upon the current best evidence and discussed in detail in Chapter 2, emphasize the clinical value of IAP monitoring in the detection of IAH and ACS, the importance of perfusion pressure calculations, the varying presentation of IAH and ACS by patient population, and the need for a standardized statistical approach to analyzing new monitoring techniques. These definitions should be carefully considered during the design of any future IAH / ACS research and will serve as a valuable tool in any educational effort. These consensus definitions will no doubt require periodic review and reassessment with changes being made as necessary based upon our understanding and interpretation of the available literature and research at that time. Only through the application of such consensus definitions will effective, clinically meaningful comparisons between future trials and investigations be possible.

Research

The past decade has witnessed an explosion of scientific investigation into the organ-specific physiology and clinical treatment of IAH and ACS. As a result, significant progress has been made in defining what constitutes IAH and ACS and which interventions are currently most effective in improving patient outcome. The majority of the research performed so far, however, has been either retrospective or observational in nature, or based upon animal or laboratory investigations.[12] Very few prospective, randomized trials on IAH or ACS have been performed. In addition, the majority of investigations have been published in surgery or critical care journals with a paucity of reports in the medicine literature. As a result, surgeons tend to be more cognizant of the existence of IAH and ACS than do their medical and pediatric colleagues simply due to the increased exposure in the surgical literature.

The scarcity of Level I and Level II scientific evidence is no doubt due, in part, to the emerging nature of these disease processes. Future research must be focused on performing rigorous, prospective, multi-center, human trials to answer the numerous questions that remain. Additionally, there is a need for more widespread publishing of articles on IAH and ACS throughout the scientific literature in order to more effectively educate all clinicians who may encounter these disease processes.

To promote future rigorous scientific trials in this area, the WSACS has formed a "Clinical Trials Working Group" to facilitate performance of prospective, multi-center, clinical trials using the proposed consensus definitions. Areas for future investigation include defining the true incidence and causative factors associated with IAH and ACS among various patient populations, evaluating the impact of early goal-directed therapy strategies in the resuscitation of patients with IAH and ACS, comparing the efficacy of various abdominal wall closure methods, and evaluating new organ-specific monitoring techniques.

Technology Development

As a result of the frequently insidious nature of IAH and the poor clinical sensitivity of physical examination in its detection, bedside monitoring techniques are crucial to the detection and management of the patient with IAH and/or ACS. The importance of IAP monitoring and APP calculations has previously been emphasized. There remains significant debate over the optimal technique for measuring IAP. Dr. Malbrain has eloquently discussed the advantages and disadvantages of each methodology in Chapter 3. Most of the current methods for IAP measurement, while effective, are cobbled together from pieces of stray pressure tubing and plastic connectors, or use devices that are intended for unrelated purposes. New techniques for IAP measurement that overcome the pitfalls of the current methods must be developed. Specifically, there is currently a great need for adoption of a standardized technique and infusion volume (for intravesicular pressure monitoring) to ensure reproducibility and accuracy of the measurements obtained. Proactive development of such monitoring techniques by bedside physicians, in collaboration with industry, will be essential.

The trend in monitoring in recent years has been a move towards continuous rather than intermittent monitoring techniques. Such is no less the case in the management of IAH and ACS. Methods for measuring IAP and APP continuously and using this real-time information to guide patient resuscitation have recently been reported.[9,13] As is currently being widely adopted for hemodynamic monitoring, continuous assessment of IAP and APP with early goal-directed resuscitation based upon these parameters will no doubt become the standard of care over the next decade.

Monitoring of IAP and APP, while state-of-the-art at this point in the evolution of IAH and ACS resuscitation, will not be sufficient for the future. Each organ system possesses a differential sensitivity and response to elevations in IAP. The kidneys and liver demonstrate significant reductions in regional perfusion at an IAP of 10-15 mm Hg while the lungs and heart appear to retain sufficient reserve to maintain adequate perfusion at higher levels. Development of organ specific monitoring techniques should be aggressively pursued to allow accurate assessment of regional as opposed to global perfusion adequacy. For some organ systems, simple markers of regional resuscitation adequacy already exist. Intracranial pressure (ICP) monitoring for the brain, intramucosal pH (pHi) and gastric luminal carbon dioxide tension ($PrCO_2$) for the stomach, and indocyanine green (ICG) clearance for the liver have all been proposed as useful markers of regional perfusion adequacy in the management of the patient with IAH and ACS. Clearly, additional regional markers are needed and should be a primary focus of future research in IAH and ACS.

World Society on Abdominal Compartment Syndrome (WSACS)

Unlike many commonly encountered disease processes which remain within the purview of a given discipline, IAH and ACS readily cross the usual barriers and may occur in any patient population regardless of age, illness, or injury. As a result, no one scientific society or association can represent the wide variety of physicians, nurses, respiratory therapists, and other allied healthcare personnel who might encounter patients with IAH and/or ACS in their daily practice. To fill this void, the World Society on Abdominal Compartment Syndrome (WSACS) has been founded to serve as a peer-reviewed forum and educational resource for all healthcare providers as well as industry who have an interest in IAH and ACS. Launched at the Inaugural World Congress on Abdominal Compartment Syndrome in December 2004, the mission of the WSACS is to foster education, promote research, and thereby improve the survival of patients with IAH and ACS by bringing together physicians, nurses, and others from throughout the world and from a variety of clinical disciplines.

Effective communication and discussion between clinicians and researchers worldwide will be essential to accomplish these goals. The widespread availability, speed, and visual nature of Internet-based education and communication will significantly increase the rapidity and efficiency with which these objectives are achieved. The WSACS website (www.wsacs.org) is intended to serve as the definitive resource on IAH and ACS. Multimedia educational modules, lectures, case discussions, and other resources and links will be utilized to promote IAH / ACS education worldwide. Further, an electronic mail discussion list will facilitate discussion of patient problems, questions, and ideas among experienced clinicians, researchers, and others in near real-time fashion.

The WSACS Clinical Trials Working Group (CTWG) will promote prospective, multi-center scientific trials to study pertinent research questions and hypotheses based upon the foundation of the proposed consensus definitions. The CTWG, in collaboration with other scientific organizations worldwide, will set the standard and lead the effort towards performing the prospective, multi-center clinical trials that are necessary to more fully define the nature and appropriate management of IAH and ACS.

Modeled on the highly successful Inaugural World Congress meeting, biannual international scientific symposia will be organized to promote face-to-face discussion of research findings and the current state-of-the-art in IAH and ACS treatment and management. During these sessions, the consensus definitions will be reassessed and revised as necessary. In addition, joint collaborative educational opportunities will be pursued with other scientific organizations worldwide.

Conclusions

IAH and ACS are significant causes of organ failure, increased resource utilization, decreased economic productivity, and increased mortality among a wide variety of patient populations. Considerable progress has been made in the field of IAH and ACS over the past decade, but there is significant work yet to be done. We must study and learn from the past and, at the same time, proactively "invent" the future. As aptly described by Dr. Ivatury, IAH / ACS is "...a clinical entity that had been ignored for far too long".[14] The future of IAH and ACS is in our hands. It is time to pay attention.

References

1. Lyons WS. Uncontrolled resuscitation and "sepsis". J Trauma 2002; 52:412.
2. Balogh Z, Moore FA. Intra-abdominal hypertension: Not just a surgical critical care curiosity. Crit Care Med 2005; 33:447-449.
3. Malbrain MLNG, Chiumello D, Pelosi P et al. Incidence and prognosis of intraabdominal hypertension in a mixed population of critically ill patients: A multiple-center epidemiological study. Crit Care Med 2005; 33:315-322.
4. Dellinger RP, Carlet JM, Masur H et al. Surviving sepsis campaign guidelines for management of severe sepsis and septic shock. Crit Care Med 2004; 32:858-873.
5. Kimball EJ, Mone MC, Barton RG et al. Survey of ICU physicians on the recognition and management of abdominal compartment syndrome. Presented at the 2004 World Congress on Abdominal Compartment Syndrome. Noosa, Queensland, Australia: 2004.
6. Shem S. The House of God. Dell Publishing. ISBN: 0-440-13368-8.
7. Sugrue M, Bauman A, Jones F et al. Clinical examination is an inaccurate predictor of intraabdominal pressure. World J Surg 2002; 26:1428-1431.
8. Cheatham ML, Safcsak K. Intraabdominal pressure: A revised method for measurement. J Am Coll Surg 1998; 186:594-595.
9. Malbrain ML. Different techniques to measure intra-abdominal pressure (IAP): Time for critical reappraisal. Intensive Care Med 2004; 30:357-371.
10. Cheatham ML, White MW, Sagraves SG et al. Abdominal perfusion pressure: A superior parameter in the assessment of intra-abdominal hypertension. J Trauma 2000; 49:621-627.
11. Malbrain MLNG. Abdominal perfusion pressure as prognostic marker in intra-abdominal hypertension. In: Vincent JL, ed. Yearbook of Intensive Care and Emergency Medicine. Berlin, Heidelberg, New York: Springer, 2002:792-814.
12. Andrews K, Lynch J, Sugrue M. A scientific evaluation of the level of evidence of publications relating to abdominal compartment syndrome. Presented at the 2004 World Congress on Abdominal Compartment Syndrome. Noosa, Queensland, Australia: 2004.
13. Balogh Z, Jones F, D'Amours S et al. Continuous intra-abdominal pressure measurement technique. Am J Surg 2004; 188:679-684.
14. Ivatury RR, Sugerman HJ. Abdominal compartment syndrome: A century later, isn't it time to pay attention? Crit Care Med 2000; 28:2137-2138.

Index

I

Iberti technique 30-32, 35, 37, 56, 57
Ileus 3, 21, 76, 78, 83, 215, 233, 234, 238, 284
Incidence 10, 15, 60, 77, 78, 82-87, 100, 105, 124, 125, 138, 140, 141, 158-160, 162, 166, 167, 170, 184, 185, 193, 199, 200, 203, 204, 210, 211, 212, 215, 218, 220, 224-226, 230, 237, 238, 270, 286, 289, 293, 297-300
Incisional hernia 21, 24, 74, 189, 190, 193, 194, 215, 278
Independent predictor 20, 83, 159-161, 166, 172-174, 298
Indocyanine green clearance 112, 139
Infection 5, 21, 28, 30, 32, 35, 37, 40, 43, 45, 46, 48, 53-55, 57, 108, 114, 162, 198, 200, 210, 213-215, 218, 225, 232, 237, 269, 273, 276, 278, 283, 289, 291, 293
Inferior vena cava pressure (IVCP) 40, 54, 55
Inhalation agent 256
Interleukin (IL)-1β 134, 135, 163
Interleukin (IL)-8 134, 161
Intermittent mandatory ventilation (IMV) 112, 113
Intracranial pressure (ICP) 8, 20, 49, 70, 72, 121, 140, 144, 145-155, 191, 192, 219, 254, 255, 257, 260, 286, 301
Intramucosal pH 46, 71, 74, 83, 112, 131, 167, 301
    see also pHi
Intrarectal pressure (IRP) 57-60
Intrathoracic blood volume (ITBV) 79, 245, 246, 248, 249, 251
Intrathoracic blood volume index (ITBVI) 98, 99, 101, 141
Intrauterine pressure 54, 56
Intravital video microscopy 130
Ischemia-reperfusion injury 133, 157, 172, 219, 230, 255, 299
Isoflurane 256

K

Ketamine 255-257
Klebsiella pneumoniae 108

L

Laparoscopy 4, 8, 21, 27, 51, 55, 61, 63, 106, 122, 131, 144, 154
Laser doppler flow studies 124, 132
Left ventricular end-diastolic pressure (LVEDP) 93
Liver transplantation 14, 16, 85, 120, 138, 140-142, 224, 229
Lorazepam 256
Lung myeloperoxidase (MPO) 134, 163

M

Major trauma outcome study 153
Manometry (MANO) 28, 39, 40, 43, 56-60, 64
Mathematical coupling 96, 97
Mean arterial pressure (MAP) 9, 10, 69-77, 92, 101, 122, 124, 129, 132-134, 145-147, 149, 155, 161, 163, 165, 219, 236, 243-245, 255, 256
Mechanical ventilation 21, 64, 70, 86, 89, 93, 105-109, 111-114, 141, 155, 171, 192, 224, 229, 258, 259, 269
Meconium perforation 220
Melatonin 164, 233, 235, 236
Mesh closure 82, 170
Metoclopromide 234
Microchip Transducer Tipped Catheters 55
Midazolam 255, 256
Mitral valve regurgitation 94
Mitral valve stenosis 94
Monro-Kellie doctrine 72, 146, 148, 154
Morbid obesity 12, 23, 189, 192-194
Multiple organ dysfunction syndrome (MODS) 89, 129, 134, 135, 224, 229, 230, 283, 284, 287
Multiple organ failure (MOF) 14, 15, 86, 100, 107, 108, 122, 129, 135, 142, 157-163, 165-167, 171, 172, 198, 212, 226-229, 237, 260, 272, 294
Muscle relaxation 8, 26, 27, 63, 241, 255, 257, 258
Myeloperoxidase (MPO) 134, 135, 163-165